VAUXHALL

CAVALIER, VECTRA, CALIBRA

WORKSHOP MANUAL

by
Jim Tyler

Ring this number to find your nearest Vauxhall dealer in the UK – 01582 427 200.

CONTENTS

Detailed Contents are shown at the start of each Chapter.

FACT FILE

'LEFT' AND 'RIGHT' SIDES OF THE CAR

• Throughout this manual, we refer to the 'left' and 'right' sides of the car. They refer to the sides of the car that you would see if you were sitting in the driver's seat, looking forward.

First published in 2001 by:
Porter Publishing Ltd.
The Storehouse • Little Hereford Street
Bromyard • Hereford • HR7 4DE • England

Tel: 01885 488 800
Fax: 01885 483 012
www.portermanuals.com

British Library Cataloguing in Publication Data.

A catalogue record for this book is available from the British Library.

ISBN 1-899238-43-3

Series Editor: Lindsay Porter
Layout and Typesetting: Pineapple Publishing, Worcester
Printed in England by The Trinity Press, Worcester

CHAPTER 1: FACTS & FIGURES

*Please read **Chapter 2, Safety First** before carrying out any work on your car.*

This chapter provides you with all the information you will need about your car, especially in connection with servicing and repairing it. First, you'll need to identify the engine type. If you don't know it already, see **Chapter 6, Engine**.

Before buying parts, be sure to take your vehicle's chassis and engine numbers with you.

CONTENTS

IMPORTANT NOTE:
→ Many detail changes have taken place over the years, and there have been many different special editions and options available.
→ The following information will be true of most cases but can only be taken as a general guide.
→ Vauxhall's quoted torque settings and other information sometimes vary between publications. The figures shown here are the latest available to us.

→ If in any doubt, consult your local Vauxhall dealer for confirmation.

→ At the time of going to print, no information was available on the following engines: Y26SE (Petrol); Y17DTR; X22DTH; Y22DTR (Diesel).

Part A: Model Changes

Calibra
June 1990: Calibra models introduced, based on Cavalier floorpan and drivetrain.
September 1990: Calibra 4x4 launched.
April 1992: Calibra 2.0 4x4 Turbo introduced.
October 1993: Calibra 2.5i V6 version introduced.
1998: Calibra discontinued.

Cavalier
1971: Original Cavalier introduced to UK. Rear-wheel drive. Badged Opel Ascona in rest of Europe. Discontinued 1981. NOT COVERED BY THIS MANUAL.
September 1981: Cavalier Mk 2 front-wheel drive introduced into UK in hatchback or saloon form. Still known as Opel Ascona in rest of Europe.
June 1982: Cavalier 1600 L diesel version introduced.
October 1982: Cavalier 1800 CD and 1800 SRi models introduced.
September 1983: Cavalier Estate versions launched.
March 1984: Diesel version of Cavalier Estate introduced.
August 1987: Cavalier 2.0 SRi version launched.
October 1988: Cavalier Mk 3 introduced, replacing Mk 2 versions. Now known as Opel Vectra in rest of Europe ('**Vectra A'** within Opel).

January 1989: Cavalier 2.0i 4x4 launched.
October 1989: First versions of Cavalier start being fitted with catalytic converter in exhaust - certain models and options.
March 1992: Cavalier 1.7 TD turbo-diesel version launched in UK.
October 1992: Restyled Cavalier range introduced with side-impact beams in doors.
March 1993: Cavalier 2.5 V6 engine version introduced.
1995: Cavalier discontinued.

Vectra
October 1995: Vectra introduced to UK - now with same name as rest of Europe (except called 'Opel' outside UK - and known as '**Vectra B'** within Vauxhall/Opel). Extensively redeveloped version of Cavalier. Available initially in saloon and hatchback versions only.
October 1996: Vectra Estate and V6 engine versions launched.
September 1997: Vectra 2.0 TDi turbo-diesel version introduced.
April 1998 : Vectra 2.0 bi-fuel duel-fuel petrol/LPG version introduced.
April 1999: New Vectra 2.0 diesel engine introduced.
2000: New Vectra 2.2 turbo-diesel engine introduced, with balancer shaft.

Part B: Vital Statistics

Wheels and Tyres

Model	Rim width & type	Tyre size	Tyre pressures Normal Front	Rear	Laden Front	Rear
Cavalier 1981-1988						
Cavalier	5J x 13	155 TR 13	2.0 (29)	1.8 (26)	2.2 (32)	2.4 (35)
Cavalier L	5J x 13	165 TR 13	1.9 (28)	1.7 (25)	2.1 (30)	2.3 (33)
Cavalier GL	51/2J x 13	165 TR 13	1.9 (28)	1.7 (25)	2.1 (30)	2.3 (33)
Cavalier SRi	51/2J x 14	195/60HR 14	2.1 (30)	1.9 (28)	2.2 (32)	2.4 (35)
Cavalier CD and GLi GLSi	51/2J x 13	185/70HR 13 (Manual)	2.0 (30)	1.8 (26)	2.2 (32)	2.4 (35)
Convertible		185/70TR 13 (Auto)				
Cavalier 1988-1995						
Cavalier 1.4,1.6	51/2J x 13	165 R 13-82 T	1.8 (26)	1.6 (23)	2.0 (29)	2.2 (32)
16NZ, 1.7D TC4EE1, 1.8	5/12J x 14	175/70 R 14-84 T	2.0 (29)	1.8 (26)	2.2 (32)	2.4 (35)
Cavalier GL 4 x 4	51/2J x 14	195/60 R 14-85V	2.2 (32)	2.0 (29)	2.4 (35)	3.0 (43)
C20NE, 20SHE	6J x 15	195/60 R 15-87V	2.2 (32)	2.0 (29)	2.4 (35)	3.0 (43)
		205/55 R 15-87V	2.2 (32)	2.0 (29)	2.4 (35)	3.0 (43)
Cavalier 2000	6J x 15	195/60 R 15-87V	2.3 (33)	2.1 (30)	2.5 (36)	3.2 (46)
C20XE, 20XEJ	6J x 15	205/55 R 15-87V	2.3 (33)	2.1 (30)	2.5 (36)	3.2 (46)
Cavalier 2000 4 x 4	51/2J x 14	175/70 R 14 TM+S	2.3(33)	2.1(30)	2.5(36)	3.2(46)
Calibra 1990-1998						
	51/2J x 14	195/60 R 14-85H or V	2.3 (33)	2.1 (30)	2.5 (36)	3.2 (46)
	51/2J x 14	175/70 R 14-84Q M+S	2.3 (33)	2.1 (30)	2.5 (36)	3.2 (46)
	6J x 15	195/60 R 15-87H or V	2.4 (34)	2.2 (32)	2.5 (36)	3.2 (46)
Vectra						
X16SZR	51/2J x 14	175/70 R 14-84Q M+S	2.3 (33)	2.3 (33)	2.4 (34)	3.0 (43)
		175/70 R 14-84T	2.3 (33)	2.3 (33)	2.4 (34)	3.0 (43)
16LZ2	51/2J x 14	185/70 R 14-88T	2.0 (29)	2.0 (29)	2.1 (30)	2.6 (38)
	6J x 15	195/65 R 15-91T	2.0 (29)	2.0 (29)	2.1 (30)	2.6 (38)
X16XEL	51/2J x 14	175/70 R 14-84Q M+S	2.0 (29)	2.0 (29)	2.1 (30)	2.6 (38)
		185/70 R 14-88H	2.0 (29)	2.0 (29)	2.1 (30)	2.6 (38)
	6J x 15	195/65 R 15-91H	2.0 (29)	2.0 (29)	2.1 (30)	2.6 (38)
	6J x 16	205/55 R 16-91V	2.0 (29)	2.0 (29)	2.1 (30)	2.6 (38)
Y16XE	6J x 15	195/65 R 15-91H	2.0 (29)	2.0 (29)	2.1 (30)	2.6 (38)
X17DT	51/2J x 14	175/70 R 14-84Q M+S	2.5 (36)	2.5 (36)	2.5 (36)	3.0 (44)
		185/70 R 14-88T	2.5 (36)	2.5 (36)	2.5 (36)	3.0 (44)
	6J x 15	195/65 R 15-91T	2.5 (36)	2.5 (36)	2.5 (36)	3.0 (44)
X20DTL	51/2J x 14	175/70 R 14-84Q M+S	2.3 (33)	2.3 (33)	2.4 (34)	3.0 (43)
		185/70 R 14-88T	2.3 (33)	2.3 (33)	2.4 (34)	3.0 (43)
	6J x 15	195/65 R 15-91T	2.3 (33)	2.3 (33)	2.4 (34)	3.0 (43)
X18XE, X18XE1, 20NEJ	51/2J x 14	175/70 R 14-84Q M+S	2.1 (31)	2.1 (31)	2.3 (33)	2.8 (41)
		185/70 R 14-88H	2.1 (31)	2.1 (31)	2.3 (33)	2.8 (41)
	6J x 15	195/65 R 15-91H	2.1 (31)	2.1 (31)	2.3 (33)	2.8 (41)
	6J x 16	205/55 R 16-91V	2.1 (31)	2.1 (31)	2.3 (33)	2.8 (41)
C20SEL, C22SEL, X20XEV	6J x 15	195/65 R 15091V	2.1 (30)	2.1 (30)	2.2 (31)	2.7 (40)
		205/60 R 15-91V	2.1 (30)	2.1 (30)	2.2 (31)	2.7 (40)
X25XE	6J x 15	195/65 R 15-91V	2.4 (34)	2.2 (31)	2.5 (35)	2.8 (41)
		205/60 R 15-91V	2.4 (34)	2.2 (31)	2.5 (35)	2.8 (41)
	6J x 16	205/60 R 16-91V	2.4 (34)	2.2 (31)	2.5 (35)	2.8 (41)
Space saving wheel	4J x 15	T 125/85 R 15	4.2 (61)	4.2 (61)	4.2 (61)	4.2 (61)

Weights and Dimensions

All weights in kg. All sizes in mm. NB Maximum load capacity = (Maximum Laden Weight) minus (Unladen Weight). Trailer weights are maximum loaded weight.

Weights

MODEL	Unladen Weight Manual	Unladen Weight Automatic	Maximum Laden Weight Manual	Maximum Laden Weight Automatic	Unbraked Trailer	Braked Trailer Manual	Braked Trailer Automatic
Cavalier 1980-1988							
1.3 engine							
4-door saloon	965	991	1460	1500	475	1000	650
Hatchback	995	1021	1480	1520	475	1000	650
1.6 engine							
4-door saloon	1015	1005	1515	1515	500	1100	1100
Hatchback	1045	1035	1535	1535	500	1100	1100
Estate	1062	1060	1585	1585	500	1100	1100

MODEL	Unladen Weight Manual	Unladen Weight Automatic	Maximum Laden Weight Manual	Maximum Laden Weight Automatic	Unbraked Trailer	Braked Trailer Manual	Braked Trailer Automatic
1.6 diesel engine							
4-door saloon	1044		1545		515	750	
Hatchback	1070		1565		515	750	
Estate	1115		1585		515	750	
1.8i engine							
4-door saloon	1030	1064	1530	1560	500	1300	1100
Hatchback	1060	1094	1550	1580	500	1300	1100
2.0i engine (CD)							
4-door saloon	1073	1103	1545	1575	500	1300	1300
Hatchback	1115	1145	1565	1595	500	1300	1300
2.0i engine (SRi)							
4-door saloon	1050		1545		500	1300	
Hatchback	1080		1565		500	1300	

Cavalier 1988-1995 (models with air conditioning add 30Kgs)

MODEL	Unladen Weight Manual	Unladen Weight Automatic	Maximum Laden Weight Manual	Maximum Laden Weight Automatic	Unbraked Trailer	Braked Trailer Manual	Braked Trailer Automatic
4-door							
14NV	990 to 1013		1530		500	1000	
C16NZ	1020 to 1041		1550		500	1250	
16SV	1005 to 1036	1050 to 1081	1550		500	1200	850
17D	1080 to 1100		1610		500	700	
TC4 EE1	1135 to 1163		1660		500	1000	
C18NZ	1075 to 1111	1110 to 1146	1610	1650	500	1300	1250
18SV	1060 to 1079	1095 to 1114	1590	1630	500	1350	1250
2.0 engines	1115 to 1144	1150 to 1179	1645	1685	600	1350	1300
5-door							
14NV	1005 to 1028		1530		500	1000	
C16NZ	1035 to 1056		1550		500	1250	
16SV	1020 to 1051	1065 to 1096	1550	1590	500	1200	850
17D	1095 to 1115		1610		500	700	
TC4 EE1	1150 to 1178		1660		500	1000	
C18NZ	1090 to 1126	1125 to 1161	1610	1650	500	1300	1250
18SV	1075 to 1078	1110 to 1113	1590	1630	500	1350	1250
2.0 engines	1130 to 1159	1165 to 1194	1645	1685	600	1350	1300

GSi 4-door

MODEL	Unladen Weight Manual	Unladen Weight Automatic	Maximum Laden Weight Manual	Maximum Laden Weight Automatic	Unbraked Trailer	Braked Trailer Manual	Braked Trailer Automatic
C18NZ	1085		1600				
18SV	1075		1600				
C20NE, 20NE, 20SHE	1125		1645				

GSi 5-door

MODEL	Unladen Weight Manual	Unladen Weight Automatic	Maximum Laden Weight Manual	Maximum Laden Weight Automatic	Unbraked Trailer	Braked Trailer Manual	Braked Trailer Automatic
C18NZ	1100		1600				
18SV	1090		1600				
C20NE, 20NE, 20SHE	1140		1645				
Cavalier GL 4 x 4	1245		1750				
Cavalier 2000	1215		1720				
Cavalier 2000 4 x 4	1310		1805				

Calibra 1990-1998

MODEL	Unladen Weight Manual	Unladen Weight Automatic	Maximum Laden Weight Manual	Maximum Laden Weight Automatic	Unbraked Trailer	Braked Trailer Manual	Braked Trailer Automatic
C20NE	1215	1245	1635	1665	660	1350	1350
X20XEV	1250	1280	1670	1700	650	1350	1350
C20XE	1245		1665		580	1300	
C25XE	1325	1345	1745	1765	660	1350	1350

MODEL	Unladen Weight Manual/auto	Unladen Weight Manual/auto	Unladen Weight Manual/auto	Maximum Laden Weight Manual/auto	Maximum Laden Weight Manual/auto	Maximum Laden Weight Manual/auto	Braked Trailer Manual/auto	Braked Trailer Manual/auto	Unbraked Trailer Manual/auto	Unbraked Trailer Manual/auto

Where only one figure is given, this relates to manual only.

Vectra	Saloon	Hatch	Estate	Saloon	Hatch	Estate	Hatch/Saloon	Estate	Hatch/Saloon	Estate
Envoy X16SZR	1210	1216		1745	1760					
X18XE1			1315			1935				
LS, Club, Sxi										
X16XEL	1254/1284	1269/1299		1815/1830	1830/1845		1300/1300		665/665	
X18XE1	1307/1322	1322/1337	1352/1367	1830/1845	1845/1860	1935/1950	1400/1450	1300/1400	670/670	695/695
X20DTL	1369	1384	1432	1910	1925	2015	1300	1200	725	745
X20DTH	1369	1384	1437	1920	1935	2020	1300	1200	725	745
X20XEV	1337/1369	1352/1384	1387/1417	1860/1890	1875/1905	1960/1990				
GLS										
X18XE1	1325/1340	1340/1355	1370/1385	1830/1845	1845/1860	1935/1950	1400/1450	1300/1400	670/670	695/695
X20DTL	1387	1402		1450	1910	1925	1300	1200	725	745
X20DTH	1387	1402		1455	1920	1935	1300	1200	725	745
X20XEV	1355/1387	1370/1402	1387/1417	1860/1890	1875/1905	1960/1990	1400/1450	1300/1400	670/670	695/695

MODEL	Unladen Weight Manual/auto	Unladen Weight Manual/auto	Unladen Weight Manual/auto	Maximum Laden Weight Manual/auto	Maximum Laden Weight Manual/auto	Maximum Laden Weight Manual/auto	Braked Trailer Manual/auto	Braked Trailer Manual/auto	Unbraked Trailer Manual/auto	Unbraked Trailer Manual/auto
Sri, Gsi.	Saloon	Hatch	Estate	Saloon	Hatch	Estate	Hatch /Saloon	Estate	Hatch /Saloon	Estate
X18XE1	1318	1333	1365	1830	1845	1935	1250	1150	670	695
X20XEV	1348	1363	1382	1860	1875	1960	1400/1450	1300/1400	670/670	695/695
X25XE	1379	1405	1425	1895	1910	1995	1500	1500	695	740
CD										
X20DTH	1399	1414	1468	1920	1935	2020	1300	1200	725	745
X20XEV	1367/1399	1382/1414	1400/1430	1860/1890	1875/1905	1960/1990	1400/1450	1300/1400	670/670	695/695
X25XE	1398/1424	1424/1439	1443/1458	1895/1910	1910/1925	1995/2010	1500/1500	1500/1500	695/695	740/740
CDX										
X20DTH	1400	1421	1474	1920	1935	2020	1300	1200	725	745
X20XEV	1374/1406	1389/1421	1406/1436	1860/1890	1875/1905	1960/1990	1400/1450	1300/1400	670/670	695/695
X25XE	1405/1431	1431/1446	1449/1464	1895/1910	1910/1925	1995/2010	1500	1500	695	740

Dimensions

	Overall Length	Overall Width	Overall Height	Wheelbase	Track - front 1.3litre	Track - front All other	Track - rear
Cavalier 1981-1988							
Saloon	4366	1668	1395	2573	1400	1406	1406
Hatchback	4264	1668	1385	2573	1400	1406	1406
Estate	4366	1668	1368	2573	1400	1406	1406
Cavalier 1988-1995							
4-door	4430	1700	1425	2600		1426	1423
4-door							
2000 4X4	4430	1700	1425	2600		1426	1444
5-door	4350	1700	1418	2600		1426	1423
Calibra 1990-1998							
	4492	1688	1320	2600		1426	1446
Vectra 1995-on							
Saloon	4495	1707	1425	2637		1463/1483*	1450/1478*
Hatch	4495	1707	1425	2637		1463/1483*	1450/1478*
Estate	4490	1707	1490	2637		1463/1483*	1450/1478* * Depending on model

Part C: Capacities

All fluid figures are given in litres.
ENGINE OIL: SAE 10W/40 to 20W/50 to API SF/CC,SF/CD,SG/CC or SG/CD
MANUAL GEARBOX AND FINAL DRIVE: SAE 80EP API-GL3/GL4
AUTOMATIC GEARBOX: Dexron II ATF all engines
POWER ASSISTED STEERING: DOT 4 or SAE J1703
BRAKE FLUID: DOT 4 or SAE J1703
COOLING SYSTEM: Ethylene Glycol based antifreeze

Cavalier 1981 - 1988

Engine Code	13S	16S	16D	18i	2.0i
COOLANT CAPACITY INC. HEATER:					
Manual	6.7	7.7	7.5	7.5	6.9
Automatic	6.7	7.7	–	7.5	6.9
ENGINE OIL CAPACITY – WITH (WITHOUT) FILTER CHANGE:					
	3.0 (2.75)	3.5 (3.25)	5.0 (4.75)	4.0 (3.75)	4.0 (3.75)
TRANSMISSION:					
Manual: 4/5 speed	1.75/1.85	1.75/1.85	2.0/2.1	2.1	2.1
Automatic:	7.0	7.0	–	7.0	7.0
STEERING GEAR:					
CV JOINT CAVITIES AND BOOTS:					
BRAKE FLUID:	0.5	0.5	0.5	0.5	0.5

Cavalier 1988 - 1995 and Calibra 1990 - 1998

FUEL TANK: 61 litres (except Cavalier 2000, Cavalier 4 x 4 and Calibra - 63 Litres)

COOLING SYSTEM:	14NV	E16NZ, C16NZ 16SV	17D	TC4 EE1	E18NVR	18SV C18NZ	20SEH 20NE C20NE	20XEJ C20XE	X20 XEV
	5.6	5.8	9.1	6.9	6.7	7.2	7.2	7.2	7.2
ENGINE OIL CAPACITY - WITH(WITHOUT) FILTER CHANGE									
	3.0 (2.75)	3.5 (3.25)	4.75 (4.5)	4.5 (4.25)	4.0 (3.75)	4.0 (3.75)	4.0 (3.75)	4.5 (4.25)	4.0 (3.75)
TRANSMISSION: Manual									
	1.6	1.6		1.9	1.9	2.2	1.9	1.9	
TRANSMISSION:Automatic									
	3.3	–	–	3.25	3.25	3.25	3.25	3.25	

	14NV	E16NZ, C16NZ 16SV	17D	TC4 EE1	E18NVR	18SV C18NZ	20SEH 20NE C20NE	20XEJ C20XE	X20 XEV
STEERING GEAR:	1.0	1.0	1.0	1.0	1.0	1.0	1.0	1.0	1.0
CV JOINT CAVITIES AND BOOTS:									
BRAKE FLUID:	0.4	0.4	0.4	0.4	0.4	0.4	0.4	0.4	0.4

Vectra 1995 on	X16SZR	X16XEL	X18XE1	X20XEV	X25XE	X20DTL	X20DTH
FUEL TANK:	60	60	60	60	60	60	60
COOLANT CAPACITY:							
Manual Trans.	6.1	6.7	6.8	7.2	7.7	7.4	7.2
with air con.	6.4	7	7.1	7.2	7.7	7.4	7.2
Automatic Trans.		6.9	7.0	7.1	7.6		
with air con.		6.9	7.0	7.1	7.6		
ENGINE OIL CAPACITY - WITH (WITHOUT) OIL FILTER:	3.5 (3.25)	3.25 (3.0)	4.25 (4.0)	4.25 (4.0)	4.5 (4.25)	5.5 (5.25)	5.5 (5.25)
TRANSMISSION:							
Manual:	1.6	1.6	1.6	1.55	1.55	1.55	1.55
Automatic:	3.25	3.25	3.25	3.25	3.25	3.25	3.25
STEERING GEAR:	1.4	1.4	1.4	1.4	1.4	1.4	1.4
CV JOINT CAVITIES AND BOOTS:							
BRAKE FLUID:	0.4	0.4	0.4	0.4	0.4	0.4	0.4

Part D: Service Data

All settings in mm. unless stated otherwise.

Engine

FIRING ORDER:
4 cylinder petrol engines: 1-3-4-2
V6 petrol engines: 1-2-3-4-5-6
INJECTION ORDER:
Diesel engines: 1-3-4-2
IGNITION TIMING in degrees Before Top Dead Centre - BTDC:
Petrol engines:
Cavalier 1981-1988: 10 (Crankshaft pulley notch and oil pump housing in alignment.)
Cavalier 1988-1995: 14NV - 5 deg (Mark on pulley aligns with mark on crank case.) All other engines electronically controlled no adjustment possible.
Vectra 1995-on and Calibra: Electronically controlled no adjustment possible.
Diesel Engines: 0 degrees BTDC

SPARK PLUG TYPES AND GAPS

Engine	NGK	Gap (mm)	Champion	Gap (mm)
1.3, 1.4, 1.6, 1.8 and 2.0 (8 valve engines)				
	BPR6ES	0.7	RN7YCC	0.8
All 16 valve engines (up to 1.8)				
	BKR5EK	0.7	RC10DMC	Not adjustable
Above 1.8	BCPR6ES	0.7	RN9YCC	0.8

Diesel Engines (Glow Plugs)

	Lucas/Cav	Champion
16D and 16DA		CH68
17D and 17DR		CH68
17DT	HDS346	CH158
20DTL/H	HDS360	

IDLE SPEED: (RPM) NA = Not adjustable
Petrol Engines

1279cc 8v SOHC	13S	900 - 950
1389cc 8v SOHC	14NV	900 - 950
1598cc 8v SOHC	16S, 16SH, 16SV	900 - 950
	C16NZ	720 - 880 NA
	C16SZR, 16NZ2	770 _ 930 NA
1598cc 16v DOHC	X16XEL, Y16XE	770 - 930 NA
1796cc 8v SOHC	18E	900 - 950
	18SE	800 - 900
	C18NZ	820 - 980 NA
1796cc 16v DOHC	X18XE,	670 - 1030 NA
	X18XE1	780 NA
1998cc 8v SOHC	20SHE	720 - 780
	20NE	720 - 880 NA
	C20NE	820 - 980 NA
1998cc 16v DOHC	20XE	860 - 1020
	C20XE, 20XEJ	860 - 1020 NA
	X20XEV	670 - 1030 NA
2498cc V6 DOHC	C25XE, X25XE	600 - 800 NA

Diesel Engines

1598cc 8v	16D, 16DA	800 - 900
1686cc 8v	17DT	780 - 880
1699cc 8v	17D,17DR	720 - 920
1994cc 8v	X20DTL, X20DTH	750 - 850

EMISSIONS

	CO	CO$_2$	O$_2$	HCppm
Carburated engines	1 - 1.5	13 - 16	0.5 - 2	300
Cat + central fuel injection	0.4 non-adj.	14.5 - 16	0.1 - 0.5	100
Cat + Multi-point fuel injection	0.5 non-adj.	14.5 - 16	0.1 - 0.5	100

VALVE CLEARANCES (mm)
(Check when engine cold)

	Inlet	Exhaust
All petrol engines automatic by hydraulic valve lifters.		
Diesel engines		
16D, 16DA, 17D, X17D, X17DT	0.15	0.25
All other diesel engines	Hydraulic	Hydraulic

Other Settings

CLUTCH ADJUSTMENT:
Cavalier 81-88: 138 mm
Cavalier 88 - 95 and Calibra: 134-141 mm
Vectra 1995-on: 135–140 mm

BRAKE DISC PAD MINIMUM THICKNESS:
Cavaliers 1981–1995: 7.0 mm including back plate
Vectra 1995–on: 7.5 mm including back plate

BRAKE SHOE FRICTION LINING MINIMUM THICKNESS:
Cavalier 1981-88: 0.5 mm above rivets
All other models: 1 mm.

Part E: Repair Data 1297cc to 1796cc Petrol Engines

Dimensions in mm unless stated otherwise

Engine 'bottom end'

	13S 1297cc	14NV 1389cc	16S,16SH 1598cc	16SV,C16NZ E16NZ,X16SZ 1598cc	C16NZ2 1599cc	X16SZR 1598cc	X16XEL,Y16XE 1589cc	18E,18SE 1796cc	18SV,C18NZ, E18NVR 1796cc	X18XE 1799cc	X18XE1 1796cc
BORE:	75.0	77.6	80.0	79.0	80.0	79.0	79.0	84.8	84.8	81.6	80.5
STROKE:	73.4	73.4	79.5	81.5	79.5	81.5	81.5	79.5	79.5	86.0	88.2
PISTON SIZES:											
Grade 1:	74.93 to 74.96	77.54 to 77.56	79.935 to 79.975	78.93 to 78.96	79.955 to 79.965	78.935 to 78.945	78.955 to 78.965	84.735 to 84.775	84.76	81.555 to 81.565	80.455 to 80.465
Grade 2:	74.97 to 75.0	77.57 to 77.60	79.975 to 80.015	78.97 to 79.00	79.965 to 79.975	78.945 to 78.955	78.965 to 78.975	84.775 to 84.815	84.77 to 84.80	81.565 to 81.575	80.465 to 80.475
Grade 3:	75.01 to 75.04		80.015 to 80.055	79.01 to 79.04	79.975 to 79.985	78.955 to 78.965	78.975 to 78.985	84.815 to 84.855		81.575 to 81.585	80.475 to 80.485
Grade 4:	75.05 to 75.08		80.055 to 80.095	79.05 to 79.08	79.985 to 79.995		78.985 to 78.995	84.855 to 84.895		81.585 to 81.595	80.515 to 80.525
					79.995 to 80.005		78.995 to 79.005			81.595 to 81.605	
0.5 mm oversize:	75.45 to 75.48	78.05	80.455 to 80.495	79.45	80.445 to 80.455	79.445 to 79.455	79.445 to 79.455	85.255 to 85.295	84.25	82.045 to 82.055	80.965 to 80.975
PISTON CLEARANCES IN BORE:	0.01 to 0.03	0.02 to 0.04	0.02 to 0.04	0.02 to 0.04	0.01 to 0.03	0.01 to 0.03	0.02 to 0.04	0.01 to 0.04	0.02 to 0.04	0.02 to 0.04	
PISTON RING THICKNESS:											
TOP:	1.50	1.50	1.5	1.20	1.20	1.20	1.20	1.5	1.50	0.02 to 0.04	1.175 to 1.19
SECOND:	1.50	1.50	1.75	1.50	1.50	1.50	1.50	1.5	1.50	1.50	1.175 to 1.19
BOTTOM:	4.0	3.0	4.0	3.0	3.0	3.0	2.5	3	3.0	3.0	1.88 - 1.96
PISTON RING CLEARANCES - RING-TO-GROOVE:											
TOP:					0.02 to 0.04	0.02 to 0.04	0.02 to 0.04			0.02 to 0.04	
SECOND:					0.02 to 0.04	0.04 to 0.06	0.04 to 0.06			0.04 to 0.06	
BOTTOM:					0.02 to 0.03	0.01 to 0.03	0.01 to 0.03			0.01 to 0.03	
PISTON PROJECTION ABOVE BLOCK:				0.4	0.4	0.4				0.4	0.4
PISTON RING END GAP:											
TOP:	0.3 to 0.5	0.3 to 0.5	0.3 to 0.5	0.3 to 0.5	0.3 to 0.5	0.3 to 0.5	0.3 to 0.5	0.3 to 0.5	0.3 to 0.5	0.3 to 0.5	0.20 to 0.40
SECOND:	0.3 to 0.5	0.3 to 0.5	0.3 to 0.5	0.3 to 0.5	0.3 to 0.5	0.3 to 0.5	0.3 to 0.5	0.3 to 0.5	0.3 to 0.5	0.3 to 0.5	0.40 to 0.60
BOTTOM:	0.4 to 1.4	0.4 to 1.4	0.4 to 1.4	0.4 to 1.4	0.4 to 1.4	0.4 to 1.4	0.4 to 1.4	0.4 to 1.4	0.4 to 1.4	0.4 to 1.4	0.25 to 0.75
PISTON RING OVERSIZES: All models 0.25 and 0.5											
CRANK MAIN JOURNAL DIAMETER:											
Standard:	54.972 to 54.985	54.980 to 54.997	57.974 to 57.995	54.980 to 54.997	57.974 to 57.995	54.980 to 54.997	54.980 to 54.997	57.974 - 57.995	57.974 to 57.995	57.974 - 57.995	54.980 to 54.997
0.25 mm undersize:	54.722 to 54.735	54.730 to 54.747	57.732 to 57.745	54.730 to 54.747	57.732 to 57.745	54.730 to 54.747	54.730 to 54.747	57.732 - 57.745	57.732 to 57.745	57.732 - 57.745	54.730 to 54.747
0.50 mm undersize:	54.472 to 54.485	54.482 to 54.495	57.482 to 57.495	54.482 to 54 495	57.482 to 57.495	54.482 to 54.495	54.482 to 54 495	57.482 - 57.495	57.482 to 57.495	57.482 - 57.495	54.482 to 54 495
CRANK, BIG-END DIAMETER:											
Standard:	42.971 to 42.987	42.971 to 42.987	48.971 to 48.987	42.971 to 42.987	48.970 to 48.988	42.971 to 42.987	42.971 to 42.987	48.971 to 48.987	48.970 to 48.988	48.970 to 48.988	42.971 to 42.987
Undersizes:	0.25 and 0.50	0.25 and 0.50	0.25 and 0.50	0.25 and 0.50	0.25 and 0.50	0.25 and 0.50	0.25 and 0.50	0.25 and 0.50	0.25 and 0.50	0.25 and 0.50	0.25 and 0.50
CENTRE (THRUST) MAIN BEARING WIDTH:											
Standard:	26.00 to 26.052	26.00 to 26.052	26.00 to 26.052	26.00 to 26.052	25.950 to 26.002	26.00 to 26.052	26.00 to 26.052	26.00 to 26.052	25.950 to 26.002	25.950 to 26.002	26.00 to 26.052
0.25 mm undersize:	26.20 to 26.252	26.20 to 26.252	26.20 to 26.252	26.20 to 26.252	26.150 to 26.202	26.20 to 26.252	26.20 to 26.252	26.20 to 26.252	26.150 to 26.202	26.150 to 26.202	26.20 to 26.252
0.50 mm undersize:	26.40 to 26.452	26.40 to 26.452	26.40 to 26.452	26.40 to 26.452	26.350 to 26.402	26.40 to 26.452	26.40 to 26.452	26.40 to 26.452	26.350 to 26.402	26.350 to 26.402	26.40 to 26.452
MAIN BEARING CLEARANCE:											
Size 1:	0.025 to 0.05	0.017 to 0.047	0.015 to 0.04	0.017 to 0.047	0.015 to 0.040	0.017 to 0.047	0.013 to 0.043	0.015 to 0.04	0.015 to 0.040	0.015 to 0.040	0.013 to 0.043
Size 2:											
MAIN BEARING SHELL COLOUR CODES:											
Standard:											
Top:	Brown	Brown	Brown	Brown	Brown	Brown	Brown	Brown	Brown	Brown	Brown
Bottom:	Green	Green	Green	Green	Green	Green	Green	Green	Green	Green	Green
0.25mm undersize:											
Top:	Brown/blue	Brown/blue	Brown/blue	Brown/blue	Brown/blue	Brown/blue	Brown/blue	Brown/blue	Brown/blue	Brown/blue	Brown/blue
Bottom:	Green/blue	Green/blue	Green/blue	Green/blue	Green/blue	Green/blue	Green/blue	Green/blue	Green/blue	Green/blue	Green/blue
0.50 mm undersize:											
Top:	Brown/white	Brown/white	Brown/white	Brown/white	Brown/white	Brown/white	Brown/white	Brown/white	Brown/white	Brown/white	Brown/white
Bottom:	Green/white	Green/white	Green/white	Green/white	Green/white	Green/white	Green/white	Green/white	Green/white	Green/white	Green/white
MAIN BEARING UNDERSIZES:											
1:	0.25	0.25	0.25	0.25	0.25	0.25	0.25	0.25	0.25	0.25	0.25
2:	0.5	0.5	0.5	0.5	0.5	0.5	0.5	0.5	0.5	0.5	0.5
BIG-END BEARING SHELL COLOUR CODES: All models											
Standard: None											
0.25mm undersize: Blue											
0.50mm undersize: White											
BIG-END BEARING CLEARANCE:	0.019 to 0.071	0.019 to 0.071	0.019 to 0.063	0.019 to 0.071	0.006 to 0.031	0.019 to 0.071	0.019 to 0.071	0.019 to 0.063	0.006 to 0.031	0.015 to 0.043	0.019 to 0.071
BIG-END BEARING UNDERSIZES:											
1:	0.25	0.25	0.25	0.25	0.25	0.25	0.25	0.25	0.25	0.25	0.25
2:	0.50	0.50	0.50	0.50	0.50	0.50	0.50	0.50	0.50	0.50	0.50
CRANKSHAFT END FLOAT:	0.1 to 0.2	0.1 to 0.2	0.07 to 0.30	0.1 to 0.2	0.05 to 0.152	0.07 to 0.24	0.07 to 0.24	0.07 to 0.30	0.05 to 0.152	0.05 to 0.152	0.07 to 0.24

Engine 'top end' and valve gear

	3S 1297cc	14NV 1389cc	16S,16SH 1598cc	16SV,C16NZ E16NZ,X16SZ 1598cc	C16NZ2 1599cc	X16SZR 1598cc	X16XEL,Y16XE 1589cc	18E,18SE 1796cc	18SV,C18NZ, E18NVR 1796cc	X18XE 1799cc	X18XE1 1796cc
CAMSHAFT BEARING OUTER DIAMETER:											
1:	39.435 to 39.455	39.435 to 39.455	42.500 to 42.525	42.455 to 42.470	42.455 to 42.470	n/a	n/a	42.500 to 42.525	42.455 to 42.470	27.939 to 27 960	n/a
2:	39.685 to 39.705	39.685 to 39.705	42.750 to 42.775	42.705 to 42.720	42.705 to 42.720	n/a	n/a	42.750 to 42.775	42.705 to 42.720		n/a
3:	39.935 to 39.955	39.935 to 39.955	43.000 to 43.025	42.955 to 42.970	42.955 to 42.970	n/a	n/a	43.000 to 43.025	42.955 to 42.970		n/a
4:	40.185 to 40.205	40.185 to 40.205	43.250 to 43.275	43.205 to 43.220	43.205 to 43.220	n/a	n/a	43.250 to 43.275	43.205 to 43.220		n/a
5:	40.435 to 40.455	40.435 to 40.455	43.500 to 43.525	43.455 to 43 470	43.455 to 43 470	n/a	n/a	43.500 to 43.525	43.455 to 43 470		n/a
CAM SHAFT END FLOAT:	0.09 to 0.21	0.09 to 0.21	0.09 to 0.21	0.09 to 0.21	0.09 to 0.21	0.09 to 0.21	0.04 to 0.144	0.09 to 0.21	0.09 to 0.21	0.04 to 0.144	0.04 to 0.144
INLET VALVE HEAD SIZE:	33.0	33.0	35.0	38.0	38.0	38.0	n/a	41.0	41.8	32.0	n/a
EXHAUST VALVE HEAD SIZE:	29.0	29.0	32.0	31.0	31.0	31.0	n/a	35.0	36.5	29.0	n/a
VALVE FACE RE-CUTTING ANGLE:	45 deg	45 deg	44 deg	44 deg	44 deg	44 deg 40 min	44 deg 40 min	44 deg			44 deg 40 min

VALVE CLEARANCES: All Models - Automatic adjustment by hydraulic lifters.

Cooling System

All figures in degrees Celsius unless stated otherwise.

	3S 1297cc	14NV 1389cc	16S,16SH 1598cc	16SV,C16NZ E16NZ,X16SZ 1598cc	C16NZ2 1599cc	X16SZR 1598cc	X16XEL,Y16XE 1589cc	18E,18SE 1796cc	18SV,C18NZ, E18NVR 1796cc	X18XE 1799cc	X18XE1 1796cc
THERMOSTAT:											
Starts to open:	91.0	92.0		92.0	92.0	92.0	92.0	92.0	92.0	92.0	n/a
Fully open:	103.0	107.0		107.0	107.0	107.0	107.0	107.0	107.0	107.0	n/a
PRESSURE CAP RATING: (bar)											
Blue	1.2 to 1.35		1.2 to 1.35	1.2 to 1.35	1.2 to 1.35	1.2 to 1.35	1.2 to 1.50	1.2 to 1.35	1.2 to 1.35	1.2 to 1.35	1.2 to 1.50
Yellow	1.02 to 1.15		1.02 to 1.15	1.02 to 1.15					1.02 to 1.15		
COOLING FAN:											
Switches on:	97.0	100.0	97.0	97.0	100.0	100.0	100.0	97.0	100.0	100.0	n/a
Switches off:	93.0	95.0	93.0	93.0	95.0	95.0	95.0	93.0	95.0	95.0	n/a

Clutch

	3S 1297cc	14NV 1389cc	16S,16SH 1598cc	16SV,C16NZ E16NZ,X16SZ 1598cc	C16NZ2 1599cc (C16NZ2,18SV)	X16SZR 1598cc	X16XEL,Y16XE 1589cc	18E,18SE 1796cc	18SV,C18NZ, E18NVR 1796cc	X18XE 1799cc	X18XE1 1796cc
LINING, OUTER DIA. (mm):	190.0	190.0	203.0	200.0	216.0	200.0	200.0	216.0	204 (except C18NZ)	216.0	216.0
LINING, INNER DIA. (mm):	134.0	134.0	131.0	134.0	134.0	134.0	134.0	144.0	131 (except C18NZ)	144.0	144.0

Brakes

	3S 1297cc	14NV 1389cc	16S,16SH 1598cc	16SV,C16NZ E16NZ,X16SZ 1598cc	C16NZ2 1599cc	X16SZR 1598cc	X16XEL,Y16XE 1589cc	18E,18SE 1796cc	18SV,C18NZ, E18NVR 1796cc	X18XE 1799cc	X18XE1 1796cc
MINIMUM ALLOWED FRONT DISC THICKNESS (mm):(Up to 10/91)	10.7	10.7	10.7	10.7	10.7			18			
10/91-on		18		18	18	22	22		18	22	22
MINIMUM ALLOWED REAR DISC THICKNESS - when applicable:					8		8			8	8
MAXIMUM ALLOWED BRAKE DRUM INTERNAL DIAMETER - when applicable-(mm): (Diameter when in new in brackets)											
Saloon and Hatch:	201	201	201	201	201			201	201	231	231
Estate:	231		231	231				231			

Steering

TOE: Laden (mm)

Front:

Up to 1985: 0.5 to 2.5 toe out

1985-on: 0.0 +/- 1.0

1988-95

Toe in: + 15 min +/- 10 min

Toe out on turns: 1 deg 30 min +/- 45 min

1995-on:

Toe: + 0 deg to 20 min

Toe in on turns: 1 deg +/- 45 deg

Rear:

1988-95

DOHC - Toe in: +25 min +/- 30 min/ - 20 min

SOHC - Toe in: +10 min +/- 30 min/ -20 min

1995-on:

Toe in: Up to 01/96 +16 min +/- 10 min

01/96 to 1997: + 25 min +/- 10 min

1997-on:

Saloon: + 27 min +/- 10 min

Estate: + 26 min +/- 10 min

Part E: Repair Data
1998cc to 2498cc Petrol Engines and 1598cc to 1994cc Diesel Engines

Dimensions in mm unless stated otherwise

Engine 'bottom end'

	20NE, 20SEH C20NEF, C20NE 1998cc	C20XE,20XE, 20XEJ, C20LET 1998cc	X20XEV 1998cc	C22SEL 2198cc	C25XE, X25XE, 2498cc V6	16D, 16DA 1598cc diesel	17D, 17DR 1699cc diesel	TC4EE1 (17DT) X17DT 1686cc	X20DTL 1994cc diesel	X20DTH 1994cc diesel
BORE:	86.0	86.0	86.0	86.0	81.6	80	82.5	79.0	84.0	84.0
STROKE:	86.0	86.0	86.0	94.0	79.6	79.5	79.5	86.0	90.0	90.0
PISTON SIZES:										
Grade 1:	85.945 to 85.995	85.945 to 85.995	85.945 to 85.995	85.945 to 85.995	81.540 to 81.550	80.035 to 80.045	82.415 to 82.445	78.975 to 78.984	83.905 to 83.915	83.885 to 83.895
Grade 2:	85.995 to 85.965	85.995 to 85.965	85.995 to 85.965	85.995 to 85.965	81.550 to 81.560	80.045 to 80.055	82.445 to 82.505	78.985 to 78.994	83.915 to 83.925	83.895 to 83.905
Grade 3:	85.965 to 85.975	85.965 to 85.975	85.965 to 85.975	85.965 to 85.975	81.560 to 81.570	80.055 to 80.065	82.505 to 82.575	78.995 to 79.004	83.925 to 83.935	83.905 to 83.915
Grade 4:	85.975 to 85.985	85.975 to 85.985	85.975 to 85.985	85.975 to 85.985	81.570 to 81.580	80.065 to 80.075			83.935 to 83.945	83.915 to 83.925
	85.985 to 85.995	85.985 to 85.995	85.985 to 85.995	85.985 to 85.995	81.580 to 81.590	80.075 to 80.085			83.945 to 83.995	83.925 to 83.935
0.5 mm oversize:	86.450	86.450	86.450	86.450	82.030 to 82.040		82.935 to 82.975		84.405 to 84.415	84.385 to 84.395
PISTON CLEARANCES IN BORE:										
	0.02 to 0.04	0.02 to 0.04	0.02 to 0.04	0.02 to 0.04	0.025 to 0.045	0.015 - 0.04	0.020 to 0.040	0.016 to 0.034	0.06 to 0.08	0.06 to 0.08
PISTON RING THICKNESS:										
TOP:	1.50	1.50	1.50	1.50	1.50	2.00	2.0	2.0	2.0	2.5
SECOND:	1.50	1.50	1.50	1.50	1.50	2.00	2.0	2.0	1.75	1.75
BOTTOM:	3.0	3.0	3.0	3.0	3.0	3.0	3.0	3.0	3.0	3
PISTON RING CLEARANCES - RING-TO-GROOVE:										
TOP:		0.02 to 0.04	0.02 to 0.04	0.02 to 0.04	0.02 to 0.04	n.a	0.12 to 0.18	0.12 to 0.18	0.11 to 0.15	0.02 to 0.04
SECOND:		0.02 to 0.04	0.02 to 0.04	0.02 to 0.04	0.02 to 0.04	n.a	0.05 to 0.15	0.05 to 0.15	0.05 to 0.09	0.05 to 0.04
BOTTOM:		0.01 to 0.03	0.01 to 0.03	0.01 to 0.03	0.01 to 0.03	n.a	0.025 to 0.15	0.025 to 0.15	0.03 to 0.07	0.03 to 0.07
PISTON PROJECTION ABOVE BLOCK:										
	0.4	0.4	0.4	0.4	0.45	0.4		0.58 to 0.78	0.40 to 0.70	0.4 to 0.70
PISTON RING END GAP:										
TOP:	0.3 to 0.5	0.3 to 0.5	0.3 to 0.5	0.3 to 0.5	0.3 to 0.5	0.20 to 0.40	0.20 to 0.40	0.25 to 0.80	0.30 to 0.50	0.30 to 0.50
SECOND:	0.3 to 0.5	0.3 to 0.5	0.3 to 0.5	0.3 to 0.5	0.3 to 0.5	0.20 to 0.40	0.20 to 0.40	0.20 to 0.80	0.30 to 0.50	0.30 to 0.50
BOTTOM:	0.4 to 1.4	0.4 to 1.4	0.4 to 1.4	0.4 to 1.4	0.4 to 1.4	0.25 to 0.50	0.25 to 0.50	0.20 to 0.80	0.40 to 1.40	0.40 to 1.40
PISTON RING OVERSIZES: All models 0.25 and 0.5										
CRANK MAIN JOURNAL DIAMETER:										
Standard	57.974 - 57.995	57.974 - 57.995	57.974 - 57.995	57.974 - 57.995	67.980 to 67.996	57.982 - 57.995	57.974 - 57.995	55.992 to 56.000	67.966 to 67.982	67.966 to 67.982
0.25 mm undersize:	57.732 - 57.745	57.732 - 57.745	57.732 - 57.745	57.732 - 57.745	67.730 to 67.746	57.732 - 57.745	55.984 to 55.992	67.716 to 67.732
67.716 to 67.732										
0.50 mm undersize:	57.482 - 57.495	57.482 - 57.495	57.482 - 57.495	57.482 - 57.495	67.480 to 67.496	57.482 - 57.495	55.976 to 55.984	67.466 to 67.482
67.466 to 67.482										
CRANK, BIG-END DIAMETER:										
Standard:	48.970 to 48.988	48.970 to 48.988	48.970 to 48.988	48.970 to 48.988	48.971 to 48.990	48.971 to 48.990	48.970 to 48.988	51.928 to 51.938	48.971 to 48 990	48.971 to 48 990
Undersizes:	0.25 and 0.50	0.25 and 0.50	0.25 and 0.50	0.25 and 0.50	0.25 and 0.50	0.25 and 0.50	0.25 and 0.50	0.25 and 0.50	0.25 and 0.50	0.25 and 0.50
CENTRE (THRUST) MAIN BEARING WIDTH:										
Standard:	26.00 to 26.052	26.00 to 26.052	26.00 to 26.052	26.00 to 26.052	24.500 to 24.552		26.00 to 26.052	26.995 to 27.000	25.950 to 26.002	25.950 to 26.002
0.25:mm undersize:	26.20 to 26.252	26.20 to 26.252	26.20 to 26.252	26.20 to 26.252	24.700 to 24.752		26.20 to 26.252		26.150 to 26.202	26.150 to 26.202
0.50 mm undersize:	26.40 to 26.452	26.40 to 26.452	26.40 to 26.452	26.40 to 26.452	24.900 to 24.952		26.40 to 26.452		26.350 to 26.402	26.350 to 26.402
MAIN BEARING CLEARANCE:										
Size 1:	0.015 to 0.040	0.015 to 0.040	0.015 to 0.040	0.015 to 0.040	0.014 to 0.043	0.015 to 0.040	0.015 to 0.040	0.03 to 0.06	0.016 to 0.069	0.016 to 0.069
Size 2:										
MAIN BEARING SHELL COLOUR CODES:										
Standard:										
Top:	Brown	Brown	Brown	Brown	Brown	Brown	Brown	Brown	Brown	Brown
Bottom:	Green	Green	Green	Green	Green	Green	Green	Green	Green	Green
0.25 mm undersize:										
Top:	Brown/blue	Brown/blue	Brown/blue	Brown/blue	Brown/blue	Brown/blue	Brown/blue	Brown/blue	Brown/blue	Brown/blue
Bottom:	Green/blue	Green/blue	Green/blue	Green/blue	Green/blue	Green/blue	Green/blue	Green/blue	Green/blue	Green/blue
0.50 mm undersize:										
Top:	Brown/white	Brown/white	Brown/white	Brown/white	Brown/white	Brown/white	Brown/white	Brown/white	Brown\purple	Brown\purple
Bottom:	Green/white	Green/white	Green/white	Green/white	Green/white	Green/white	Green/white	Green/white	Green\Purple	Green\Purple
MAIN BEARING UNDERSIZES:										
1:	0.25	0.25	0.25	0.25	0.25	0.25	0.25	0.25	0.25	0.25
2:	0.50	0.50	0.50	0.50	0.50	0.50	0.50	0.50	0.50	0.50
BIG-END BEARING SHELL COLOUR CODES: All models										
Standard: None										
0.25 mm undersize: Blue										
0.50 mm undersize: White										
BIG-END BEARING CLEARANCE:										
	0.006 to 0.031	0.006 to 0.031	0.006 to 0.031	0.006 to 0.031	0.010 to 0.061	0.019 to 0.063	0.019 to 0.063	0.02 to 0.06	0.01 to 0.02	0.01 to 0.02
BIG-END BEARING UNDERSIZES:										
1:	0.25	0.25	0.25	0.25	0.25	0.25	0.25	0.25	0.25	0.25
2:	0.50	0.50	0.50	0.50	0.50	0.50	0.50	0.50	0.50	0.50
CRANKSHAFT END FLOAT:										
	0.050 to 0.152	0.050 to 0.152	0.050 to 0.152	0.050 to 0.152	0.010 to 0.061	0.070 to 0.242	0.070 to 0.242	0.03 to 0.06	0.01 to 0.02	0.01 to 0.02

Engine 'top end' and valve gear

	20NE, 20SEH C20NEF, C20NE 1998cc	C20XE,20XE, 20XEJ, C20LET 1998cc	X20XEV 1998cc	C22SEL 2198cc	C25XE, X25XE, 2498cc V6	16D, 16DA 1598cc diesel	17D, 17DR 1699cc diesel	TC4EE1 (17DT) X17DT 1686cc	X20DTL 1994cc diesel	X20DTH 1994cc diesel
CAMSHAFT BEARING OUTER DIAMETER:										
1:	42.455 to 42.470	27.939 to 27 960	27.939 to 27 960	27.939 to 27 960	27.939 to 27 960	42.500 to 42.525				
2:	42.705 to 42.720					42.750 to 42.775				
3:	42.955 to 42.970					43.000 to 42.025				
4:	43.205 to 43.220					43.250 to 43.275				
5:	43.455 to 43 470					43.500 to 43.525				
CAM SHAFT END FLOAT:	0.09 to 0.21	0.04 to 0.144	0.04 to 0.144	0.04 to 0.144	0.04 to 0.144	0.04 to 0.144	0.09 to 0.21	0.05 to 0.20	0.04 to 0.144	0.04 to 0.144
INLET VALVE HEAD SIZE:	41.8	33.0	32.0	32.0	32.0	36.0	36.0	34.6	29.0	29.0
EXHAUST VALVE HEAD SIZE:	36.5	29.0	29.0	29.0	29.0	32.0	32.0	30.6	26.0	26.0
VALVE FACE RE-CUTTING ANGLE:	44 deg	44 deg 40 min	44 deg 40 min	44 deg 40 min	44 deg 40 min	44 deg	44 deg	45 deg		

VALVE CLEARANCES: All Models - Automatic adjustment by hydraulic lifters.

Cooling System

All figures in degees Celsius unless stated otherwise.

THERMOSTAT:										
Starts to open:	92.0	92.0	92.0	92.0	92.0	91.0	92.0	86 to 90	92.0	92.0
Fully open:	107.0	107.0	107.0	107.0	107.0	106.0	107.0	100.0	107.0	107.0
PRESSURE CAP RATING: (bar)										
Blue	1.2 to 1.35	1.2 to 1.35	1.2 to 1.35	1.2 to 1.35	1.2 to 1.35	1.2 to 1.35	1.4 to 1.5	1.4 to 1.5	1.4 to 1.5	1.4 to 1.5
Yellow						1.02 to 1.15				
COOLING FAN:										
Switches on:	100.0	100.0	100.0	100.0	100.0	100.0	100.0	100.0	100.0	100.0
Switches off:	95.0	95.0	95.0	95.0	95.0	95.0	95.0	95.0	95.0	95.0

Clutch

LINING, OUTER DIA. (mm):	216.0 (+ C18NZ)	228.0	216.0	216.0	228.0	203.0	204(17D)200(17DR)	200.0	228.0	228.0
LINING, INNER DIA. (mm):	144.0(+ C18NZ)	154.0	144.0	144.0	155.0	131.0	131(17D)134(17DR)	134.0	155.0	155.0

Brakes

MINIMUM ALLOWED FRONT DISC THICKNESS (mm): (Up to 10/91)										
	22					10.7				
10/91-on	22	22	23	23	22	18	22	22	22	22
MINIMUM ALLOWED REAR DISC THICKNESS - when applicable:										
	8	8	8	8	8	7	8	8	8	8

MAXIMUM ALLOWED BRAKE DRUM INTERNAL DIAMETER - when applicable (mm): (Diameter when in new in brackets)

Saloon and Hatch:	201	231	231	231	231	201	231	231	231	231
Estate:	231					231				

Steering

TOE: Laden (mm)

Front:

Up to 1985: 0.5 to 2.5 toe out

1985-on: 0.0 +/- 1.0

1988-95

Toe in: + 15 min +/- 10 min

Toe out on turns: 1 deg 30 min +/- 45 min

1995-on:

Toe: + 0 deg to 20 min

Toe in on turns: 1 deg +/- 45 deg

Rear:

1988-95

DOHC - Toe in: +25 min +/- 30 min/ - 20 min

SOHC - Toe in: +10 min +/- 30 min/ -20 min

1995-on:

Toe in: Up to 01/96 +16 min +/- 10 min

01/96 to 1997: + 25 min +/- 10 min

1997-on:

Saloon: + 27 min +/- 10 min

Estate: + 26 min +/- 10 min

Part F: Torque Wrench Settings

Key to Engine Types and Sizes

A=13S,
B=14NV, 16SV, C16NZ, X16SZ
C=C16NZ2, 18SV, C18NZ, C20NE
D=X18XE, 20XEJ, C20XE, X20XEV
C20LET, C20SEL,C22SEL

E= X16XEL
F=16S, 18E, 20NE, 20SEH
G=C25XE, X25XE
H=16D, 16DA
I=17DT, X17DT

J=X20DTL, X20DTH
K=17D, 17DR, 17DTL

ENGINE	A	B	C	D	E	F	G	H	I	J	K	Torque (Nm)
Main bearing cap, bolt	•					•						65
Main bearing cap, bolt		•	•	•	•			•			•	50 + 45 deg + 15 deg
Main bearing cap, bolt						•						50 + 60 deg + 15 deg
Main bearing cap, bolt									•			88
Main bearing cap, bolt										•		90 + 60 deg + 15 deg
Cylinder head to crankcase fixing, bolt, see chapter 6 part C job 6												
Engine support to crankcase fixing, bolt	•					•						50
Engine bracket to cylinder block		•	•	•	•		•		•	•		60
Engine bracket to transmission block		•	•	•	•				•	•		60
Mounting to bracket/subframe							•		•	•		45
Camshaft cap fixing bolt and nut				•								10
Camshaft cap fixing bolt and nut									•			19
Camshaft cap fixing bolt and nut				•						•		20
Camshaft cap fixing bolt and nut				•			•					8 (X20XEV)
Camshaft cap fixing bolt and nut					•							
Camshaft cover to cylinder head fixing	•	•	•	•	•	•	•	•	•	•	•	8
Flywheel to crankshaft fixing, bolt	•											60
Flywheel to crankshaft fixing, bolt						•						50 + 25 to 35 deg
Flywheel to crankshaft fixing, bolt		•			•							35 + 30 deg + 15 deg
Flywheel to crankshaft fixing, bolt										•		45 + 30 deg + 15 deg
Flywheel to crankshaft fixing, bolt			•	•			•					65 + 30 deg + 15 deg
Flywheel to crankshaft fixing, bolt									•			30 + 45 to 60 deg
Flywheel to crankshaft fixing, bolt								•			•	50 + 30 deg + 5 deg
Crankshaft pulley bolt	•											55
Crankshaft pulley bolt						•			•		•	20
Crankshaft pulley bolt				•								50 + 60 deg + 15deg
Crankshaft pulley bolt		•	•									55 + 45deg + 15 deg
Crankshaft pulley bolt					•							95 + 30 deg + 15 deg
Crankshaft pulley bolt										•		150 + 45 deg + 15 deg
Crankshaft pulley bolt							•					250 + 45 deg + 15 deg
Crankshaft sprocket bolt											•	145 + 30 deg + 10 deg
Crankshaft sprocket bolt			•	•	•		•					130 + 40 to 50 deg
Crankshaft sprocket bolt									•			196
Starter motor bolts	•	•			•							25
Starter motor bolts							•		•			38
Starter motor bolts(M10)			•	•		•				•	•	45
Starter motor bolts(M12)			•	•							•	60
Toothed belt drive gear to crankshaft				•			•					250 + 40 to 50 deg
Big end bearing cap fixing, bolt	•											28
Big end bearing cap fixing, bolt		•	•		•							25 + 30 deg
Big end bearing cap fixing, bolt							•					35 + 30 deg
Big end bearing cap fixing, bolt				•		•	•			•	•	35 + 45 deg + 15 deg
Big end bearing cap fixing, bolt												35 + 45 to 60 deg
Big end bearing cap fixing, bolt									•			25 + 100 deg + 15 deg
Fuel Pump to camshaft housing	•	•	•	•		•						18
Inlet manifold to cylinder head fixing, nut							•					20
Inlet manifold to cylinder head fixing, nut		•	•	•	•			•				22
Exhaust manifold to cylinder head							•					20
(M8)		•	•	•	•						•	22 (17DTL = 38)
Exhaust manifold to pipe						•					•	25
Driveplate to crankshaft	•				•	•						60
Driveplate to crankshaft							•					65 + 30 deg + 15 deg

ENGINE	A	B	C	D	E	F	G	H	I	J	K	Torque (Nm)
Camshaft sprocket to camshaft,bolt	●	●	●			●						45
Camshaft sprocket to camshaft,bolt				●	●		●					50 + 60 deg + 15 deg
Camshaft sprocket to camshaft,bolt								●			●	75 + 60 deg + 5 deg
Camshaft sprocket to camshaft,bolt										●		90 + 60 deg + 30 deg
Camshaft sensor to cylinder head		●	●	●	●							6
Camshaft sensor to camshaft bearing cap							●					8
Timing belt tensioner indicator bolt		●		●	●				●			20
Timing belt idler roller bolt (DOHC)				●	●							25
Timing belt idler roller bolt									●			76
Timing chain tensioner blade pivot bolt										●		20
Timing chain tensioner cap										●		60
Injection pump fixing, nut and bolt (M8)									●			69
Injection pump sprocket bolt										●		20
Injection pump front mounting								●	●	●	●	23
Injection pump drive gear									●			25
M8 bolts									●			25
M10 bolts									●			40
Injector holder to cylinder head									●			50
Injector holder to cylinder head								●			●	70
Injector holder assembly									●			45
Injector holder assembly								●				59 to 68
Injector holder assembly											●	80
Complete injector									●	●		50
Heater plugs										●		10
Heater plugs									●		●	20
Heater plugs								●				15
Vacuum pump to camshaft housing								●	●	●	●	28
Turbocharger to exhaust manifold									●			30
Turbocharger to exhaust pipe									●			65
Turbocharge support to block									●			51
Water pump to crankcase fixing, bolt	●	●	●		●							8
Water pump to crankcase fixing, bolt									●	●		20
Water pump to crankcase fixing, bolt				●		●	●	●			●	25
Alternator bracket to block	●				●		●				●	40
Alternator mounting		●	●	●								25
M8 bolts									●			24
M10									●			48
Alternator mounting				●	●					●	●	35
Oil sump to crankcase fixing, bolt		●	●	●								15
Oil sump to crankcase fixing, bolt					●							10
Oil sump to crankcase fixing, bolt											●	8
Oil sump to covers fixing, nut (M6)		●			●		●	●				8
Oil sump to covers fixing, bolt (M6)			●	●								15
Oil sump to covers fixing, bolt (M8)							●			●		20
Oil pump to crankcase fixing, bolt	●	●	●	●	●	●		●			●	6
Oil pump to crankcase fixing, bolt (M6)							●					8
Oil pump to crankcase fixing, bolt (M8)							●					12
Oil pressure switch									●	●		20
Oil pressure switch	●	●			●	●		●			●	30
Oil pressure switch			●	●				●				40
Oil filter to engine	●	●	●	●	●	●	●	●			●	15
Oil drain plug	●				●	●		●			●	45
Oil drain plug		●	●	●								55
Oil drain plug										●		18
Oil drain plug									●			20 to 25
Sump bolts	●					●		●				5
Oil temp. sender unit		●	●	●	●							20
Nut fixing exhaust manifold to cylinder head		●	●	●	●							25
Inlet manifold fixing adjustment screw (M6)		●		●					●	●	●	22
Inlet manifold (upper) nuts and bolts				●								8
Inlet manifold (lower) nuts and bolts				●								20
Turbo to manifold & cyl. head fixing, nut (M8)									●	●	●	29
EGR valve to inlet or exhaust manifold, nut (M8)			●	●	●				●	●	●	20

ENGINE

	A	B	C	D	E	F	G	H	I	J	K	Torque (Nm)
Pipe to exhaust or inlet manifold (M6)	●					●		●				22
Spark plugs	●					●		●				20
			●	●	●	●	●	●				25
Thermostat housing to cylinder block	●	●										10
Thermostat housing to cylinder block			●	●		●		●			●	15
Thermostat housing to cylinder block									●			24
Thermostat housing to cylinder block					●		●			●		20
Thermostat housing to cylinder head										●		8
Coolant temp. sender unit in manifold	●					●		●		●		10
Coolant temperature sender in thermostat housing	●	●	●	●		●		●	●			8
Coolant temperature sender in thermostat housing											●	11
Coolant temperature sender in thermostat housing												20

Engine Exhaust

	A	B	C	D	E	F	G	H	I	J	K	Torque (Nm)
Exhaust downpipe-to-manifold bolts		●	●	●	●	●		●	●	●		25
Lambda sensor		●	●	●			●					30
Knock sensor to crankcase		●	●	●	●		●		●	●		20
Knock sensor to crankcase		●										13 (X16XZ only)
Rear exhaust to catalytic converter, nut (M8)									●		●	25
Front exhaust to catalytic converter, nut (M8)									●		●	25

Transmission

	A	B	C	D	E	F	G	H	I	J	K	Torque (Nm)
Selector rod to gear lever fixing, self-lock nut				●	●		●		●	●	●	12 + 180 deg
Engine to body shell bolt	●					●		●			●	40
Engine to crankcase mounting bracket	●					●		●				50
Engine/transmission left-hand mounting bolts				●	●				●	●	●	60
Flexible mounting to side member											●	65
Rear mounting bracket to cylinder block											●	60
Rear flexible mounting to bracket											●	35
Rear flexible mounting to crossmember											●	40
Lever to gear selector and engine shaft, bolt		●	●	●								15
Gear control to support fixing, nut (M6)		●	●	●								6
F18 transmission unit					●				●	●		40
Alloy cover plate		●	●	●								18
Steel cover plate		●	●	●								30
Endplate to transmission												
(M7)		●	●	●								15
(M8)		●	●	●								20
Oil drain plug (Automatics)				●	●		●					35
Engine to body frame.	●					●		●				40
Flywheel housing to engine bolts	●					●		●				75
Left-hand transmission mounting to body		●	●	●								75
Left-hand transmission mounting to transmission		●	●	●								65
Rear transmission mounting to front sub frame		●	●	●								40
Transmission to engine unit, bolts												
M8					●				●	●		20
M10					●				●	●		40
M12					●		●		●	●		60
Transmission to engine, bolt	●					●		●				30
Transmission to engine, bolt (M10)		●	●	●								45
Transmission to engine (M12)		●	●	●								60

Part F: Torque Wrench Settings

Key to Engine Types and Sizes

A = Cavalier 1981 to 1988,
B = Cavalier 1989 to 1995 (SOHC)
C = Cavalier 1989 to 1995 (DOHC),
D = Vectra 1995 - on,
E = Calibra 1990 to 1998

ENGINE TYPES

	A	B	C	D	E	
Front Suspension						
Front cross member to bodyshell front fixing, bolt with wide flange		●	●			100 + 60 to 75 deg
Front cross member to bodyshell rear fixing, bolt with normal flange for nut		●	●			100 + 75 to 90 deg
Front axle body to floor crossmember		●	●			170
Front axle body to radiator crossmember		●	●			115
Bolts front sub frame (except engine mount nuts)				●		100 + 45 deg + 15 deg
Engine mounting, nuts				●		45
Support to damper, nut	●					20
Anti-rollbar links				●		65
Anti-rollbar to sub frame				●	●	20
Suspension control arm pivot, bolts	●					110
Smaller bolts	●					110
Subframe front to lower arm					●	100 + 60 deg + 15 deg
Larger centre bolts	●					130
Suspension control arm support (longer type) bolts	●					110
Upper shock absorber to mounting fixing, nut		●	●	●	●	55
Suspension lower arm to sub frame				●		90 + 75 deg + 15deg
Upper shock absorber mounting to bodyshell fixing, bolt with wide flange						
Suspension strut ring	●	●	●		●	200
Strut to hub carrier				●		50 + 90 + 45 deg + 15 deg
Lower arm to ball joint				●		60
Tie rod ball joint to steering arm	●	●	●	●	●	60
Ball joint to stub axle carrier	●				●	70
Ball joint to stub axle carrier				●		100
Lower ball joint to lower arm				●		35
Suspension strut top mounting, nuts	●					20
Strut piston rod, nut	●					55
Shock absorber piston rod to support bearing				●		70
Anti-roll bar U-clamps	●					20
Driveshaft to hub carrier castellated, nut	●					100 then slacken + 20 Nm + 90 deg
Driveshaft to hub carrier castellated, nut		●	●	●		130 then slacken then 20 Nm + 80 deg.
Subframe to underbody, front bolts					●	115
Subframe to underbody, centre bolts					●	170
Subframe to underbody rear bolts					●	100 + 75 deg + 15 deg
Wheel studs	●					90
Wheel studs		●	●	●	●	110
Rear Suspension						
Shock absorber upper mounting		●	●	●	●	20
Shock absorber lower mounting		●				70
Shock absorber lower mounting			●		●	110
Lower damper, bolt - flange (Estate)	●					55
Upper damper, bolt - flange (Estate)	●					20
Lower damper, bolt - flange (Saloon - Hatch)	●					70
Crossmember (front) to underbody			●		●	125
Crossmember mounting to underbody			●		●	65
Rear hub unit		●				50 + 30 to 45 deg
Rear axle to underbody pivot, bolts	●					100
Rear anti-roll bar, clamps.	●					18
Rear anti-roll bar		●				30
Rear anti-roll bar			●		●	22
Rear anti-roll bar				●		55
Rear hub fixing stub axle		●				130 then slacken then 20 Nm + 80 deg
Rear hub fixing stub axle			●		●	300
Rear hub bracket to trailing arm				●		50 + 30 deg + 15 deg

	A	B	C	D	E	ENGINE TYPES
Semi-trailing arm to crossmember					●	100
Suspension strut to trailing arm				●		150 + 30 deg + 15 deg
Suspension strut upper mounting to body				●		55
Suspension upper and lower arms to subframe				●		90 + 60 deg + 15 deg
Trailing arm front mounting bracket to underbody				●		90 + 30 deg + 15 deg
Trailing arm to underbody		●	●			105
Wheel to hub fixing		●	●			25
Road wheels	●					90
Road wheels		●	●			110
Steering						
Adjustment screw locknut	●					60
Column spindle coupling clamp	●					22
Column to dash panel bolts	●	●	●		●	22
Steering box to cross member, bolt	●					15
Steering column to cross member				●	●	22
Intermediate shaft clamp, steering column, bolt				●		22
Steering wheel to steering control shaft, nut	●	●	●	●	●	25
Air bag module to steering wheel (M6)		●	●		●	10
Air bag module to steering wheel				●		8
Braking System						
Handbrake lever to floor	●	●	●		●	20
Handbrake lever bracket to fix plate (M8)				●		10
Front caliper bracket to hub carrier	●	●	●	●	●	95
Front caliper guide		●	●	●	●	30
Front brake caliper to steering knuckle, bolt		●	●			80
Front caliper mounting					●	30
Master cylinder mounting		●	●	●	●	22
Solid disc models		●	●			95
Vented disc models		●	●			30
Cylinder to rear brake back plate	●		●	●		9
Rear brake backplate/stub axle spring		●	●		●	50 + 30 deg + 15 deg
Rear brake caliper to plate, bolt (rear disc brakes only)		●	●	●	●	80
Roadwheel bolts	●					90
Roadwheel bolts		●	●	●	●	110
Pedals						
Brake pedal support bracket to bulkhead	●					20
Brake servo to bulkhead		●	●			22
Bodywork						
Seats to floor fixing, bolt (M8)		●	●	●	●	20
Belt adjuster to upper section of centre pillar		●	●	●	●	35
Seat belt reel fixing, bolts		●	●	●		35
Air Bag System						
Control unit fixing, bolt		●	●	●	●	10
Passenger air bag fixing		●	●	●	●	8
Power Steering (where fitted)						
Power steering box to chassis, bolt (M10 x 1.25)	●	●	●			40
Power steering bracket, nut	●	●	●			15

CHAPTER 2: SAFETY FIRST!

Please read this chapter before carrying out any work on your car.

You must always ensure that safety is the first consideration in any job you carry out. A slight lack of concentration, or a rush to finish the job quickly can easily result in an accident, as can failure to follow the precautions outlined in this manual.

Be sure to consult the suppliers of any materials and equipment you may use, and to obtain and read carefully any operating and health and safety instructions that may be available on packaging or from manufacturers and suppliers.

GENERAL

RAISING THE VEHICLE SAFELY

• ALWAYS ensure that the vehicle is properly supported when raised off the ground. Don't work on, around, or underneath a raised vehicle unless axle stands or hoist lifting pads are positioned under secure, load bearing underbody areas. If the vehicle is driven onto ramps, the wheels remaining on the ground must be securely chocked to prevent movement.

• NEVER work on a vehicle supported on a jack. Jacks are made for lifting the vehicle only, not for holding it off the ground while it is being worked on.

❒ 1: ALWAYS ensure that the safe working load rating of any jack, hoist or lifting gear used is sufficient for the job, and that lifting gear is used only as recommended by the manufacturer.

• NEVER attempt to loosen or tighten nuts that require a lot of force

to turn (e.g. a tight oil drain plug) with the vehicle raised, unless it is safely supported. Take care not to pull the vehicle off its supports when applying force to any part of the vehicle. Wherever possible, initially slacken tight fastenings before raising the vehicle off the ground.

• ALWAYS wear eye protection when working under the vehicle and when using power tools.

• Follow the instructions in *Chapter 4* entitled *Using a Trolley Jack*.

WORKING ON THE VEHICLE

• ALWAYS seek specialist advice from a qualified technician unless you are justifiably confident about your ability to carry out each job. Vehicle safety affects you, your passengers and other road users.

❒ 2: DON'T lean over, or work on, a running engine unless it is strictly necessary, and keep long hair and loose clothing well out of the way of moving mechanical parts.

• Note that it is theoretically possible for fluorescent striplighting to make an engine fan appear to be stationary - double check whether it is spinning or not! This is the sort of error that happens when you're really tired and not thinking straight. So...

• ...DON'T work on a vehicle when you're over tired.

• ALWAYS work in a well ventilated area and don't inhale dust - it may contain asbestos or other harmful substances.

• NEVER run an engine indoors, in a confined space or over a pit.

• REMOVE your wrist watch, rings and all other jewellery before doing any work on the vehicle - and especially when working on the electrical system.

• DON'T remove the radiator or expansion tank filler cap or other openings when the cooling system is hot, or you may get scalded by escaping coolant or steam. Let the system cool down first and even then, if the engine is not completely cold, cover the cap with a cloth and gradually release the pressure.

• NEVER drain oil, coolant or automatic transmission fluid when the engine is hot. Allow time for it to cool sufficiently to avoid scalding you.

• ALWAYS keep antifreeze, brake and clutch fluid away from vehicle paintwork. Wash off any spills immediately.

• TAKE CARE to avoid touching any engine or exhaust system component unless it is cool enough not to burn you.

RUNNING THE VEHICLE

• NEVER start the engine unless the gearbox is in neutral (or 'Park' in the case of automatic transmission) and the parking brake is fully applied.

• NEVER run a vehicle fitted with a catalytic converter without the exhaust system heat shields in place.

• TAKE CARE when parking vehicles fitted with catalytic converters. The 'cat' reaches extremely high temperatures and any combustible materials under the car, such as long dry grass, could be ignited.

PERSONAL SAFETY

• NEVER siphon fuel, antifreeze, brake fluid or other potentially harmful liquids by mouth, or allow contact with your skin. Use a suitable hand pump and wear gloves.

• BEFORE undertaking dirty jobs, use barrier cream on your hands as a protection against infection. Preferably, wear suitable gloves.

• WEAR IMPERVIOUS GLOVES - disposable types are ideal - when there is a risk of used engine oil or any other harmful substance coming into contact with your skin.

❒ 3: Wurth produce a huge range of workshop products, including the safety-related items shown here.

• WIPE UP any spilt oil, grease or water off the floor immediately.

• MAKE SURE that spanners/wrenches and all other tools are the right size for the job and are not likely to slip. Never try to 'double-up' spanners to gain more leverage.

• SEEK HELP if you need to lift something heavy which may be beyond your capability. Don't forget that when lifting a heavy weight, you should keep your back vertical and straight and bend your knees to avoid injuring your back.

• NEVER take risky short-cuts or rush to finish a job. Plan ahead and allow plenty of time.

• BE METICULOUS and keep the work area tidy - you'll avoid frustration, work better and lose less.

• KEEP children and animals well away from the work area and from unattended vehicles.

• ALWAYS tell someone what you're doing and have them regularly check that all is well, especially when working alone on, or under, the vehicle.

HAZARDS

FIRE!

• Petrol (gasoline) is a dangerous and highly flammable liquid requiring special precautions. When working on the fuel system, disconnect the vehicle battery earth (ground) terminal whenever possible and always work outside, or in a very well ventilated area. Any form of spark, such as that caused by an electrical fault, by two metal surfaces striking against each other, by a central heating boiler in the garage 'firing up', or even by static electricity built up in your clothing can, in a confined space, ignite petrol vapour causing an explosion. Take great care not to spill petrol on to the engine or exhaust system, never allow any naked flame anywhere near the work area and don't smoke.

❒ 4: There are several types of fire extinguisher. Take advice from your accredited supplier to make sure that you have the right type for workshop use. Note that water fire extinguishers are not suitable for workshop or automotive use.

PRESSURE

• DON'T disconnect any pipes on a fuel injected engine or on an ABS braking system without releasing residual pressure. The fuel or brake fluid may be under very high pressure - sufficient to cause serious injury. Remember that many systems retain high pressure for sometime after last use. If necessary seek specialist advice.

FUMES

• Vapour which is given off by petrol (gasoline) and many solvents, thinners, and adhesives is potentially very harmful and under certain conditions can lead to unconsciousness or even death, if inhaled. The risks are increased if such fluids are used in a confined space so always ensure adequate ventilation. Always read the supplier's instructions and follow them with care.

• Never drain petrol (gasoline) or use solvents, thinners, adhesives or other toxic substances in an inspection pit. It is also dangerous to park a vehicle for any length of time over an inspection pit. The fumes from even a slight fuel leak can cause an explosion when the engine is started.

MAINS ELECTRICITY

❒ **5:** Avoid the use of mains electricity when working on the vehicle, whenever possible. Use rechargeable tools and a DC inspection lamp, powered from a remote 12V battery

- both are much safer. However, if you do use mains-powered equipment, ensure that the appliance is connected correctly to its plug, that where necessary it is properly earthed (grounded), and that the fuse is of the correct rating for the appliance. Do not use any mains powered equipment in damp conditions or in the vicinity of fuel, fuel vapour or the vehicle battery. Always use an RCD (Residual Current Device) circuit breaker with mains electricity. Then, if there is a short, the RCD circuit breaker minimises the risk of electrocution by instantly cutting the power supply.

IGNITION SYSTEM

• Never work on the ignition system with the ignition switched on, or with the engine being turned over on the starter, or with the engine running.

❒ **6:** Touching certain parts of the ignition system, such as the HT leads, distributor cap, ignition coil etc., can result in a severe electric shock or physical injury as a hand is pulled sharply away. Voltages produced by electronic ignition systems are sometimes very high indeed and could prove fatal, particularly to people with cardiac pacemaker implants. Consult your vehicle's handbook or main dealer if in any doubt.

COOLING FAN

• On many vehicles, the electric cooling fan can switch itself on even with the ignition turned off. This is especially likely after driving the vehicle immediately before turning off, after which heat rises to the top of the engine and turns the fan on, suddenly and without warning. If you intend working in the engine bay, it's best to do so when the engine is cold, to disconnect the battery, or keep away from the fan, if neither of these are possible.

BATTERY

• Never cause a spark, smoke, or allow a naked light near the vehicle's battery, even in a well ventilated area. Highly explosive hydrogen gas is given off as part of the charging process.

• Battery terminals should be shielded, since a spark can be caused by any metal object touching the battery's terminals or connecting straps.

• IMPORTANT NOTE: Before disconnecting the battery earth (ground) terminal read the relevant information in **Chapter 10** regarding saving computer and radio settings. When using a battery charger, switch off the power supply before the battery charger leads are connected or disconnected. If the battery is not of the 'sealed-for-life' type, loosen the filler plugs or remove the cover before charging. For best results the battery should be given a low rate trickle charge. Do not charge at an excessive rate or the battery may burst. Always wear gloves and goggles when carrying or when topping up the battery. Acid electrolyte is extremely corrosive and must not be allowed to contact the eyes, skin or clothes. If it does, wash with copious amounts of water. Seek medical advice if necessary

BRAKES AND ASBESTOS

• Obviously, a vehicle's brakes are among its most important safety related items. ONLY work on your vehicle's braking system if you are trained and competent to do so. If you have not been trained in this work, but wish to carry out the jobs described in this manual, we strongly recommend that you have a garage or qualified mechanic check your work before using the vehicle.

• Whenever you work on the braking system: i) wear an efficient particle mask; ii) wipe off all brake dust from the brakes after spraying on a proprietary brand of brake cleaner (never blow dust off with compressed air); iii) dispose of brake dust and discarded shoes or pads in a sealed plastic bag; iv) wash your hands thoroughly after you have finished working on the brakes and certainly before you eat or smoke; v) replace shoes and pads only with asbestos-free shoes or pads. Note that asbestos brake dust can cause cancer if inhaled; vi) always replace brake pads and/or shoes in complete 'axle' sets - never replace them on one wheel only.

BRAKE FLUID

• Brake fluid absorbs moisture rapidly from the air and this can cause brake failure. Never use a previously opened container of brake fluid.

ENGINE OIL

• Always wear disposable plastic or rubber gloves when draining the oil from your engine. i) Note that the drain plug and the oil are often hotter than you expect. ii) There are very real health hazards associated with used engine oil. Use barrier cream on your hands and try not to get oil on them. Always wear impermeable gloves and wash hands with hand cleaner soon after carrying out the work. Keep oil out of the reach of children; iii) NEVER, EVER dispose of old engine oil into the ground or down a drain.

PLASTIC MATERIALS

• Be aware of dangers in the form of poisonous fumes, skin irritants, and the risk of fire and explosion. Do not allow resin or 2-pack filler or adhesive hardener to come into contact with skin or eyes. Read carefully the safety notes supplied on the can, tube or packaging.

FLUOROELASTOMERS

• Fluoroelastomers are commonly used for oil seals, wiring and cabling, bearing surfaces, gaskets, diaphragms, hoses and 'O' rings. If they are subjected to temperatures greater than 315 degrees Celcius, they will decompose and can be potentially hazardous. Some decomposition may occur when a car has been in a fire or has been dismantled with the assistance of a cutting torch.

• According to the Health and Safety Executive, "Skin contact with this liquid or decomposition residues can cause painful and penetrating burns. Permanent irreversible skin and tissue damage can occur". Damage can also be caused to eyes or by the inhalation of fumes created as fluoroelastomers are burned or heated.

• After a vehicle has been exposed to fire or high temperatures:

1. Do not touch blackened or charred seals or equipment.

2. Preferably, don't handle parts containing decomposed fluoroelastomers, but if you must do so, wear goggles and PVC (polyvinyl chloride) or neoprene protective gloves while doing so. Never handle such parts unless they are completely cool.

3. Contaminated parts, residues, materials and clothing, including protective clothing and gloves, should be disposed of by an approved contractor to currently applicable national or local regulations. Oil seals, gaskets and 'O' rings, along with contaminated material, must not be burned.

WORKSHOP

• Always have a fire extinguisher of the correct type at arm's length when working on anything flammable. If you do have a fire, DON'T PANIC. Direct the extinguisher at the base of the fire.

• NEVER use a naked flame in the workplace.

❒ 7: KEEP your inspection lamp well away from any source of flammable materials.

• NEVER use petrol (gasoline) to clean parts. Use only a proprietary degreaser.

• NO SMOKING. There's a risk of fire or of transferring dangerous substances to your mouth.

• BE METHODICAL in everything you do, use common sense, and think of safety at all times.

ENVIRONMENT FIRST!

• The used oil from the sump of just one car can cover an area of water the size of two football pitches, cutting off the oxygen supply and harming swans, ducks, fish and other river life.

❒ 8: When you drain your engine oil - don't oil the drain! Pouring oil down the drain will cause pollution. It is also an offence.

• Don't mix used oil with other materials, such as paint and solvents, because this makes recycling difficult.

• Take used oil to an oil recycling bank. Telephone FREE in the UK on 0800 663366 to find the location of your nearest oil bank, or contact your local authority recycling officer.

**OIL POLLUTES WATER
USE YOUR BRAIN-
NOT THE DRAIN!**

CHAPTER 3: GETTING THROUGH THE ANNUAL TEST

This chapter relates mostly to the UK where vehicles need to pass the 'MoT' test but also has relevance for those in other countries with a similar annual test. Obviously, you won't be able to examine your car to the same degree of thoroughness as the MoT testing station. But you can reduce the risk of being among the four out of 10 who fail the test first time!

The checks shown below are correct for the MoT Test in the UK at the time of writing but they do tend to become stricter! Your local testing station will have the latest information. DON'T BE TURNED AWAY! The vehicle, when presented for test, must be reasonably clean. Testing Stations can refuse to test vehicles that are very dirty and have excessive mud on the underside.

CONTENTS

Part A: Inside the Vehicle

STEERING WHEEL AND COLUMN

❒ 1: Try to move the steering wheel towards and away from you and then from side to side. There should be no appreciable movement or play. Check that the steering wheel is not loose on the column and that there are no breaks or loose components on the steering wheel itself.

❒ 2: Lightly grip the steering wheel between thumb and finger and turn from side to side. Vehicles with a steering rack: free play should not exceed approximately 13 mm (0.5 in.), assuming a 380 mm (15 in.) diameter steering wheel. Vehicles fitted with a steering box: free play should not exceed approximately 75 mm (3.0 in.), assuming a 380 mm (15 in.) diameter steering wheel.

A-2

❒ 3: If there is a universal joint at the bottom of the steering column inside the vehicle, check for movement. Place your hand over the joint while turning the steering wheel to-and-fro a little way with your other hand. If ANY free play can be felt, the joint must be replaced.

❒ 4: Steering security and locking devices (where fitted) must be in working order.

ELECTRICAL EQUIPMENT

❒ 5: With the ignition turned ON, ensure that the horn works okay.

❒ 6: Check that the front wipers work.

❒ 7: Check that the screen washers work.

❒ 8: Check that the internal warnings for the indicator and hazard warning lights work okay. When ABS brakes are fitted: Make sure that there is an ABS warning light that illuminates and that the lamp follows the correct sequence.

CHECKS WITH AN ASSISTANT

❒ 9: Check that the front and rear side lights and number plate lights work and that the lenses and reflectors are secure, clean and undamaged.

❏ 10: Check the operation of the headlights and check that the lenses are undamaged. The reflectors inside the headlights must not be tarnished, nor must there be condensation inside the headlight.

❏ 11: Turn on the ignition and check the direction indicators, front and rear, and the side markers.

❏ 12: Check that the hazard warning lights operate on the outside of the vehicle and at the internal warning light.

❏ 13: Check that the rear fog light/s, including the warning light inside the vehicle, all work correctly.

❏ 14: Check that the rear brake lights work correctly.

❏ 15: Operate the brake lights, side lights and each indicator in turn, then all at the same time. None should affect the operation of the others.

SAFETY FIRST!

• Follow the Safety information in *Chapter 2, Safety First!* but bear in mind that the vehicle needs to be even more stable than usual when raised off the ground.

• There must be no risk of it toppling off its stands or ramps while suspension and steering components are being pushed and pulled in order to test them.

FRONT SCREEN AND MIRRORS

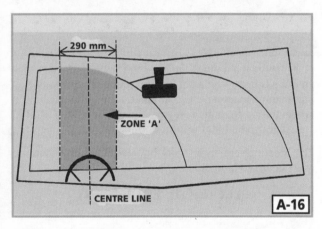

290 mm

ZONE 'A'

CENTRE LINE

A-16

❏ 16: In zone 'A' of the front screen, no items of damage larger than 10 mm in diameter will be allowed. In the rest of the area swept by the screen wipers, no damage greater than 40 mm in diameter will be allowed, nor should stickers or other obstructions encroach on this area.

❏ 17: Check that the exterior mirror on the driver's side is in good condition.

❏ 18: There must be one other mirror in good condition, either inside the vehicle or an external mirror on the passenger's side.

BRAKES

❏ 19: You cannot check the brakes properly without a rolling road brake tester but you can carry out the following checks:

❏ 20: Pull on the parking brake. It should be fully ON before the lever reaches the end of its travel.

❏ 21: Knock the parking brake lever from side to side and check that it does not then release itself.

❏ 22: Check the security of the parking brake lever mountings and check the floor around them for rust or splits.

❏ 23: Check that the front brake pedal is in good condition and that, when you take hold of it and move it from side to side, there is not too much play.

❏ 24: Push the footbrake down hard with your foot. If it creeps slowly down towards the floor, there is probably a problem with the master cylinder. Release the pedal, and after a few seconds, press down again. If the pedal feels spongy or it travels nearly to the floor, there is air in the system or another dangerous fault with the brakes.

❏ 25: Check the servo unit (when fitted) as follows: Pump the brake pedal several times then hold it down hard. Start the engine. As the engine starts, the pedal should move down slightly. If it doesn't the servo or the vacuum hose leading to it may be faulty.

SEAT BELTS AND SEATS

❏ 26: Examine all of the seat belt webbing (pull out the belts from the inertia reel if necessary) for cuts, fraying or deterioration.

❏ 27: Check that each inertia reel belt retracts correctly.

❏ 28: Fasten and unfasten each belt to ensure that the buckles work correctly.

❏ 29: Tug hard on each belt to ensure that the inertia reel locks, and inspect the mountings, as far as possible, to ensure that all are okay.

A-29

IMPORTANT NOTE: Checks apply to rear seat belts as much as to front ones.

❏ 30: Make sure that the seat runners and mountings are secure and that all back rests lock in the upright position.

DOORS AND DOOR LOCKS

❏ 31: Check that doors latch securely when closed and that they can be opened and closed from both outside and inside the vehicle.

Part B: Outside of Vehicle

ELECTRICAL EQUIPMENT

See *Part A: Inside the Vehicle* for checks on the operation of the electrical equipment.

❏ 1: Examine the wiper blades and replace those that show any damage.

VEHICLE IDENTIFICATION NUMBERS (VIN)

❏ 2: The VIN (or chassis number on older vehicles) must be clearly displayed and legible.

❏ 3: Number (licence) plates must be secure, legible and in good condition with correct spacing between letters and numbers which must be of correct size and style.

BRAKING SYSTEM

❏ 4: Inside the engine bay inspect the master cylinder, servo unit (if fitted), brake pipes and mountings. Look for corrosion, loose fitting or leaks.

STEERING AND SUSPENSION

❏ 5: While still in the engine bay, have your assistant turn the steering wheel lightly from side to side and look for play in steering universal joints or steering system mountings and any other steering connections.

❏ 6: If the vehicle is fitted with power steering, check the security and condition of the steering pump, hoses and drivebelt, in the engine bay.

❏ 7: While your assistant turns the steering wheel more vigorously from side to side, place your hand over each track rod end in turn and feel for playing. Inspect all of the steering linkages, joints and attachments for wear.

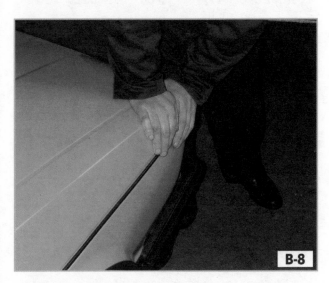

B-8

❏ 8: Go around the vehicle and 'bounce' each corner of the vehicle in turn. Release at the lowest point and the vehicle should rise and settle in its normal position without continuing to 'bounce' of its own accord. If not, a shock absorber is faulty. Always renew in 'axle' pairs or sets.

BODYWORK STRUCTURE

❏ 9: Any sharp edges on the external bodywork, caused by damage or corrosion will cause the vehicle to fail.

❏ 10: Check all load bearing areas for corrosion. Open the doors and check the sills inside and out, above and below. Any corrosion in structural metalwork within 30 cm (12 in.) of seat belt mounting, steering and suspension attachment points will cause the vehicle to fail.

WHEELS AND TYRES

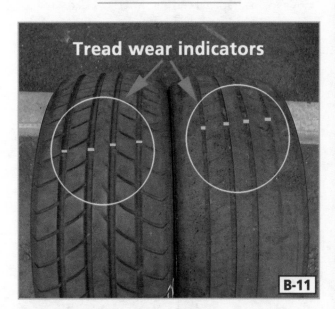

Tread wear indicators

B-11

❏ 11: To pass the test, the tread must be at least 1.6 mm deep throughout a continuous band comprising the central three-quarters of the width of the tread. The Tread Wear Indicators (TWI) will tell you when the limit has been reached, on most tyres. (They are not coloured on 'real' tyres!)

IMPORTANT NOTE: Tyres are past their best, especially in wet conditions, well before this point is reached! (Illustration courtesy of Dunlop)

❏ 12: Check that the front tyres match and that the rear tyres match each other - in terms of size and type but not necessarily make. They must be the correct size for the vehicle and the pressures must be correct.

❏ 13: With each wheel off the ground in turn, check the inside and the outside of the tyre wall for cuts, lumps and bulges and check the wheel for damage. Note that tyres deteriorate progressively over a period of time and if they have degraded noticeably, replace them.

Part C: Under the Vehicle

You will need to support the front of the vehicle off the ground with the rear wheels firmly chocked in both directions.

❏ 1: Have your helper turn the steering from lock to lock and check that the steering turns smoothly and that the brake hoses or pipes do not contact the wheel, tyre or any part of the steering or suspension.

❏ 2: Particular attention should be paid to evidence of corrosion at the steering rack or steering box fixing points.

❏ 3: Have your assistant hold down the brake pedal firmly. Check each brake flexible hose for bulges or leaks. Inspect all the rigid brake pipes underneath the front of the vehicle for corrosion, damage or leaks and also look for signs of fluid leaks at the brake calipers. Rigid fuel pipes also need to be checked for corrosion, damage or leaks.

❏ 4: At each full lock position, check the steering rack rubber gaiters for splits, leaks or loose retaining clips.

❏ 5: Check the track rod end dust covers to make sure they are in place and are not split.

❏ 6: Inspect each constant velocity joint gaiter - both inners and outers - for splits or damage. You will have to rotate each wheel to see the gaiters all the way round.

❏ 7: Check all of the suspension rubber mountings, including the anti-rollbar mountings (when fitted). Take a firm grip on each shock absorber in turn with both hands and try to twist the damper to check for deterioration in the top and bottom mounting bushes.

❏ 8: Check that the shock absorbers are not corroded, that the springs are in good condition and that there are no fluid leaks down the body of the shock absorber. Renew if necessary

❏ 9: Check the front of the exhaust for corrosion and secure fixing at manifold and mounting points.

C-10

❏ 10: Grasp each wheel at 12 o'clock and 6 o'clock positions and try rocking the wheel.
FRONT WHEELS: Look for movement at suspension ball joints, suspension and steering mountings. Repeat while grasping each wheel at 3 o'clock and 9 o'clock.
ALL WHEELS: At the wheel bearing, look for movement between the wheel and hub.

❏ 11: Spin each wheel and check for noise or roughness in the wheel bearing and binding in either the wheel bearing or the brake.

❏ 12: If you suspect wear at any of the suspension points, try levering with a screwdriver to see whether or not you can confirm any movement in that area.

❏ 13: Vehicles fitted with other suspension types such as hydraulic suspension, torsion bar suspension etc. need to be checked in a way that is relevant to the system, with the additional point that there must be no fluid leaks or damaged pipes on vehicles with hydraulic suspension.

❏ 14: Inspect the rear springs for security at their mounting points and for cracks, severe corrosion or damage.

❏ 15: Check the rear shock absorbers in the same way as the checks carried out for the fronts.

❏ 16: Check all rear suspension mounting points, including the rubbers to any locating rods or anti-roll bar that may be fitted.

❏ 17: Check all of the flexible and rigid brake pipes and the fuel pipes just as for the front of the vehicle.

❏ 18: Have your assistant press down firmly on the brake pedal while you check the rear brake flexible hoses for bulges, splits or other deterioration.

❏ 19: Check the fuel tank for leaks or corrosion. Remember also to check the fuel filler cap - a correctly sealing filler cap is a part of the MoT test.

❏ 20: Examine the parking brake mechanism. Frayed or broken cables or worn mounting points, either to the bodywork or in the linkage will all be failure points.

❏ 21: Check each of the rear wheel bearings as for the fronts.

❏ 22: Spin each rear wheel and check that neither the wheel bearings nor the brakes are binding. Pull on and let off the parking brake and check once again to make sure that the parking brake mechanism is releasing.

SAFETY FIRST!
- Only run the car out of doors.
- Beware of burning yourself on a hot exhaust system!

C-23

❏ 23: While you are out from under the vehicle, but with the rear end still raised off the ground, run the engine. Hold a rag over the end of the exhaust pipe and listen for blows or leaks in the system. You can then get back under the vehicle and investigate further if necessary.

□ 24: Check the exhaust system mountings and check for rust, corrosion or holes in the rear part of the system.

□ 25: Check the rear brake back plate or calipers (as appropriate) for any signs of fluid leakage.

Part D: Exhaust Emissions

TOP TIP!

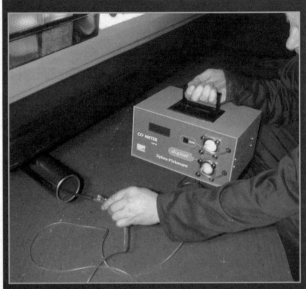

• This is a Sykes-Pickavant CO meter.
• If you don't own a CO meter, you could have your testing station carry out the emission part of the test first so that if it fails, you don't waste money on having the rest of the test carried out.

FACT FILE

FACT FILE: VEHICLE EMISSIONS

The information shown here applies, at the time of writing, to the UK. For information applicable to other territories, or for later amendments, check with the relevant local testing authorities.

PETROL/GASOLINE ENGINED VEHICLES WITHOUT CATALYSER

Vehicles first used before 1 August 1973 - visual smoke check only.

Vehicles first used between 1 August 1973 and 31 July 1986 - 4.5% carbon monoxide and 1,200 parts per million, unburned hydrocarbons.

Vehicles first used between 1 August 1986 and 31 July 1992 - 3.5% carbon monoxide and 1,200 parts per million, unburned hydrocarbons.

PETROL/GASOLINE ENGINED VEHICLES FITTED WITH CATALYTIC CONVERTERS

Vehicles first used from 1 August 1992 (K-registration - on, in the UK)

• All have to be tested at an MoT Testing Station specially equipped to handle vehicles fitted with catalytic converters whether or not the vehicle is fitted with a 'cat'.

• Required maxima are - 3.5% carbon monoxide and 1,200 parts per million, unburned hydrocarbons. There will be a further check to make sure that the catalyst is in working order.

TOP TIP!

• Because 'cats' don't work properly at lower temperatures, ensure that the engine is fully warm!

DIESEL ENGINES' EMISSIONS STANDARDS

• IMPORTANT NOTE: The diesel engine test puts a lot of stress on the engine. It is IMPERATIVE that the vehicle's engine is in good condition before you take it in for the MoT test. The tester is entitled to refuse to test the vehicle if he feels that the engine is not in serviceable condition.

Vehicles first used before 1 August, 1979
• Engine run at normal running temperature; engine speed taken to around 2,500 rpm (or half governed max. speed, if lower) and held for 20 seconds. FAILURE, if engine emits dense blue or black smoke for next 5 seconds, at tick-over.

Vehicles first used on or after 1 August, 1979
• After checking engine condition, and with the engine at normal running temperature, the engine will be run up to full revs between three and six times to see whether the engine passes the prescribed smoke density test. (2.5k for non-turbo vehicles; 3.0k for turbo diesels. An opacity meter probe will be placed in the vehicle's exhaust pipe.) Irrespective of the meter readings, the vehicle will fail if smoke or vapour obscures the view of other road users.

MULTI-FUEL VEHICLES

• Vehicles which run on more than one fuel (eg petrol and LPG) will normally be tested on the fuel they are running on when presented for test.
• There is a slight difficulty with LPG vehicles and unless the testing station analyser has the facility for conversion, the mechanic will have to do a calculation. The machine is set to measure propane, but LPG power gives out hexane. The analyser will have a 'PEF' number shown. This is used as follows: 'propane' reading ÷ PEF no. = hexane value.

CHAPTER 4: WORKSHOP TOP TIPS!

Please read Chapter 2 Safety First! before carrying out any work on your car.

Here are a few *Top Tips!* to help keep things running well in the workshop.

❒ **1: DON'T LOSE IT!** Buy sandwich bags and store small items in them, in groups, as they are removed. Keep the bags in a box or boxes, and keep the box/es in the vehicle if you have to go off and do something else. If you leave stuff lying around you'll lose some of it - right?

LOOK ON THE BRIGHT SIDE! Don't always assume the worst. That misfire - could it be the ECU? Highly unlikely, so try all the small stuff first. Engine running faults in particular are caused, 90% of the time, by failures in simple components such as spark plugs, leads, loose terminals and so on. So don't be a pessimist!

DON'T BE A BLUEBOTTLE! Work methodically; don't whizz around from one thing to another. Make a resolution to finish one thing before starting the next - even when you hit a tough patch, work through it! You'll finish jobs more quickly and you'll lose less stuff!

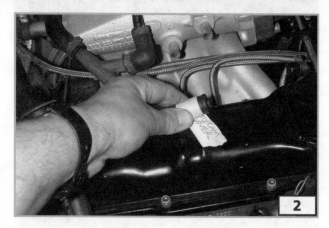

❒ **2: LABEL IT!** Even in a manual like this, it isn't possible to cover every possible variation of wiring and pipework layout. If you assume that you WON'T remember how every single part goes back together - you'll almost certainly be right! Use tags of masking tape stuck on the ends of all removed connections, and label or number the matching parts. You'll save ages when it's time for reassembly!

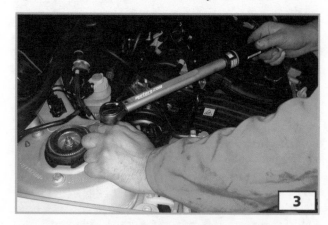

❒ **3: TIGHTEN RIGHT!** Under-torquing and over-torquing threaded fixings is all too common. Some mechanics pride themselves on being able to judge the correct torque 'by feel'. They can't! Yes, they can get closer than a raw amateur, but the demands of modern components leave no room for guessing.

→ Under-torqued fixings can come loose, or allow components to 'work' or chaff; over-torqued fixings can be even worse and can fail catastrophically or distort essential parts. Always check that threads run freely, and use a torque wrench!

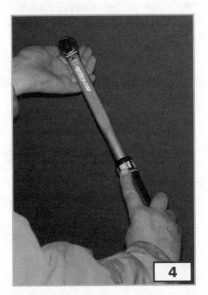

☐ 4: KEEP TORQUING! **Sykes-Pickavant** advise that their torque wrenches - and it actually applies to all makes - will read accurately for much longer if backed-off to the ZERO position before putting away, after each use.

CHOOSING AND USING A HOIST

The best way of raising a vehicle off the ground - almost essential if you intend making your living or part of your income from working on vehicles - is to use a hoist. There are several types available, and the pros and cons are explained here by leading vehicle hoist manufacturer, **Tecalemit**:

→ **FOUR POST:** This type of hoist is the least expensive, it's stable and capable of taking the greatest weights, but it's also the least versatile. With a post in each corner, the vehicle is driven onto ramps which raise the whole of the vehicle off the ground. The ramps do get in the way and the suspension is compressed by the weight of the vehicle. This restricts access in the wheel wells. On the other hand, it is possible to use a cross-beam from which you can jack specific parts of the vehicle - essential when vehicle testing. A four-post hoist is also useful if it's essential to raise a dangerously rusty vehicle off the ground.

→ **TWO POST:** A post each side of the vehicle each carries two legs. The legs are swung so that a foot on the end of each leg is positioned under each end of the vehicle's body, usually under the normal jacking points. The great thing about these hoists is that they are 'wheels free' - the wheels and suspension hang down, providing almost ideal access to the underside of the vehicle. A two-post hoist should never be used on a vehicle that is dangerously rusty, because it will be raised on body parts which may collapse if the corrosion is very severe.

→ **SINGLE POST:** This type has a single post, and swing-out legs reaching right under the vehicle. The advantage gained from the 'loss' of a post is offset by the intrusion of the extra-long support legs. The legs impede under-car access; the second post of a two-post hoist doesn't.

→ **OUR CHOICE:** Without hesitation, we fitted a **Tecalemit** two-post hoist into the Porter Manuals workshop. Excellent service life from this famous-name manufacturer, and easy access for our mechanics, authors and photographers have made the hoist a wise choice!

☐ 5: The legs can only be swung into position when they are fully lowered. They are extended, as necessary and aligned beneath the lifting points of the main body tub of the vehicle. As soon as they begin to raise off the ground, the legs are locked into position.

☐ 6: Access to the vehicle's underside is ideal and, because there are no ramps - and no depressions in the floor, as is often the case with 4-post hoists - there is plenty of room to drive another vehicle beneath the one on the hoist for overnight storage.

☐ 7: These **Tecalemit** hoists can run from 3-phase or 1-phase electrics, and can be fitted with a converter to enable a domestic level of current supply to power the hoist. In such cases, there will be a momentary delay while the converter builds up the power to the level required.

❏ **8A:** When a vehicle is raised on a hoist, it's perfectly safe, PROVIDED that the hoist has received its regular maintenance check by the suppliers. The **Tecalemit** two-post hoist is raised by screw threads and the legs are locked immovably when the motor is not being operated.

TOP TIP!

❏ **8B:** • Put a piece of tape on the post when you've established your best working height.
• Now you can raise the vehicle with the legs lined up with this mark every time!

RAISING THE VEHICLE - SAFELY!

Read this section in conjunction with the essential safety notes in *Chapter 2, Safety First!*

For those who don't have access to a pro. hoist:
➜ NEVER work beneath a vehicle held solely on a jack, not even a trolley jack. The safest way of raising a vehicle may be to drive one end of it up onto a pair of ramps. Sometimes, however, there is no alternative but to use axle stands because of the nature of the work being carried out.
➜ Do not jack-up the vehicle with anyone on board, or when a trailer is connected (it could pull the vehicle off the jack).
➜ Pull the parking brake on and engage first (low) gear.
➜ WHEELS ON THE GROUND SHOULD BE CHOCKED AFTER THE VEHICLE HAS BEEN RAISED, SO THAT THE VEHICLE CANNOT MOVE.

USING RAMPS

Make absolutely certain that the ramps are parallel to the wheels of the vehicle and that the wheels are exactly central on each ramp. Always have a helper watch both sides of the vehicle as you drive up.
➜ Wrap a strip of carpet into a loop around the first 'rung' of the ramps and drive over the doubled-up piece of carpet on the approach to the ramps. This prevents the ramps from skidding away, as they are inclined to do, as the vehicle is driven on to them.

➜ Drive up to the end 'stops' on the ramps but never over them!
➜ Apply the parking brake firmly and put the vehicle in first or reverse gear (or 'P' in the case of auto).
➜ Chock both wheels remaining on the ground, both in front and behind so that the vehicle can't move in either direction.

USING A TROLLEY JACK

On many occasions, you will need to raise the vehicle with a trolley jack - invest in one if you don't already own one. Ensure that the floor is sufficiently clear and smooth for the trolley jack wheels to roll as the vehicle is raised and lowered, otherwise it could slip off the jack.
➜ Before raising the vehicle, ENSURE THAT THE PARKING BRAKE IS OFF AND THE TRANSMISSION IS IN NEUTRAL. This is so that the vehicle can move as the jack is raised.
➜ Reapply brake and place in gear after the raising is complete and chock each wheel to prevent vehicle movement.
➜ Always remember to release brake and gear and remove chocks before lowering again.

❏ **9:** Axle stands also need to be man enough for the job. These inexpensive **Clarke** stands have an SWL of 3 tonnes. Make sure that the axle stands will each be placed beneath a reinforced part of the body, suitable for jacking from, or a main suspension mounting. Never place the jack or axle stand under a moving suspension part.

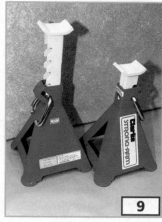

SAFETY FIRST!

• Whenever you're working beneath a vehicle, have someone primed to keep an eye on you!

• If someone pops out to see how you are getting on at regular intervals, it could be enough to save your life!

• Be especially careful when applying force to a spanner or when pulling hard on anything, when the vehicle is supported off the ground.

• It is all too easy to move the vehicle so that it topples off the axle stand or stands.

TOOLS AND EQUIPMENT

This section shows some of the tools and equipment that we have used while working on the vehicles that have been photographed for this manual.

You'll never have a complete set of tools; there will always be something else that you need! But over the years, if you buy equipment a little at a time, as you need it, you will accumulate a surprisingly large range.

When buying tools, it certainly pays to shop around. Tools that you won't need to use regularly, such as an impact screwdriver or a rubbing block for use with abrasive paper can be picked up for a song.

When it comes to larger and more expensive and specialised items, it pays to stick to a known maker rather than to take a chance with an apparently cheap tool whose make you may never have heard of.

☐ **10:** The **Clarke** 'Strong-Arm' engine hoist has the added advantage of being able to be folded into a really small space for storage.

☐ **11:** This engine stand, from the same manufacturer, is remarkably inexpensive. The engine is held at a comfortable working height and can be turned through 360 degrees. Recommended!

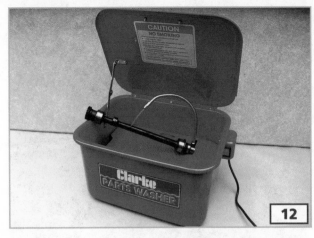

☐ **12:** When you've stripped components down, the most effective way of getting them clean is with a parts washer, this one from **Clarke** again.

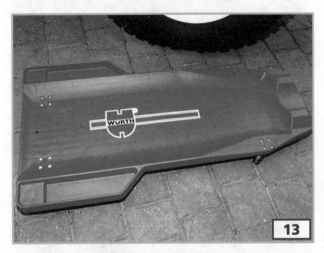

☐ **13:** Sliding beneath the vehicle will be a hundred times easier with a good quality car crawler, such as this plastic moulded crawler from **Wurth**.

☐ **14:** Another tool that you can scarcely do without is a compressor. At the bottom end of the range, both in terms of price and performance, is a compressor such as the **Clarke** Monza. This tiny compressor will power a spray gun sufficiently for 'blowing-in' a panel and you'll also be able to inflate tyres and carry out all sorts of other lightweight jobs.

15

17

❐ **17:** Another use to which you will be able to put your compressor is spraying cavity protection wax. This **Wurth** injection gun is dual-purpose. It takes disposable **Wurth** screw-on canisters and also has its own large separate canister for injecting any protection wax that you may want to use 'loose'. Hand-powered and cheap-and-cheerful injectors simply don't atomise the protection wax or blast it far enough into nooks and crannies to be useful.

❐ **15:** A compressor such as this 60 c.f.m. unit is the smallest needed by the serious amateur or semi-pro.. It won't run larger air tools, except in shorter bursts, but it's fine for the air wrench, for instance.

16

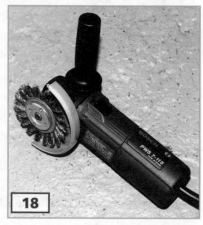

18

❐ **18:** Another invaluable tool is an angle grinder. This is the **Bosch** PWS7-115. This piece of equipment is perfect for using with grinding and cutting discs but, when used with this twisted-wire brush (available from bodyshop suppliers), scours paint and rust off steel in seconds. Always wear goggles and gloves with this tool.

❐ **16:** The Air Kit 400 provides a very useful and remarkably low-cost set of basic air tools capable of being powered by even the smaller compressors. Clockwise from top-left:
➜ The engine cleaner gun works much better than a brush.
➜ The spray gun is basic but effective.
➜ Air hose is suitable for all smaller compressors.
➜ Wear goggles when using the invaluable air duster.
➜ Double-check tyre pressures with a hand-held gauge when using this tyre inflator and gauge.

❐ **19:** Another power tool which has a lot of domestic uses as well as being invaluable when working on a vehicle is something like the Jet 3000 Power Washer. It's a marvellous tool for removing mud, oil, grease and grim before stripping down body or mechanical parts

19

and it is also extremely useful around the outside of the house.

❏ 20: If your budget – or workshop space – won't run to a stand-alone pillar drill, the **Bosch** drill stand will turn your mains-power drill (this is the **Bosch** PSB with powerful 750W motor) into a perfectly adequate light-user

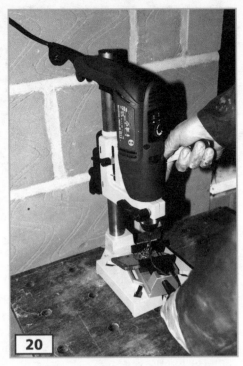

version. The same company also offer the hand vice which is an essential piece of equipment for gripping small pieces.

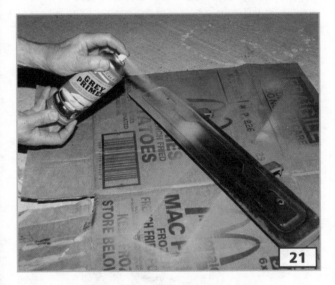

❏ 21: Aerosol cans of paint are extremely useful for small items, such as this number plate backing plate, mainly because there's no cleaning up to do afterwards. For large areas, aerosol is prohibitively expensive and you won't find the depth of paint or the quality as good as you would get from a spray gun. There's always a place for aerosol, however, and the **Hycote** range includes all the various types of primer and finish coats that you could want, as well as offering a range of mix-and-match aerosol paints which are mixed to the shade you need.

❏ 22: There is a wide range of tool boxes and chests available from **Sykes-Pickavant**. They're made of tough heavy gauge steel, are lockable, and contain separate 'filing cabinet' type drawers for tool storage. Most of the units are stackable.

❏ 23. Increasingly, the kind of work described in this manual requires the use of special tools. **Sykes-Pickavant** produce a complete range of regular workshop tools and equipment and also special tools for most purposes.

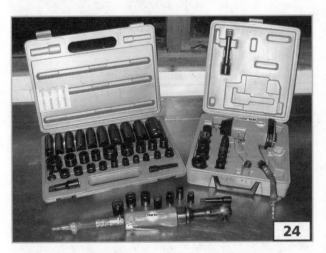

❏ 24. An air wrench can save a heck of a lot of time on dismantling and reassembly – although you should always finish off with a torque wrench, where appropriate. The **Clarke** 3/8 in. drive is 'wieldy' enough for engine bay work while the 1/2 in. drive wrench and sockets will cope with most heavy duty jobs. Note the flexible 'tail' we add to each tool to protect the female connector on the air line.

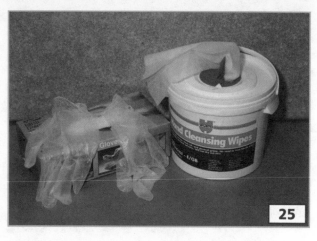

25

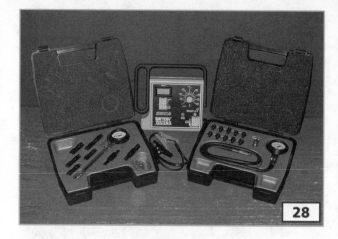

28

❏ **25:** You'll need hand cleaner - the sort with granules shifts heavy grease best. **Wurth** also produce these hand wipes - useful if you need to touch upholstery in the middle of a job - and packs of disposable gloves.

❏ **26: Wurth** produce a huge range of workshop products including electrical connectors and that wonderful 'shrink-fit' wire insulation tubing – slide it on, heat it up, and it 'shrinks' into place and can't come unwrapped.

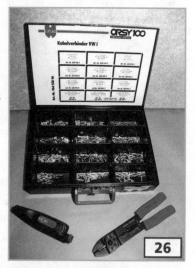

26

❏ **28:** With these three **Sykes-Pickavant** kits, you can check (from left to right) many of the engine's most basic functions:

➜ Cylinder compression tester - essential for a whole range of mechanical diagnostics, without the need for engine dismantling. Different testers are available for Diesel engines with their very high compression pressures.

➜ Battery tester - essential for eliminating or confirming the battery as a source of problems.

➜ Oil pressure tester. When combined with the cylinder compressions tester, this is capable of providing virtually a complete picture of the 'inner health' of any engine.

❏ **29:** Last in this Chapter, but first job before carrying out vehicle dismantling: disconnect the battery! Beware of radio sets, alarms and ECUs that need a continuous electricity supply. See *Chapter 10, Electrical, Dash, Instruments* on preserving a battery feed.

29

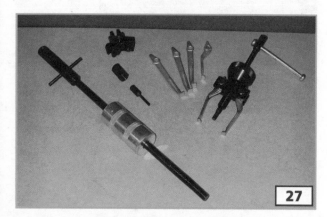

27

❏ **27:** It's sometimes necessary to use pullers to remove 'stuck' bearings or other interference fit items. This **Sykes-Pickavant** set includes a variety of arm sizes and types, and a slide hammer to supplement the usual screw-type puller.

CHAPTER 5: SERVICING

*Please read **Chapter 2 Safety First** before carrying out any work on your car.*

SAFETY FIRST!

• Please read the whole of *Chapter 2, Safety First!* before carrying out any work described here.

HOW TO USE THIS CHAPTER

Note that:
→ Each letter code tells you the Service Interval.
→ Look the code up in the Service Intervals Key.
→ Each Service Job has a Job number.
→ IMPORTANT NOTE: Each service should be carried out at EITHER the recommended mileage OR the recommended time interval, whichever comes first.

SERVICE INTERVALS: KEY

A - Every week, or before every long journey.
B - Every 6 months, or 5,000 miles.
C - Every year or 10,000 miles.
D - Every 2 years, or 20,000 miles.
E - Every 3 years, or 30,000 miles.
F - Every 4 years, or 40,000 miles
G - Every 5 years or 50,000 miles
H - Every 6 years or 60,000 miles

CONTENTS

FACT FILE
ENGINE BAY LAYOUTS

Annotations:
a – oil filler
b – engine oil dipstick
c – coolant reservoir
d – brake fluid reservoir
e – battery

f – air cleaner
g – washer fluid filler
h – power steering fluid reservoir
(when fitted)
i – diesel fuel filter

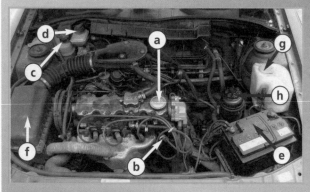

1. CAVALIER MK 3 PETROL.

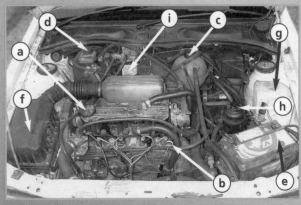

2. CAVALIER MK 3 DIESEL.

3. VECTRA PETROL.

4. VECTRA 2.2 DIESEL.

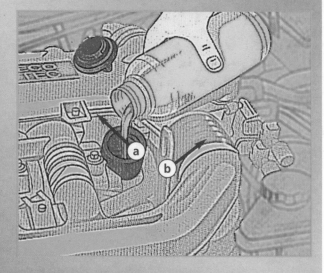

5. VECTRA V6 AND SOME DIESELS.

→ The oil filler is located here (**a**) on V6 models.
→ The power steering fluid reservoir is here (**b**) on V6 and some diesel models.

Part A: Regular Checks

JOB 1: ENGINE OIL - *check level.*

❏ **1:** On the V6 engine the dipstick is just in front of the oil filler cap.
→ On some models, the dipstick goes right into a hole in the block, rather than a tube.

1

JOB 2: COOLING SYSTEM - *check level.*

❏ **2:** If necessary, top up the system with a mixture of 50% anti-freeze and water to the level mark on the expansion bottle.

JOB 3: HYDRAULIC FLUID - *check level.*

❏ **3:** Check/top-up brake fluid level as required.
➔ Note that on all disc-braked cars a gradual drop in the fluid level over a lengthy period is natural.
➔ This is because, as the disc pads' friction material wears, the fluid level drops naturally - though slowly, of course.
➔ If the fluid level drops quickly, something is wrong!

JOB 4: BATTERY - *check electrolyte level.*

Check the fluid level, and for corrosion around the terminals.
➔ Wash white 'fur' off with hot water.
➔ Protect terminals with petroleum jelly or (better still) proprietary terminal fluid. It's best not to use grease!

TOP TIP!

• A so-called 'maintenance-free' battery may have flush-fitting strips over its cells which can be prised up for the addition of electrolyte, perhaps prolonging its life beyond general expectation!
• This does not apply to batteries supplied as original equipment on new Vauxhalls - they cannot be opened up.

❏ **4:** All the vehicles covered by this manual are fitted with a 'maintenance-free' battery when new.
➔ These are sealed-for-life units that do not require topping-up and no provision is made for doing so.
➔ They are provided with a 'condition indicator' on the top surface, as shown here.
➔ If the indicator is darkened and shows a green dot, it is charged and in good condition.
➔ No green dot indicates the battery is part-charged, although this may not be noticeable in normal use.
➔ If the indicator shows yellow then the battery is in poor condition and may need replacement: DO NOT

ATTEMPT to jump-start - using jump-leads - if the battery is in this state.

| JOB 5: | SCREEN/HEADLIGHT WASHER FLUID AND WASHERS - *check level*. |

☐ 5: Check the screenwash fluid level reservoir and top-up if necessary.
➜ Don't forget the rear washer bottle on pre-1988 models.

The windscreen washer jets are non-adjustable on some models.
➜ On others, the screen washer jets can be aimed with a pin.
➜ If one of the jets is blocked, it can usually be freed with a piece of thin wire.

| JOB 6: | TYRES - *check pressures and condition (road wheels)*. |

Pressures should be checked when the tyres are cold - they warm up when a vehicle is driven.

☐ 6: Check the inner and outer sidewalls for bulges and splits. Check the tread depth using a gauge.

TOP TIP!

• If a tyre is worn more on one side than another, it probably means that the suspension or steering is out of adjustment but could be symptomatic of suspension damage.
• If a tyre is worn more in the centre or on both edges, it could mean that tyre pressures are wrong.
• See *Chapter 3: The Annual Test* for a more detailed explanation of tyre wear and legal limits.

JOB 7: LIGHTS - *check/change bulbs*.

See *Chapter 10, Electrical and Instruments* for bulb replacement information.

TOP TIP!

• In many cases, the bulb is a push-fit in its holder and is recessed quite deeply, which makes removal difficult.
• To remove the bulb more easily wrap a short piece of masking tape around the protruding glass and squeeze the excess into a small 'handle'.

Part B: Engine, Cooling System

| JOB 8: | ENGINE OIL AND FILTER – *change*. |

SAFETY FIRST!

• Refer to the section on ENGINE OILS in *Chapter 2, Safety First*, before carrying out the following work.

• It is essential to wear rubber or plastic gloves since used engine oil can be carcinogenic.

• Take care that the spanner does not slip causing injury to the hand or head.

• Take great care that the effort needed to undo the drain plug does not cause the vehicle to fall or slide off the ramps or axle stands.

☐ 8A: This is the approximate location of the oil drain plug on the great majority of these engines.

☐ 8B: Be ready to reposition your bowl - the angle of 'spurt' changes as the oil flows out of the sump!

8C: Remember to use a new drain plug washer - not applicable to tapered-plug types.

8C

8D: Use a strap or chain wrench to remove the filter.

8D

8E: Apply clean engine oil to the rubber sealing ring to prevent it buckling as the filter is screwed home.

8E

8F: When the sealing ring contacts the face on the engine, tighten it a further three-quarters of a turn - by hand only.

8G: Pour in fresh oil - slowly so that it doesn't spurt back.

8F

Finally after the vehicle has been lowered to the ground, run the engine for a minute or so then re-check the oil level on the dipstick, topping-up if necessary.

8G

➔ Take a look beneath the car to check for any leaks around the sump plug or filter, tightening if required.

JOB 9: VALVE CLEARANCES – check/adjust.

See *Chapter 9, Ignition, Fuel, Exhaust.*

JOB 10: CAMSHAFT BELT - check.

Check the timing belt for damage, cracking or fraying. Renew it if any is found - see *Chapter 6, Part A.*

JOB 11: CAMSHAFT DRIVE BELT - renew.

See *Chapter 6, Engine.*

JOB 12: COOLING SYSTEM HOSES - check.

Check all the coolant and heater hoses for security and leaks.
➔ Squeeze the larger hoses, and with the smaller hoses, bend the straights and straighten the bends, listening for cracking sounds, which tell you that the hose is brittle and needs replacing.

JOB 13: COOLANT - replace.

See *Chapter 8: Cooling System.*

JOB 14: RADIATOR - clean.

TOP TIP!

• As the radiator picks up most of its debris from the air flowing in from the front, the best way to clear any which is inside the fins is to blow it out from the back.
• We don't recommend blasting the core with high pressure compressed air as this could damage it.

JOB 15: WATER PUMP - *check*.

Examine the engine block around the water pump mountings for any sign of a leak.

➔ Leaks are characterised by a white, powdery deposit, sometimes tinged with the colour of the antifreeze, and may be seen in 'runs' down the side of the block.
➔ If such evidence is present, check coolant hoses, connections and thermostat housing.
➔ Replace the pump, if necessary, before it fails.

Part C: Transmission

JOB 16: TRANSMISSION OIL - *check*.

This check requires the vehicle to be on level ground, so it will be necessary to raise both the front and rear of the vehicle at the same level.

SAFETY FIRST!

• The following checks have to be carried out with the engine hot and running. The vehicle must be out of doors because of dangerous exhaust fumes.

• Make sure that hair, jewellry and loose clothing cannot become caught by moving parts.

• Do not touch any part of the electrical system.

• Take care not to be burnt by the hot engine.

MANUAL TRANSMISSION

TOP TIP!

• To check the oil level, you will have to add a quantity of new oil.
• The level plug is located on the differential casing which, while an integral part of the gearbox, contains oil that is kept at a higher level than that in the main gearbox.
• As a result, oil may be present in the differential casing, but not in the gearbox adjacent to it, thereby giving a false indication of the level.

❏ **16A:** The level plug will be in one of these two positions (depending on type) adjacent to a driveshaft on the differential casing, either on the right or left-hand side of the casing.

❏ **16B:** Looking down into the engine bay right-hand side, the oil filler also doubles as a vent for the gearbox, and as such may not readily be recognisable.

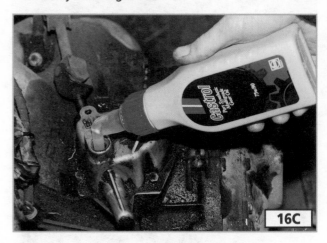

❏ **16C:** Add oil into the gearbox, while a helper watches the level outlet.
➔ When oil appears, the level is now correct.
➔ Tighten the level plug and replace the vent/filler.

AUTOMATIC TRANSMISSION

The auto. transmission fluid is checked with the engine running and the drive selector in the 'P' or PARK position.
➔ On a cold transmission, the level is checked using the side of the dipstick marked '+20 C'.
➔ At normal operating temperature the reading is taken from the side marked '+94 C'.
➔ Normal operating temperature is only reached after ten to fifteen miles driving, so allowing the car to 'warm-up' for five or ten minutes is not enough to warm the fluid sufficiently.
➔ The handle of the dipstick is coloured red to distinguish it from the engine oil dipstick.
➔ Fluid is added by pouring into the mouth of the dipstick tube.

JOB 17: TRANSMISSION OIL - *change*.

This is not part of Vauxhall's service schedule but common sense suggests that fresh lubricant will allow components to last longer.

MANUAL TRANSMISSION

❑ **17:** There is no drain plug!
➜ Unbolt this plate from the gearbox differential housing. Remove all the bolts, starting with

the lowest one and working up.
➜ This will allow the plate to maintain its seal with the gasket until a drip tray can be positioned beneath it, when a sharp tap from a soft-faced hammer or block of wood, struck sideways, will free the plate and allow the oil to drain.
➜ Always use a new gasket when refitting the plate, after draining is complete.

AUTOMATIC TRANSMISSION

Follow the approach for manual transmission.
➜ When you've removed the sump pan (very similar to the manual pan but with 18 bolts instead of 10) the next thing you come to, once the fluid is drained, is a large flat filter screen.
➜ Wash the pan and screen meticulously in white spirit before reassembling with a new gasket.

JOB 18: DRIVESHAFT GAITERS - *check*.

❑ **18:** If any splits are found, replace the gaiters - see *Chapter 11, Steering and Suspension*.

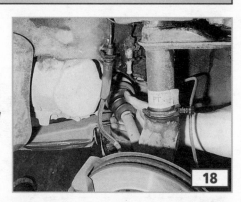

JOB 19: CLUTCH - *check, adjust*.

If the pedal action feels jerky, 'dry' or heavy, try lubricating the pivots and self-adjusting mechanism located at the top of the pedal with Castrol DWF or releasing fluid.
➜ If the fault persists, the clutch cable probably needs replacing.
➜ For cable replacement and adjustment, see *Chapter 7: Transmission, Clutch*.

TOP TIP!

• Note that the correctly-adjusted clutch pedal will come to rest at a higher position than the brake pedal - this is normal. If the pedals are level, then adjustment may be required, otherwise difficulty may be encountered in gear selection.

Part D: Ignition, Electrics

JOB 20: SPARK PLUGS - *check*.

TOP TIP!

• If a plug gets tighter as you turn it, there's every possibility that it is cross threaded. Once out, it probably won't go back in again.
• If the threads can't be cleaned up with a thread chaser, you will have to add a thread insert to the cylinder head.

Before replacing the plug leads, clean them by use of a maintenance spray and a piece of rag or tissue. Also clean the exterior of the distributor cap (into which the plug leads fit) using the same method.

JOB 21: SPARK PLUGS - *renew*.

Renew the spark plugs irrespective of apparent condition.

JOB 22: CONTACT BREAKER POINTS/DWELL ANGLE - *check/set*.

See *Chapter 9: Ignition, Fuel, Exhaust*.

TOP TIP!

• Remove the spark plugs.
• Apply the parking brake, select second gear and support the right-hand front wheel off the ground - use an axle stand.
• Now, you can easily turn the engine by turning the road wheel by hand.

JOB 23: CONTACT BREAKER POINTS - *renew*.

See *Chapter 9: Ignition, Fuel, Exhaust*.

JOB 24: DISTRIBUTOR - *lubricate*.

Apply a smear of grease to the sides of the cam rotor - be sparing with the grease so that none can transfer to the points themselves.
➔ Where a distributor features a small felt pad in the centre of the cam rotor, apply a little oil to it, just enough to soak in without depositing oil on the points, which could prevent them working.

On re-assembly, clean the rotor arm of any dust or grime, also the inside of the distributor cap, with a soft cloth. In winter, spray the outside of the cap with a very small amount of water repellent, such as Castrol DWF maintenance spray which helps prevent condensation forming and possibly 'shorting' the spark to earth.

JOB 25: IGNITION TIMING - *check*.

See *Chapter 9: Ignition, Fuel, Exhaust*.

JOB 26: AUXILIARY DRIVE BELT/S - *check*.

See *Chapter 6, Part A* for information on checking and adjusting belts

JOB 27: IDLE SPEED AND EMISSIONS - *adjust*.

See *Chapter 9, Ignition, Fuel, Exhaust* for all engine types

27

Part E: Fuel, Exhaust

JOB 28: FUEL PIPES – *check*.

Check fuel lines and hose clips from fuel tank into engine compartment.

JOB 29: EXHAUST SYSTEM AND MOUNTINGS - *check*.

If exhaust mountings break or come loose, the extra stresses on the exhaust system will soon cause fracturing.

SAFETY FIRST!

• **Never run the vehicle's engine in an enclosed space – only ever out of doors.**

• **Exhaust gases are toxic!**

TOP TIP!

• If you suspect a leak but it's location isn't obvious, hold a piece of board so that it blocks off the tailpipe. Under pressure, the leak should be more noisy, enabling you to track down its position.

JOB 30: AIR CLEANER ELEMENT - *renew*.

❏ **30A: CARBURETOR ENGINED MODELS:** The air filter housing is secured by a combination of clips and cross-head screws.

30A

30B: FUEL INJECTION AND LATER CARBURETOR MODELS: Remove the clips and screws (if fitted).
→ Lift the top cover as far as the air intake hose will allow, as much as is necessary.

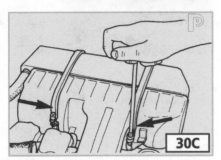

30C: CAVALIER DIESEL: Undo the two cross-head screws or clips, and remove two further screws that secure the air scoop to the front engine-bay panel.

Place the new filter in position, ensuring the latex seal on the upper edge of the filter seats properly in the groove on the lower housing - the upper casing will not fit properly if the filter is misplaced.
→ Refit the top cover, clips and screw (if fitted).

JOB 31: AIR INTAKE CONTROL VALVE - *check*.

31: CARBURETOR MODELS: Check the operation of the valve by observing the flap through the end of the intake, first with the

engine cold, then when it has reached operating temperature.
→ If the valve fails to open when the engine is started, check the vacuum pipe from the inlet manifold to the vacuum capsule for signs of splitting or other damage.
→ Otherwise, you may need to replace the capsule.

JOB 32: DIESEL ENGINES FUEL FILTER - *drain*.

The filter is located at the back of the engine bay.
→ Drain it by partly unscrewing the plastic drain-plug fitted to the base of the filter unit, after placing a number of rags beneath the filter to catch any water

and fuel that may escape.
→ See *Chapter 9, Part B* for full information.

JOB 33: FUEL FILTER - *renew*.

DIESEL ENGINES

See *Chapter 9, Ignition, Fuel, Exhaust*.
→ Use an oil filter strap to release the disposable type.

TOP TIP!

• Before fitting, fill the diesel filter with fuel - the engine has to suck all the air out of the system before it is fed with diesel fuel, and this will save a lot of engine churning.

Before screwing on the new filter, lubricate the rubber sealing ring with a smear of clean diesel fuel and ensure the retainer ring in the centre of the filter is in place.
→ Fit the new filter firmly by hand, close the drain valve and vent screw, then run the engine at a fast idle for a minute or so to vent the system. Check for leaks.

PETROL ENGINES

33A: CARBURETOR ENGINES UP TO 1985: The fuel filter is contained within the fuel pump and consists of a fine screen through which

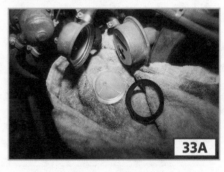

the fuel is drawn from the tank.
→ To remove the filter for cleaning, remove the screw in the centre of the fuel pump and lift away the lid, revealing the screen filter and sealing rubber.

33B: CARBURETOR ENGINES AFTER 1985: The in-line fuel filter is located at the top of the engine, just behind the cam/valve

cover. It must be renewed.
→ The filter is held in place on a small bracket and secured by two screws - remove them and discard the filter. After replacing, run the engine and check for leaks.

➜ Ensure that the arrow on the filter points in the direction the fuel will flow through it. Some filters work in either direction, in which case there will be no arrow.

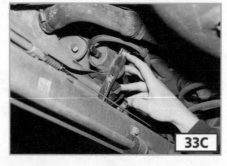

❑ **33C: FUEL INJECTION ENGINES:** The fuel filter is positioned at the rear of the fuel tank beneath the car.

33C

➜ On models up to 1992 the filter is free-hanging, while on later vehicles it is fixed by a bracket, held by a single screw.

➜ When fitting a new filter, make sure its arrow points the same way as the original - with the arrow in the direction of fuel flow.

JOB 34:	DIESEL GLOW PLUGS - *check/clean.*

Unscrew each glow plug in turn - see *Chapter 10: Electrical, Dash, Instruments* and clean the carbon off the end with proprietary carburetor cleaner.

JOB 35:	ACCELERATOR CONTROLS - *lubricate.*

35

❑ **35:** Apply a smear of grease to the sides of the cam rotor - be sparing with the grease so that none can transfer to the points themselves.

➜ Lubricate the accelerator control linkage at the injection system in the engine bay.

➜ Also lubricate, the throttle pedal pivot in the footwell recess. Use a spray lubricant like Castrol DWF or white silicone grease in the footwell, so as not to spoil your shoes with dripping oil - it stains leather!

➜ There are rather more lubrication points on a carburetor. With air filter removed, oil every spring, moving part and joint in the linkages.

Part F: Steering, Suspension

JOB 36:	FRONT AND REAR WHEEL BEARINGS – *check.*

In order to check for wear, rock the wheel about its centre, feeling for excess bearing play. Also, try spinning each wheel - as far as you can, with driven wheels - feeling for rough rotation.

TOP TIP!

• If a wheel bearing is worn, you will normally hear a noise on the outer, loaded bearing when cornering.

JOB 37:	STEERING AND SUSPENSION - *check.*

❑ **37:** Raise each front wheel just clear of the ground and support with an axle stand.

➜ Put a pry bar under the wheel and try to lever it upwards.

➜ Free play suggests wear in suspension bushes or strut.

➜ Check the springs for obvious breaks or sagging.

➜ Check tie bar bushes for cracking, softness or wear.

37

JOB 38: STEERING RACK GAITERS - *check.*

☐ **38:** If any splits are found, replace the gaiters - see *Chapter 11, Steering, Suspension*.

JOB 39: STEERING BALL JOINTS - *check.*

See *Chapter 11, Steering, Suspension*.

JOB 40: SUSPENSION JOINTS AND BUSHES - *check.*

Test the inner suspension arm bush by levering. There should be only a just-perceptible amount of movement.
→ Check all rubber bushes and mountings for distortion, splits or perishing.

JOB 41: SHOCK ABSORBER ACTION - *check.*

See *Chapter 3: The Annual Test* for shock absorber tests.

JOB 42: POWER STEERING FLUID - *check*

PRE-1988 CAVALIER: With the engine at operating temperature the fluid level should be at the FULL mark, while on a cold engine the level shouldn't fall below the ADD mark.

☐ **42: LATER MODELS:** Check the level by unscrewing the cap which has a short dipstick attached to it.
→ There are two marks on the dipstick: The lower '20 deg. C' mark is used when the fluid is cold
→ The upper '80 deg. C' when at operating temperature.
→ Top-up using Castrol Dexron R III, or as specified in *Chapter 1, Facts and Figures*.

JOB 43: WHEEL NUTS/BOLTS - *check.*

☐ **43:** To 'torque' the wheel nuts/bolts correctly, first slacken each nut or bolt and check that the threads aren't stiff or corroded, then tighten with the wrench.

Part G: Braking System

SAFETY FIRST!

• Do NOT attempt any work on the braking system unless you are fully competent to do so.

• See *Chapter 2, Safety First*.

• Brake fluid is poisonous and must never be allowed to enter the mouth!

• Clean off brake dust with proprietary brake cleaner – brake drums removed where appropriate – never use compressed air.

• Only replace brake pads and/or shoes in 'axle' sets of four.

• After fitting new brake shoes or pads, avoid heavy braking for 150 to 200 miles (250 to 300 km), except in an emergency.

JOB 44: FRONT BRAKE PADS - *check/renew.*

See *Chapter 12, Brakes*.

JOB 45: REAR BRAKE SHOES - *check/renew.*

See *Chapter 12, Brakes*.

JOB 46: PARKING BRAKE ADJUSTMENT - *check.*

Apply the parking brake. It should lock at between 2 and 4 clicks. If more or less than these figures, the parking brake must be adjusted as described in *Chapter 12, Brakes*.

JOB 47: PARKING BRAKE LINKAGE - *lubricate.*

☐ **47:** Apply grease liberally to versions with a brake cable equaliser and place a blob of grease where the cable enters each 'outer'.

JOB 48: BRAKE LINES - *check.*

☐ **48:** Make a physical check of all pipework and connections.
➜ Bend all flexible brake hoses to show up signs of cracking rubber - if any are found, the hose should be replaced as soon as possible.
➜ Hoses should also be free from bulges or chafing marks.

JOB 49: BRAKE FLUID - *renew.*

If this job is not carried out at the recommended interval, it can result in brakes which fail without warning. See *Chapter 12, Brakes*.

Part H: Bodywork, Interior

JOB 50: SEATBELTS AND SEATBELT MOUNTINGS - *check operation and security*.

➜ Examine seat belts for signs of damage, such as abrasion, cuts, contamination or frayed stitching.
➜ Where inertia-reel seatbelts are fitted, pull the belts out as far as they will go and make sure they fully retract.
➜ Make sure the seatbelt anchorage points are secure and the retaining bolts are tight.
➜ On older vehicles you should make sure that there is no body corrosion which would weaken the seatbelt mountings.

SAFETY FIRST!

• Some vehicles are fitted with seatbelt pre-tensioners. Special safety precautions are need when this type of seatbelt is being removed or refitted - see *Chapter 14, Interior, Trim*.

JOB 51: SEAT MOUNTINGS - *check.*

Check the seats:
→ For secure mounting, by trying to rock them.
→ The seat adjustment mechanism.
→ That the folding seat backs lock securely in the upright position.

JOB 52: SPARE TYRE - *check.*

TOP TIP!

• You should inflate the spare tyre to the maximum pressure recommended for high speed or load running.
• Then, if you have a puncture while on a journey, you'll be okay.
• Just carry a tyre pressure gauge with you and let some air out if necessary.

JOB 53: FRONT SCREEN WIPERS - *check.*

Check the wiper blades for splitting, hardening of the rubber blade, and that no metal part of the blade carrier is contacting the screen.

❒ 53A: 'SHEETING' – **Causes:**
→ Dirty blade lip.
→ Worn or torn blade lip – needs replacing.

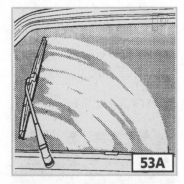

❒ 53B: DROPLETS – **Causes:**
→ Glass and rubbers contaminated by traffic film, oil or grease or wax from washing.
→ Use panel wipe to clean all traces away.

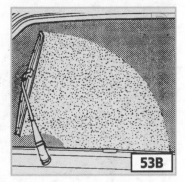

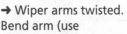
❒ 53C: 'CHATTER' – **Causes:**
→ Deformed or heavily contaminated blade lips – replace.
→ Glass dirty, contaminated by grease, was or traffic film or not enough water on screen.
→ Wiper arms twisted. Bend arm (use adjustable wrenches) to provide proper blade contact – blade must 'flip' at end of each stroke.
→ Loose or incorrect connection between blade holder and arm. Replace.

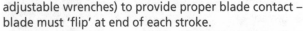

JOB 54: HORN - *check.*

FACT FILE
HORN LOCATION

CAVALIER UP TO 1988: The horn is positioned between the grille and radiator, with a second horn on CD and SRi models positioned on a bracket below the radiator.

❒ 54: **CAVALIER FROM1988:** Either a single horn is fitted between grille and radiator as above, or twin horns are fitted to a bracket found below the screenwash reservoir at the base of the passenger-side front wing, behind the lower splash guard.

VECTRA: The horn is located behind the front bumper, on the right-hand side.

For removal of grille and/or bumper, see *Chapter 13, Bodywork.*

JOB 55: LOCKS, CHECK STRAPS AND HINGES (NOT HIGH SECURITY LOCKS) - *lubricate*.

❑ **55A:** Oil or grease hinges and check straps at each door.
➔ Apply silicone grease or aerosol spray to the jaws of the lock catches in the edges of the doors and tailgate.

55A

❑ **55B:** Apply graphite lock lube to the keyholes of locks.

55B

JOB 56: ALARM SENDER UNIT BATTERIES - *renew*.

If the vehicle is fitted with a remote control alarm system renew the batteries.

JOB 57: CLEAR DRAIN HOLES.

Check and clear drain holes in doors, sills, etc.
➔ Use a thin probe to poke the holes clear, although for a lasting job a vacuum and nozzle will clear most debris.
➔ Check also the bulkhead well at the rear of the engine bay, especially for leaves in winter.

Part I: Road Test and Final Check – *after every service.*

JOB 58: ROAD TEST AND FINAL CHECK - *after every service.*

Before you can claim to have 'finished' working on the vehicle, you must check it and test it. If you are not a qualified mechanic, we strongly recommend having someone who is a properly qualified mechanic inspect all of the vehicle's safety-related items before using it on the road.

➔ **WARM-UP:** Run the car for several minutes before setting out then turn off, check fluid levels and check underneath for leaks.
➔ **STEERING:** Check that the steering moves freely in both directions and that the car does not 'pull' one way or the other when driving in a straight line - but do bear in mind the effect of the camber on the road.
➔ **BRAKES:** Make sure that the brakes work effectively, smoothly and without the need for 'pumping'. There should be no juddering or squealing.
➔ **'PULLING'** Check that the car does not 'pull' from one side to the other when you brake firmly from around 40 mph. (Don't cause a skid and don't try this if there is any risk of having an accident.)

➔ **CLUTCH OPERATION:** . Make sure the clutch pedal operates smoothly and the clutch engages without juddering.
➔ **ENGINE PERFORMANCE:** Make sure the engine is operating satisfactorily under load and on the over-run.
➔ **TRANSMISSION:** Make sure the gear changes are smooth, and the transmission appears to be working normally
➔ **INSTRUMENTS:** Must work and indicate correctly.

58

CHAPTER 6: ENGINE

Please read **Chapter 2 Safety First** before carrying out any work on your car.

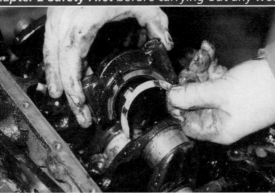

Part A: General Procedures

CONTENTS

Many of the operations when working on an engine are common to all motors. So, to save you time and help things go smoothly, here they are!

Before dismantling, disconnect the battery's negative terminal. See *Chapter 10, Electrical, Dash, Instruments, Fact File: Disconnecting the Battery* BEFORE doing so!

JOB 1: SAFETY FIRST!

You must follow the advice in *Chapter 2, Safety First!* before doing any work on your vehicle. Pay special attention to the notes on fuel safety. Petrol/gasoline is highly explosive and, as any job on or near fuel systems could well result in leaks, care must always be taken.

Quite a few vehicle components are heavy enough to cause injury if not lifted or supported properly. Check *Job 11* for information on moving the engine.

JOB 2: TOP DEAD CENTRE (TDC) - *explained*.

FACT FILE

TOP DEAD CENTRE.

• Top Dead Centre (TDC) is the highest point in a piston's travel.
• When carrying out a range of repairs and maintenance tasks it is necessary to set No. 1 cylinder (on the right hand side of an in-line engine) to TDC on the firing stroke when both valves are closed.
• TDC is determined according to the engine type.
• **ALL ENGINES:** Crankshaft rotation is clockwise.
• To find TDC the crankshaft has to be rotated. This is most easily achieved using a spanner or socket on the crankshaft pulley nut.

☐ **STEP 1A: ALL ENGINES: EITHER:** Turn the crankshaft until the crankshaft pulley notch aligns with a timing pointer…

2-1A

☐ **STEP 1B: …OR:** the crankshaft mark (**a**) aligns with the crankcase lip mark (**b**)…
➜ …**AND** one of the following.

2-1B

☐ **STEP 2A: ALL 4-CYLINDER PETROL AND DIESEL ENGINES WITH CAMSHAFT COVER REMOVED:** Both valves of No. 1 cylinder can be seen to be closed (the camshaft lobes point upward).

2-2A

➜ This is the only method for finding TDC on diesel engines.

☐ **STEP 2B:** On DOHC units, all four camshaft lobes (arrowed) of No. 1 cylinder should be pointing upward.

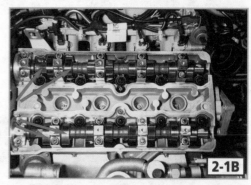

2-1B

☐ **STEP 3:** All three types mentioned here are similar to this illustration when viewed with their upper cambelt covers removed – except that there is only one cambelt pulley on SOHC engines, of course!

2-3

➜ **SOHC PETROL ENGINES:** The camshaft sprocket mark aligns with its timing mark – usually at the top of the cam belt inner cover.
➜ **DOHC PETROL ENGINES:** The timing marks on the camshaft sprockets are…
➜ 1.6 Litre; aligned with the camshaft inner cover marks (**a**).
➜ 1.8 and 2.0 Litre; adjacent to each other (**b**).
➜ **V6 ENGINE:** The rear cylinder bank camshaft timing marks align with the timing marks on the toothed belt cover.

JOB 3: CYLINDER HEAD BOLTS – *tightening, undoing*.

Be warned: if you get this wrong, you risk distorting the cylinder head. Aluminium heads are especially vulnerable to this. NEVER remove the head until the engine has cooled down completely.

☐ **STEP 1:** Generally, it is accepted practice to undo the outer bolts first, working inwards in a spiral pattern. Tightening begins with the inner bolts,

3-1

working outwards. HOWEVER - there are specific patterns for different Vauxhall engines, listed in the relevant sections of this manual.

JOB 4: CYLINDER HEAD GASKETS.

❏ **STEP 1:** NEVER re-use a cylinder head gasket, as it will almost certainly blow. The replacement should be fitted so that the word TOP or OBEN is facing **upwards**.

❏ **STEP 2: DIESEL ENGINES:** It is critical that the correct gasket thickness is used. There are three. Use one with the same number of notches in the gasket edge as the one you removed.

JOB 5: GASKET SEALANT.

➜ NEVER use gasket sealant on a cylinder head gasket.
➜ Neoprene (compressible 'rubber') gaskets do not normally need gasket paste. If necessary, use silicone sealant.

❏ **STEP 1:** All joints to be sealed must be cleaned of all the old sealant, but as gently as the materials allow. A pressed steel component can be wire

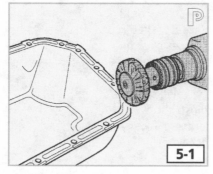

5-1

brushed, and a cast iron item scraped clean. Aluminium, however, is much softer, and should be cleaned with extra care. NEVER use power tools to clean joints.

❏ **STEP 2:** Apply silicone sealant as follows:
➜ Cut the nozzle on the sealant tube to the size of bead you need - usually 2-3 mm.

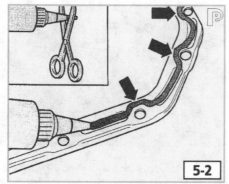

5-2

➜ Apply an unbroken bead around the *inner* sides of bolt holes.

JOB 6: 'STUCK' CYLINDER HEAD - *freeing*.

It is sometimes difficult to free a cylinder head, even after all the bolts have been undone. NEVER lever between the mating faces of the head and block. Instead:
➜ Look for strong brackets or protrusions on the head or block against which you might be able to use a lever.
➜ Turn the engine over on the starter motor, with the spark plugs in place but not connected. The compression may shift the head.
➜ Use a large, soft-faced mallet on a solid protrusion from the head. **Do not** hammer anywhere you could damage.

JOB 7: CYLINDER HEAD – *lifting, fitting*.

When the engine is still installed, lifting a cylinder head may (depending on type) be a dangerously heavy component to lift, because of the need to reach into the engine bay. It's best to have someone help you to lift it away.

TOP TIP!

❏ **STEP 1:** •
When refitting a cylinder head held down with bolts rather than studs fitted in the block – and if there are no alignment

7-1

dowels fitted by the manufacturer - you may have difficulty in aligning the bolt holes in the gasket with those in the head - and if they're not aligned, you won't be able to screw the bolts into the block.
➜ Make a pair of guides out of two old cylinder head bolts, or two pieces of plain steel bar of a size that will just slide into the threads in the block.
➜ If using old bolts, cut the heads off with an angle grinder and slot the ends so that you can use a screwdriver to remove them.
➜ Fit the bolts, or the bars, slide the gasket over them, followed by the cylinder head, so that all holes are aligned.
➜ You can now fit some of the 'proper' bolts before removing the guides.

JOB 8: CYLINDER HEAD AND COMBUSTION CHAMBERS.

❏ **STEP 1:** The cylinder head can be checked for distortion by use of a straight edge and feeler

8-1

gauge. At the same time, check for excessive corrosion. If you are in doubt, or the head gasket had blown, have the cylinder head refaced ('skimmed') by your Vauxhall agent or engine specialist.

Clean excessive carbon deposits from the crowns of the pistons without damaging the surface of the aluminium.

TOP TIP!

• If the engine is worn and you don't intend overhauling it at this stage, we strongly advise you don't scrape the carbon from the piston crowns. It can help preserve compression pressures.

JOB 9: EXHAUST AND INLET VALVES AND GUIDES – *removing, replacing*.

VALVE REPLACEMENT

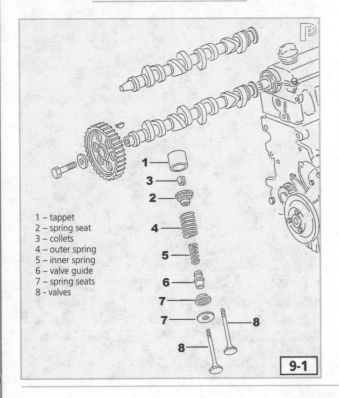

1 – tappet
2 – spring seat
3 – collets
4 – outer spring
5 – inner spring
6 – valve guide
7 – spring seats
8 - valves

9-1

❏ **STEP 1:** Remove the cam followers and valve clearance adjusting shims (not used on finger-tappet or hydraulic tappet types), taking care to maintain their original positions for reassembly.

❏ **STEP 2:** With many engines, you will need to use a valve-spring compressor with an extension jaw to reach into the recessed valve spring

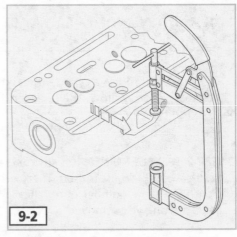

9-2

area in the head. (Illustration, courtesy Sykes-Pickavant.)

❏ **STEP 3:** Give each valve a sharp tap with a hammer to free the top spring plate from the valve. You could place a socket spanner over the plate to avoid striking the end of the valve stem. (Illustration, courtesy Sykes-Pickavant.)

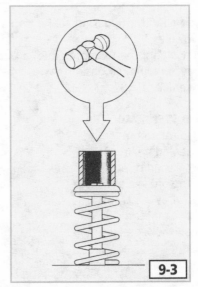

9-3

9-4

❏ **STEP 4:** Compress the valve springs so that you can reach in and remove the collets from around the heads of the valves.

❒ **STEP 5:** Slowly and carefully open the spring compressor tool to release the pressure of the valve spring, and remove the upper spring seat followed by the springs.

9-5

❒ **STEP 6:** Slide out the valves, removing the valve seals and discard the old seals; new ones must be used on re-assembly.

9-6

TOP TIP!

• The valves should slide freely out of their guides.
• Any resistance may be caused by a build up of carbon, or a slight burr on the stem where the collets engage.
• This can usually be removed by careful use of fine wet-or-dry paper, allowing you to withdraw the valves without scoring their guides.
• Keep the valves in their correct order by wrapping a numbered piece of masking tape around each stem.

❒ **STEP 7:** Once all the valves have been removed, clean them up ready for inspection. Remove the carbon deposits with a wire brush

9-7

and degrease the rest. Exhaust valves are prone to burning at their heads, as are their valve seats in the cylinder head.

TOP TIP!

• Check the height of the valve springs against new ones if possible, but if not, compare them with each other.
• If any are shorter than the others, play safe and replace the complete set. They are bound to have suffered fatigue which could cause premature valve failure.

❒ **STEP 8:** Check for valve guide wear:
➜ Lift each valve until the end of the stem is level with the top of the valve guide.

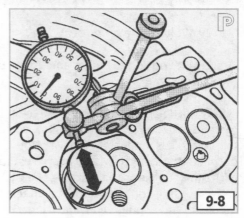

9-8

➜ Attempt to move the head from side to side.
➜ If you feel any noticeable play then you may need new guides.
➜ This is a job that has to be carried out by your Vauxhall dealer or engine rebuild specialist, who will also have the experience to confirm whether or not wear is acceptable, as well as the special tools needed to replace the valve guides.

❒ **STEP 9:** To install valves, start from one end. Lubricate a valve stem with fresh engine oil and slide it in to its guide.

9-9

❒ **STEP 10:** Locate a new valve stem seal over the stem of the valve and push down into contact with the guide. Push the seal onto its seat using a suitable metal tube.

❏ **STEP 11:** • Wrap a short length of insulating tape around the collet grooves at the end of each valve stem.
• This will protect the new valve seals from damage as you slide them over the valve stems.
• When you have pushed the new seal firmly onto the top of the valve guide, remove the tape.

❏ **STEP 12:** Refit the spring seat.

❏ **STEP 13:** Position the inner and outer springs and the spring cap.

❏ **STEP 14:** Re-apply the valve spring compressor and compress the springs enough to allow you to engage the split collets in the stem grooves. Note that the type of collets, the spring caps and the valves must match each other. Unlike the earlier type, the later type has a valve with three grooves, a collet with three single shoulders and a spring cap with an inner chamfer and a wider outer chamfer. Make sure that you buy the correct type of valves, if any have to be replaced.

• Grease the grooves so that the collets will 'stick' in place.
• Collets are easily fitted by 'sticking' the backs of them onto the end of a screwdriver with some grease and feeding them into position.

❏ **STEP 15:** Carefully release the spring compressor and check that the collets are correctly located. Tap the end of each stem with a hammer to bed them all in.

❏ **STEP 16:** Fit the remaining valves.

JOB 10: EXHAUST AND INLET VALVES
- *grinding in.*

VALVE GRINDING

• Modern engines, with hardened valve seats, are not capable of having valves ground in the traditional way, except to remove the smallest of blemishes.
• If anything more is needed, take the valves and cylinder head to a main dealer or engine specialist and have the valves and seats recut to the correct angle by machine.

❏ **STEP 1:** A power-operated valve grinder, such as this Wurth tool, attaches to the electric drill and makes valve grinding on engines with hardened valve seats a more practical proposition.

10-1

• Before grinding-in the valves, clean the tops of the valve heads back to shiny metal.
• Now the sucker on your valve grinding stick won't fall off as you use it!

❏ **STEP 2:** Apply a small quantity of coarse grinding paste evenly around the valve seat.
➜ Use a valve grinding stick tool with a suction cup slightly smaller than the valve face.

10-2

❏ **STEP 3:** Apply a dab of moisture and press the suction pad firmly onto the valve head.
➜ Lower the valve stem into the guide.
➜ Holding the grinding stick firmly between your palms, rub back and forth.
➜ Create a rotary grinding motion, while pressing gently down into the valve seat
➜ IMPORTANT NOTE: Absolutely NO paste must get into the guide, as this will rapidly create wear.

10-3

→ Lift the valve regularly to redistribute the cutting paste around the contact area.

→ When you feel the paste wearing thin remove the valve, wipe the surfaces clean and check the contact surface on both valve and seat.

→ You are aiming to create a complete, narrow grey ring around both of these. If there are any blemishes left, more coarse paste is needed.

→ Once a complete ring has been achieved, clean off the coarse paste and finish off with fine paste.

→ Finally, thoroughly remove all traces of cutting paste from the valves and the head.

TOP TIP!

• A narrow contact band means high pressure on the seat and longer contact life.

• A wide contact band reduces the contact pressure and induces early valve seat burning.

❒ **STEP 4:** Repeat this operation on the operating valves.

❒ **STEP 5:** Wash the whole cylinder head using paraffin (kerosene) and an old brush, making sure all traces of grinding paste are removed, then dry it off. Use compressed air if available.

SAFETY FIRST!

• Treat compressed air with respect. Always wear goggles to protect your eyes.

• Never allow the airline nozzle near your skin.

JOB 11: ENGINE – *lifting, moving*.

❒ **STEP 1:** Always use suitable lifting equipment, such as the Sykes-Pickavant tool we show being used. Hire one from a tool hire specialist if you don't have access to one. Most domestic garage roof structures are not strong enough to hang lifting gear from them.

11-1

The weakest part of engine lifting gear is often the connection to the engine itself. On large engines, it's best not to use ropes, which can stretch and slip, although a rope can be used on the models covered by this manual. Ideally, make up a solid lifting eye which can be attached to the block at one or more points. Alternatively, make a lifting bar up which fits to the lifting eyes on your engine, if it has them. A length of strong, closed-link chain can also be fixed to the engine. The hook on your lifting gear should be the type that snaps closed.

If you intend to remove the engine leaving the transmission in place, you should make up or buy a suitable bridge, supported on each side of the engine bay and with a chain or rod holding the weight of the engine end of the transmission.

When moving major engine and/or transmission parts around the workshop, use a strong trolley. Make sure that the components are stable, and cannot fall onto hands, feet or any other parts of the body.

JOB 12: CRANKSHAFT BEARINGS, CONNECTING ROD BEARINGS - *removing*.

❒ **STEP 1:** Check that all the connecting rods and their big-end bearing caps are marked with matching numbers, starting from the timing cover end.

→ Make sure that the marks make it clear which way round they go.

→ If there are no marks there, use typists' correction fluid on clean metal, or a centre punch to make some.

12-1

❒ **STEP 2:** If you are refitting the same pistons, mark them to show their position and which way round they face. Some Vauxhall pistons are

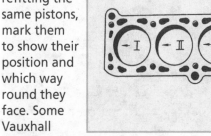
12-2

marked with arrows, which point towards the timing gear end of the engine.

STEP 3: Undo the securing bolts and remove the caps, keeping them in their correct order.

12-3

TOP TIP!

12-4

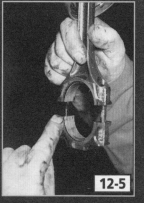

12-5

STEP 4: • Inspect the top of each cylinder bore - there may be a small ring of carbon build-up which can make it difficult to remove the pistons. If so, scrape it carefully away.
• Use a hammer handle to tap the piston/connecting rod assemblies carefully out of the bores...

STEP 5: ...keeping them in the correct order. Keep the matching conrods and bearing caps together.

STEP 6: Check that the five crankshaft main bearing caps are correctly marked, starting from the timing cover end. Undo and remove them, keeping them in the correct order.

TOP TIP!

STEP 7:
• If any of the caps are difficult to remove, lever the bolt holes with a bar, a pair

12-7

of bars or a pair of fixing bolts, and tap carefully with a hammer.
• Bearing shells are best removed by sliding them out with your thumbs, pushing the tab end out first.
• DON'T try to lever them out - it won't work.

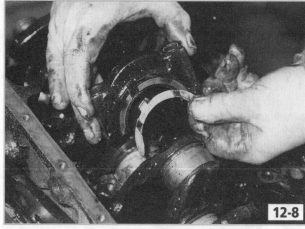

12-8

STEP 8: Retrieve the thrust washers from each side of the centre main bearing cap.

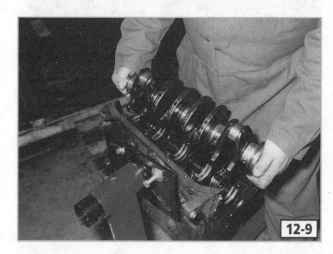

12-9

STEP 9: When lifting the crankshaft clear of the cylinder block, look out for 'stray' bearing shells or thrust washers falling into the block.

JOB 13: PISTONS – *removing, refitting.*

Pistons can be held to connecting rods in different ways - check the relevant section of this manual. See *Job 12, Step 4* for piston removal.

TOP TIP!

• Wrap insulation tape, or a short piece of plastic tube, over each conrod thread, so it cannot damage the crank as it goes in.

STEP 1: Locate a ring clamp over the piston rings and tighten enough to close the ring gaps, but not too tight. Lubricate the rings so they compress and slide easily within the clamp.

STEP 2: Position the piston/conrod assembly in its correct bore with the connecting rod identification marks facing each other, and also so that the complete assemblies face the correct way.

STEP 3: With the ring clamp touching the cylinder block, use a hammer shaft to carefully tap the piston through and into the bore.

TOP TIP!

• Turn the crankshaft so that the journal for the conrod you are working on is at bottom dead centre. Now the conrod will line up with its crank journal.

STEP 4: Locate the upper half of the big-end shell bearing in the conrod, making sure that the mating surfaces are clean.

→ Lubricate the crankpin and the big-end shell and draw the conrod down the bore so the big-end locates with the crank pin.

→ Fit the other half of the big-end shell to the bearing cap and lubricate.
→ Offer the cap to the connecting rod and make sure that the numbers match.
→ Screw in the fixing bolts and tighten progressively to the correct torque.

JOB 14: PISTON RINGS - *fitting.*

STEP 1: Make sure that the bores and pistons are clean, then the rings, preferably using a piston ring spreader. Make sure the rings are fitted with the word 'TOP' facing upwards.

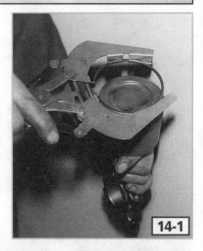

STEP 2: If you don't have access to a piston ring spreader, work the rings down a little at a time, using a feeler gauge or gauges to bridge the gaps. Piston rings are very brittle, and highly expensive to replace.

STEP 3: Fit the piston gaps at equal intervals round the pistons circumference and lubricate them well.

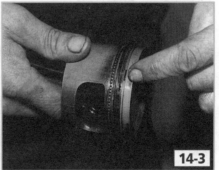

JOB 15: ENGINE COMPONENTS – *checking, measuring for wear.*

GENERAL

All parts must be thoroughly cleaned before inspection. Keep them in the right order for reassembly in case they are to be used. Check component as follows:

CYLINDER BLOCK

☐ **STEP 1:** Look for any cracks and evidence of gas or water blow-by on both sides of the gasket, on the cylinder

15-1

head and in the block casing, particularly at bolt holes and between cylinders.

☐ **STEP 2:** Check the bores as follows:
➜ Check for score marks, caused by burned pistons or broken rings.
➜ Check for a wear ridge just

15-2

below the top of the bore where the top piston ring ends its travel.
➜ If you have access to a suitable internal micrometer, measure the bores at the points shown. Otherwise, ask your engine specialist to measure the bores for wear if there is any evidence of a wear ridge.

☐ **STEP 3:** Assuming the bores to be in reasonable condition, it is sometimes possible to 'glaze bust' the bores and fit new piston rings. If not, the cylinders will have to be rebored.

15-3

☐ **STEP 4:** You can't check for bore ovality like this, but you can gain a good idea of overall wear:
• Push each piston ring squarely into the cylinder until it is about 15 mm

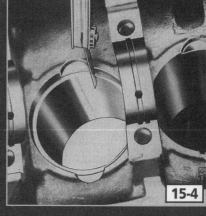

15-4

from the bottom edge where no wear will have taken place.
• Measure the ring gap with a feeler gauge.
• Now carry out the same check on the most badly worn parts of the bore and see how much wear has taken place.

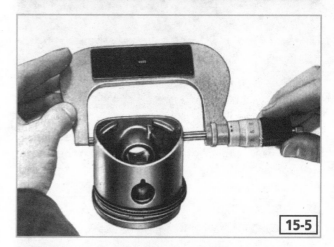

15-5

☐ **STEP 5:** You will need a suitably large external micrometer to measure the pistons. Check about 15 mm from the skirt.

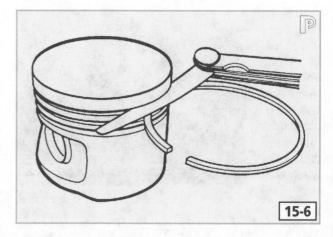

15-6

☐ **STEP 6:** Check the piston ring clearances with feeler gauges.

CRANKSHAFT

15-7

❏ **STEP 7:** Check the main journals and crankpins:
➜ for any signs of wear ridges round the circumference or scoring of the surface.
➜ for ovality, using a suitable micrometer, although the precision Vernier gauge shown here will give an excellent guide.

❏ **STEP 8:** Check the shell bearings, which should have an even, dull grey finish, like the ones shown here.

15-8

❏ **STEP 9:** If the leaded layer is scored or has worn through to the copper coloured backing, or if the crankshaft has any of the previously

15-9

mentioned faults, the crankshaft should be reground by a specialist, who will also supply new shell bearings and thrust washers.

❏ **STEP 10:** Check the crankshaft end float by using a feeler gauge between the thrust washer and the crankshaft.

❏ **STEP 11:** If an engine is being rebuilt, a new oil pump should be fitted as a matter of course. However, if you are checking a stripped engine to see how badly worn it is, include the oil pump in this inspection.

❏ **STEP 12:** Use feeler gauges to measure the backlash between the gears - it should be no more than 0.20 mm.

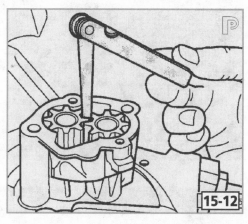

15-12

❏ **STEP 13:** Use a straight edge and feeler gauges to measure the axial clearance - maximum 0.10-0.15 mm. Check that the

15-13

bearing face of the cover is not worn. If in doubt, replace the pump, as it is the 'heart' of the engine.

CAMSHAFT

Check the following:

❏ **STEP 14:** Check each cam lobe for wear, which can progress rapidly once it starts. If you replace the camshaft, fit new followers too.

❏ **STEP 15:** Cam followers (bucket tappets) should also be checked, especially where they touch the cam lobe.

❏ **STEP 16:** Check the five camshaft bearings and their corresponding surfaces in the housing for a smooth, shiny surface without wear ridges.

JOB 16: RE-ASSEMBLING ENGINE COMPONENTS - *lubrication.*

TOP TIP!

• If you start a rebuilt engine up 'dry', severe damage can be caused well before the oil has started to circulate.
• Metal-to-metal surfaces will 'pick-up' and rubber seals will be torn out.
• You should keep an oil can full of fresh oil handy as you re-assemble an engine. Alternatively, there is purpose-made assembly lubricant available.
• DON'T use a friction-reducer meant to be added to engine oil. It could prevent the engine from bedding in properly.

16-1

❏ **STEP 1:** Apply copious amounts of lubricant to every rotating, rubbing and oil seal surface as assembly takes place. You CAN'T over-lubricate!

❏ **STEP 2:** If the engine won't be run straight away, wipe grease on seal surfaces, so they cannot dry out.

16-2

16-3

❏ **STEP 3:** Fill the oil pump housing with as much fresh oil is it can take, so the pump is both lubricated and delivering fresh oil to the engine as quickly as possible.

Before starting the engine, fit a fully charged battery and crank the engine over with spark plugs (or diesel pre-heaters) removed, so that the engine spins rapidly without starting and putting strain on the bearings. Turn the engine for 30 seconds or so to circulate oil to the bearings, before starting up.

SAFETY FIRST!

• Make sure it is safe to spin your engine with HT leads disconnected.

• Some ignition systems may suffer damage if not first completely disconnected.

• Take care not to cause injury from uncontrolled high tension sparks, or from diesel injection spray.

JOB 17: RE-ASSEMBLING ENGINE COMPONENTS - *clearances.*

Specific assembly details are given in the relevant section of this manual, but it's well worth pointing out some general information, especially regarding fitting crankshaft bearings:

IMPORTANT NOTES: These points are essential, not optional!
➔ You must change the oil pump as part of an engine overhaul.
➔ All bearings, shells, piston rings and ALL seals that bear on moving parts MUST be copiously lubricated with fresh engine oil as the engine is being reassembled.
➔ Work ONLY in clean conditions, with clean components and clean hands.
➔ Re-assemble in the reverse order of dismantling and take note of the following steps, which will help you proceed smoothly.

17-1

STEP 1: Make sure you have all the necessary gaskets, available from your Vauxhall dealer or parts specialist in complete sets.

STEP 2: Make sure all bearing sets are perfectly clean and locate the shells so their tabs engage with the slots. Once lubricated, a shell can be placed on a journal and pushed around into its correct position.

17-2

STEP 3: Screw bolts in finger tight and check that the crankshaft rotates freely and smoothly.

STEP 4: Tighten each bolt evenly and progressively until the specified torque setting is reached. Check after tightening EACH bearing that the crankshaft rotates smoothly. If it doesn't, remove the bearing cap and shells and investigate as follows:
➜ Check there is no dirt or debris under a bearing shell - the most likely cause.
➜ The next most likely cause is that there has been a build up of carbon on the rim of a seating and/or cap. Scrape off any that is present.
➜ Check that the shells supplied are the right size for any machining that may have been carried out.

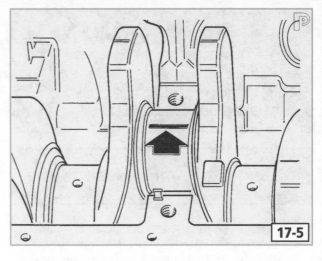

17-5

STEP 5: Put a strip of Plastigage on one of the main journals. Fit the cap and bearing shells and tighten down to the recommended torque, then remove the cap again, *all without moving the crankshaft*.

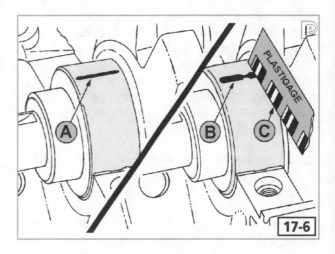

17-6

STEP 6: Use the gauge supplied with the Plastigage kit (available from vehicle parts sellers) to measure the width of the flattened-out strip.
➜ The width of the strip tells you the amount of clearance – read the information on the Plastigage gauge.
➜ If too tight, the engine will overheat; if too loose, it will fail rapidly.
➜ Check for dirt or carbon build-up causing tight-spots, inaccurate machining, or incorrectly supplied bearing shells.

Part B: Which Engine Is Which?

CONTENTS

JOB 1: UNDERSTAND ENGINE TYPES.

The range of engines used in the Cavalier, Vectra and Calibra is deceptively simple. However, one particular engine could appear in several Vauxhall models, so it's the engine, rather than the vehicle you're working on, that is your most important piece of information.

All of the engines excepting the two V6 2.5L units found in models covered by this manual are four cylinder in-line units. Although they appear in a variety of sizes, and underwent many detail specification changes, they can conveniently be divided into three distinct engine types, for the purpose of describing repair procedures. The changes materially affect neither the basic principles of each type's operation, nor the approaches required to work on them.

The three types of in-line engine are single overhead camshaft (SOHC), double overhead camshaft (DOHC) and chain-driven camshaft (2.0 and 2.2 litre diesels). The bottom end is fundamentally the same in all. Petrol DOHC variants use a separate tensioner for the timing belt (rather than the coolant pump pulley), with an additional free-wheeling pulley to complete the timing belt's run.

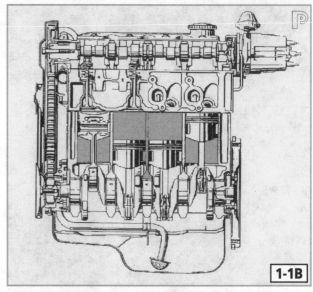

❒ **TYPE 1B:** This is a sectional view of the SOHC petrol/ gasoline engine.

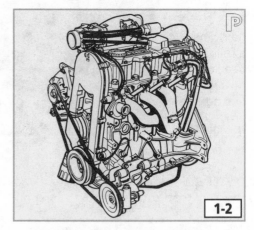

❒ **TYPE 2:** This is a 2-litre, electronic ignition/ injection engine.

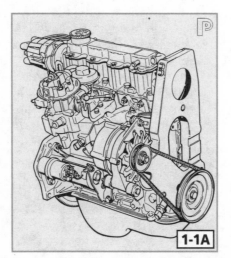

❒ **TYPE 1A:** The SOHC engine appears in 1.3, 1.4, 1.6, 1.8 and 2.0-litre versions, throughout the Cavalier range and as a 1.6 in the Vectra. It has eight valves, operated by a single overhead camshaft. It has carburetor fed and both single and multi-point fuel injected variants.

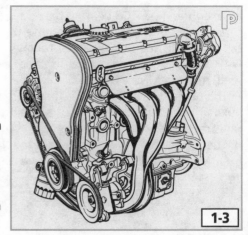

❒ **TYPE 3:** The DOHC engine appears in 1.6, 1.7, 1.8 and 2.0-litre versions, in the Cavalier and Vectra ranges. It has sixteen valves, operated by twin overhead camshafts. All versions are multi-point fuel injected.

❑ **TYPE 4A:** The diesel engines fitted to the models covered by this manual are SOHC. This is the 1.7 litre, with a timing belt. Models from 2001 have 16-valves and a timing chain. There are both naturally aspirated and turbocharged versions – this one a 1.7L turbo-version.

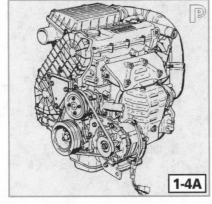

1-4A

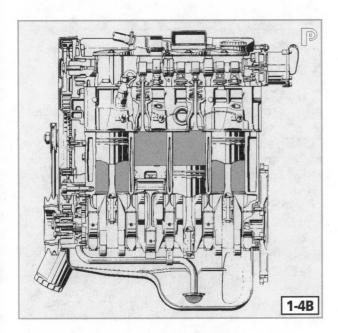

1-4B

❑ **TYPE 4B:** This is a sectional view of the 1.7 litre diesel engine.

❑ **TYPE 5:** This is the later diesel with a timing chain.

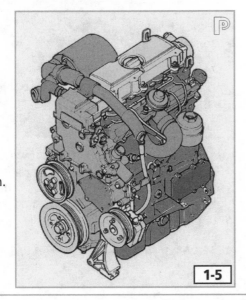

1-5

JOB 2: ENGINE DESIGNATION CODES.

FACT FILE

ENGINE CODES

Within the three basic engine types, evolutionary changes were denoted by a change to the designation code. This is stamped onto the front of the block, near the head gasket. Quote it when obtaining parts, to be sure of always getting the correct one.

ENGINE CODES

SOHC	Capacity	Used in...
13S	1.3	Cavalier
14NV	1.4	Cavalier
16LZ2	1.6	Vectra 1996-on
16S	1.6	Cavalier to 1987
16SV	1.6	Cavalier 1988-1992
X16SZ	1.6 Ecotec DIS	Cavalier 1993-1995
16SH	1.6	Cavalier
X16SZR	1.6 Ecotec	Vectra1996-2001
C16NZ	1.6 Cat	Cavalier 1990-95
C16NZ2	1.6 Cat	Cavalier 1993-94
18S	1.8	Cavalier to 1987
18SE	1.8	Cavalier 1986 to 1988
18SV	1.8	Cavalier to 1991
E18NVR	1.8	Cavalier 1988-1992
C18NZ	1.8 Cat	Cavalier to 1995
20NE	2.0	Cavalier 1989-1993, Calibra
20NEJ	2.0	Vectra 1996-2001
20SEH	2.0 SRi	Cavalier 1987-1990
20SEH	2.0	Cavalier 1990-1993
C20NE	2.0 Cat	Cavalier 1990-1994
DOHC	**Capacity**	**Used in...**
X16XEL	1.6	Vectra 1995-1997
X18XE	1.8	Vectra 1996-on
X18XE1	1.8	Vectra 1998-on
20XEJ	2.0	Cavalier 1989-1991
C20LET	2.0	Calibra 1992-1997
C20XE	2.0 Cat 16V	Cavalier 1993-on, Calibra
X20XEV	2.0 Cat Ecotec	Cavalier Vectra 1995-1998
X20XEV	2.0 Cat Ecotec	Cavalier Vectra 1998-on
X25XEV	2.5	Calibra
C20SEL	2.0	Vectra 1997-on
C22SEL	2.2	Vectra 1998-1999
Y26SE	2.6	Vectra 2001
DIESEL	**Capacity**	**Used in...**
16D	1.6	Cavalier 1982-1986
16DA	1.6	Cavalier 1986-1988
17D	1.7	Cavalier 1989-1993
17DR	1.7	Cavalier 1993-1995
17DT	1.7	Cavalier 1995-1996, Vectra 1996 on
X17DT	1.7	Cavalier 1994-1995, Vectra 1996 on
X20DTL	2.0	low pressure turbo Vectra 1995-1996
X20DTH	2.0	high pressure turbo Vectra 1997-on
Y17DTR	1.7	Vectra 2001
X22DTH	2.2	Vectra Y22DTR 2.2 Vectra 2001
V6	**Capacity**	**Used in...**
X25XE	2.5	Cavalier Vectra Calibra
C25XE	2.5	Calibra 1994-1997

Part C: Repair Procedures

CONTENTS

JOB 1: TIMING CHAINS – *remove, refit.*

NOTE: This job applies only to 2.0 Litre-and-larger diesel engines.

FACT FILE

TIMING CHAINS

• The duplex lower timing chain connects the crankshaft sprocket and fuel pump sprockets.
• The simplex upper timing chain connects the fuel pump sprockets and camshaft sprocket.
• If either timing chain is removed the valve timing can be altered unless components are locked beforehand or marked for reassembly in the correct positions.
• Several special Vauxhall tools are required to lock the camshaft and fuel injection pump flange during timing chain removal.
• If the special tools are not available, it is recommended that work involving removal of the timing chains is entrusted to a Vauxhall dealer, although the work can be undertaken provided accurate alignment marks can be made between the camshaft, injection pump flange and sprockets.

Section A: Upper timing chain.

❏ **STEP A1:** Disconnect the battery's negative terminal. See *Chapter 10, Electrical, Dash, Instruments, Fact File: Disconnecting the Battery* BEFORE doing so!

❏ **STEP A2:** Remove the exhaust front pipe – see *Chapter 9, Part C, Job 3.*

❏ **STEP A3:** Remove the air filter box and the auxiliary drive belt – see *Chapter 10, Job 1, Section B.*

❏ **STEP A4:** Remove the camshaft cover and turn the crankshaft until number one cylinder is at TDC on the compression stroke – see *Chapter 6, Part A, Job 2.*

❏ **STEP A5:** It is necessary to raise the right hand side of the engine.
➔ The engine can be raised and supported by a jack placed underneath, by an engine hoist or a suitable bar located in the front wing runners, such as this Sykes-Pickavant tool shown here.

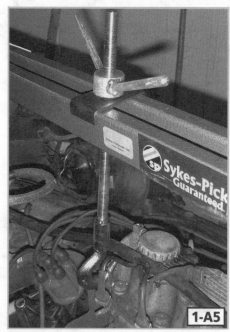

1-A5

➔ Remove the right hand engine mounting upper bolt, then raise the right hand side of the engine as high as possible without placing any strain on wiring, cables or hoses.

❏ **STEP A6:** If fitted, remove the engine undertray.

STEP A7: Remove the auxiliary drivebelt tensioning assembly.
➜ Remove the pivot and lower fastening bolts (**a**).
➜ Store the damper (**b**) the correct way up.
➜ If the damper is stored inverted or a new damper is fitted, it will have to be primed before use.

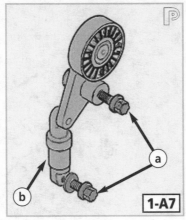

1-A7

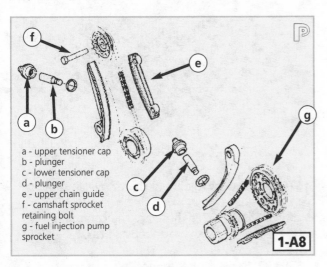

a - upper tensioner cap
b - plunger
c - lower tensioner cap
d - plunger
e - upper chain guide
f - camshaft sprocket retaining bolt
g - fuel injection pump sprocket

1-A8

STEP A8: IMPORTANT NOTE: The illustration letters in *Steps A8 to A14* refer to this illustration.
➜ Remove the upper chain tensioner cap (**a**) and withdraw the plunger (**b**).

STEP A9: Remove the lower chain tensioner cap (*1-A8, item c*) from the rear of the timing chain cover and withdraw the plunger (*1-A8, item d*).

STEP A10: Remove the Allen head bolts and injection pump sprocket cover.
➜ DO NOT use a lever to remove the pump sprocket cover because it is easily damaged.

STEP A11: In the absence of special tools to lock the camshaft and injection pump sprockets, make accurate alignment marks between pump flange and camshaft, and the chain and sprockets.

STEP A12: Remove the upper chain guide (*1-A8, item e*).
➜ The chain guide bolts are fixed with locking compound.
➜ Use a hot air gun to soften the locking compound before undoing the bolts.

STEP A13: Remove the camshaft sprocket retaining bolt (*1-A8, item f*).

➜ It is essential that the camshaft does not move.
➜ If you have the Vauxhall locking tool, remove it temporarily and refit it as soon as the camshaft sprocket bolt has been slackened.
➜ In the absence of the special tool, use a cable tie to hold the chain to the sprocket.
➜ Hold the camshaft still with a spanner on the provided flats whilst slackening the bolt.
➜ Do not remove the camshaft sprocket until the pump sprocket has been removed and the sprocket and chain can be lifted up out of the timing chain cover.

STEP A14: The fuel pump injection sprocket (*1-A8, item g*) has to be removed.
➜ If you have locked it with the Vauxhall tool, remove the tool temporarily whilst removing the sprocket bolts.
➜ In the absence of the special tool, make alignment marks on the sprocket and chain.
➜ In the absence of the special tool, make alignment marks on the sprocket and injection pump flange.
➜ Free the sprocket from the chain and remove it.

STEP A15: Lift the camshaft sprocket complete with chain up out of the upper chain cover.
➜ IMPORTANT NOTE: Renew the upper timing chain guide bolts and camshaft sprocket retaining bolt. If the upper timing chain or either sprocket are renewed, it is imperative that the alignment marks are accurately transferred. Double check that the crankshaft is still set at TDC.

STEP A16: To refit:
➜ Fit the timing chain to the camshaft sprocket, lower the chain into the chain guide and fit the sprocket.
➜ If you are not using the special Vauxhall tools, ensure that the camshaft is still in the TDC position and ensure that your alignment marks on the camshaft pulley and chain are adjacent.

STEP A17: Offer the injection pump sprocket through the aperture, engage it with the timing chain in accordance with your alignment marks.
➜ Fit the sprockets to the camshaft and fuel injection pump.
➜ Align the timing mark on the injection pump sprocket with the hole in the injection pump flange.
➜ Double check your alignment marks.
➜ Fit the camshaft pulley bolt and the injection pump retaining bolts but do not tighten them up beyond hand tightening.

STEP A18: Refit the lower timing chain tensioner.
➜ Apply fresh engine oil to the plunger and refit it ensuring that the closed end faces the chain.
➜ Renew the tensioner cap sealing ring and refit the cap. If the tensioner cap is new, press the centre pin in until it is heard to click.

STEP A19: Refit the upper timing chain guide.

STEP A20: If you have the Vauxhall special tools, fit them to check and if necessary set the valve timing.
→ If the special tools are unavailable, check your own marks yet again!

STEP A21: Using a torque wrench and angle measuring gauge, tighten the injection pump sprocket bolts to 20 Nm (15 lbs.ft)

STEP A22: Using a torque wrench and angle measuring gauge, tighten the camshaft sprocket bolt to
→ **STAGE ONE:** 90 Nm (66 lbs.ft)
→ **STAGE TWO:** Plus 45 degrees.
→ **STAGE THREE:** Plus 15 degrees.

STEP A23: Refit the upper timing chain tensioner.
→ Apply fresh engine oil to the plunger and refit it ensuring that the closed end faces the chain.
→ Renew the tensioner cap sealing ring and refit the cap. If the tensioner cap is new, press the centre pin in until it is heard to click.

STEP A24: Refit the pump sprocket and timing chain covers.
→ Clean the mating surfaces beforehand.
→ If a new gasket is not available, apply fresh sealant.

STEP A25: Further reassembly is the reverse of removal.

Section B: Lower (fuel pump) timing chain.

TOP TIP!

LOWER TIMING CHAIN.
• To remove the lower timing chain it is necessary to remove the timing chain cover. It is possible to remove the timing chain cover without removing the cylinder head, but there is a strong chance that the cylinder head gasket will be damaged in the process.
• If the cylinder head gasket is damaged, you won't know until the repair has been completed and the engine started, at which point you will have to do much of the work again in order to remove the cylinder head and renew the gasket!
• It is generally better to remove the cylinder head and fit a new gasket during lower timing chain removal.

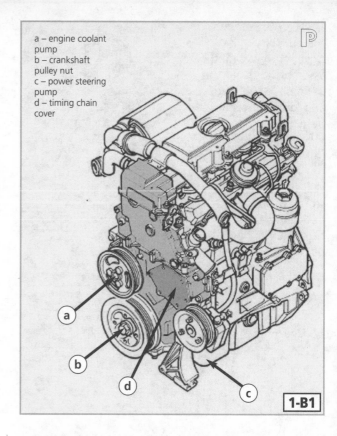

a – engine coolant pump
b – crankshaft pulley nut
c – power steering pump
d – timing chain cover

1-B1

STEP B1: The lower timing chain cover is shown here, shaded.
→ IMPORTANT NOTE: These are the components whose letters are shown in *Steps B4 to B8*.

STEP B2: Remove the upper timing chain as described in *Section A*.

STEP B3: OPTIONAL: Remove the cylinder head – see *Job 4*.

STEP B4: Remove the engine coolant pump (a) – see *Chapter 8, Job 8*.

STEP B5: Remove the crankshaft pulley nut (b), taking care not to turn the crankshaft.
→ Lock the crankshaft whilst slackening the pulley nut. With the starter motor removed, you can have a helper prevent the flywheel ring gear from turning with a large screwdriver.
→ Another way is to hold the pulley with a chain wrench, as you undo the nut.

STEP B6: Drain the engine oil and remove the sump.

STEP B7: Remove the power steering pump (c) – see *Chapter 11, Job 6*.

STEP B8: Remove the timing chain cover (d) bolts.
→ The bolts are of varying length and must be refitted in the correct positions.

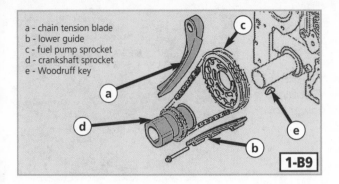

a - chain tension blade
b - lower guide
c - fuel pump sprocket
d - crankshaft sprocket
e - Woodruff key

1-B9

❏ **STEP B9:** Undo the timing chain tensioner blades' pivot bolts and remove the chain tensioner blade (**a**).
➔ Remove the timing chain lower guide bolts and the lower guide (**b**).

❏ **STEP B10:** Ease the timing chain cover off, noting the positions of locating dowels.
➔ If the locating dowels are not a tight fit, it is best to remove them and store them with the timing chain cover.
➔ The timing chain cover gasket should be renewed.

❏ **STEP B11:** Make alignment marks between the injection pump sprocket and flange, and the chain and sprockets, then pull the pump sprocket (illustration *1-B9, item c*), crankshaft sprocket (**d**) and timing chain away complete.
➔ Keep the crankshaft woodruff key (**e**) safe.

REASSEMBLY NOTE: If the timing chain or either sprocket are renewed, it is imperative that the alignment marks are accurately transferred. Double check that the crankshaft is still set at TDC.

❏ **STEP B12:** Reassemble as the reverse of removal.
➔ Ensure that the crankshaft is kept at TDC throughout.
➔ Ensure that the injection pump flange cut out aligns with the hole in the pump body.
➔ Ensure that all alignment marks line up as the rebuild progresses.

JOB 2: TIMING BELT – *inspection*.

Section A: Outer timing belt cover – remove, refit.

The timing belt cover design varies according to the engine and can be one, two or three piece.

❏ **STEP A1:** On vehicles with fuel injection, it will be necessary to remove the air filter box assembly, to gain access to the end of the engine.
➔ Where necessary, remove the auxiliary drive belt by slackening the alternator mounting bolts. If the

auxiliary drive belt is not to be renewed, mark its direction of rotation so that it can be refitted the same way around.

➔ **MODELS WITHOUT AIR CONDITIONING:** Swing the alternator towards the engine, after which the belt can be slipped off its pulleys.

2-A1

➔ **MODELS WITH AIR CONDITIONING EXCEPT 1.7 DIESEL:** On models with air conditioning, the belt travels through an engine mounting, and cannot therefore be completely removed. Simply move the loosened belt out of the way.
➔ **1.7 DIESEL:** Remove the three auxiliary drive belts.

TOP TIP!

• It is a good idea to mark the direction of the auxiliary drive belt's travel.
• If it is refitted the 'wrong' way round, its life will be shortened.

1-PIECE COVERS AND UPPER PART OF 2-PIECE COVERS

❏ **STEP A2:** Remove the bolts (arrowed) and/or clips securing the timing belt cover or upper timing belt cover, as applicable to the model you are working on – various shapes and sizes used.

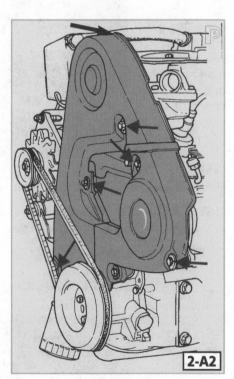

2-A2

□ STEP A3:
Withdraw the cover from the engine and through the engine bay.
➡ On models with two-part covers, unclip the upper part from the lower before lifting it out.

2-A3

LOWER PART OF 2-PIECE COVERS

□ STEP A4: To remove the lower outer cover, where present, jack the front right hand corner of the vehicle up and remove the wheel.

□ STEP A5: If necessary to make enough room to work in, remove the plastic arch liner.

□ STEP A6: Remove the crankshaft pulley, as described in *Job 3*.

□ STEP A7:
Remove the mounting bolts and, if the vehicle you are working on has one, the camshaft sensor (arrowed).

2-A7

□ STEP A8:
Unclip any wiring from the cover before removing.

□ STEP A9: Remove the lower cover by pulling it downwards and away from the engine.

□ STEP A10: Refit as the reverse of removal.

3-PIECE COVERS

□ STEP A11:
Remove the crankshaft pulley and unfasten the lower clips (arrowed).

2-A11

□ STEP A12:
Remove the lower cover.

2-A12

□ STEP A13:
Unfasten the upper cover clips and lift the cover sections away.

2-A13

V6 ENGINE

□ STEP A14: Mark the running direction of the ribbed V-belt.

□ STEP A15:
Turn the ribbed V-belt tension roller clockwise to release the tension.
➡ Remove the V-belt from the coolant pump pulley.

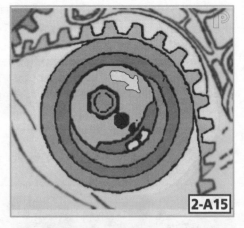

2-A15

□ STEP A16:
Remove the V-belt tensioner.

2-A16

STEP A17: Remove the coolant pump pulley (**a**) and steering hydraulic pump pulley (**b**).

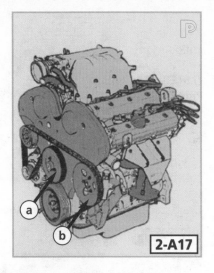

2-A17

STEP A18: Undo the four fastening bolts.
→ Remove the outer cover.

Section B: Timing belt – inspect.

> **TOP TIP!**
> • Check the tension of a used belt with the engine at its normal running temperature.
> • Check the tension of a new belt with the engine cold.

STEP B1: Remove outer timing belt cover.
→ On two-part assemblies, it is usually sufficient to remove just the upper section, although the lower cover will have to come off if adjustment or replacement of the timing belt is needed.

STEP B2: Examine the belt for wear. This inspection will only show severe damage, so if there is any cracking, or if the toothed side appears worn, or any teeth are missing, replace the belt straight away.

2-B2

STEP B3: If the belt is in good condition, it may still need re-tensioning.
→ **EITHER:** Check that you can twist the belt through approximately 90 degrees at the centre of its longest

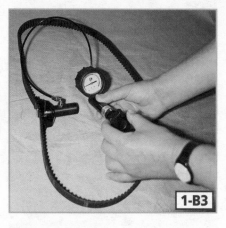

1-B3

'run', using only the strength of fingers and thumb.
→ **OR (AND PREFERABLY):** Use the correct tool, such as this Sykes-Pickavant tension checking tool – used with the belt *in situ*, of course! If necessary, re-adjust as shown in *Job 3*.

JOB 3:	TIMING BELT – *replacing, adjusting*.

On all engines, the timing or 'cam' belt – called the 'toothed drive belt' on the V6 - should be renewed at the recommended interval. If it breaks, the engine will suffer total failure and could be catastrophically damaged.

> **TOP TIP!**
> • If an engine has been standing for, say, a year or more, you should replace the belt as a matter of course, because it will be prone to early failure.

IMPORTANT NOTE: Avoid potential major problems by following this advice:
→ None of the engines covered by this manual are 'safe' engines. This means that the valves will collide with the pistons if the camshaft/s is/are turned while the engine is set to Top Dead Centre (TDC).
→ To avoid risk of damage, the camshaft/s should NEVER be replaced (or turned if already in place) while the pistons are at TDC.
→ We recommend that you turn the crankshaft/s back from TDC (the *opposite direction* to normal rotation) by a few degrees, moving the pistons safely a little way down the bore.
→ With the cylinder head fitted to the engine, the camshaft/s must now be turned to their alignment marks for when No. 1 piston is at TDC.
→ Before the timing belt is refitted, the crankshaft must be moved forward (in the *direction of* normal rotation) to the TDC position again.
→ Select 'Neutral' gear in case road wheels are turned.
→ The V6 engine is especially difficult to deal with because it has four camshafts, all of which MUST be at their correct positions when the number one

cylinder is at TDC. Vauxhall have special tools that lock the four camshaft pulleys, and it is highly recommended that the job is not tackled without them. For information on replacing/adjusting the V6 toothed belt drive, go to *Step 12*.

Section A: Timing belt replacement and adjustment
- WITH ENGINE IN VEHICLE.

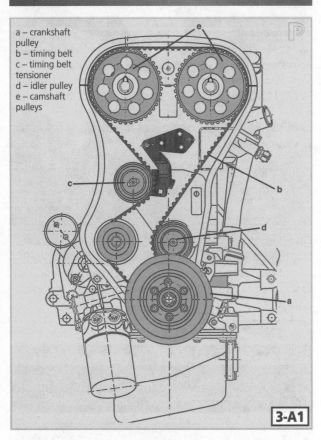

a – crankshaft pulley
b – timing belt
c – timing belt tensioner
d – idler pulley
e – camshaft pulleys

3-A1

❑ **STEP A1:** Slacken the alternator and remove the auxiliary belt.
➔ Remove the upper outer timing belt cover (see *Job 2*).

❑ **STEP A2A:** Turn the crankshaft until the TDC mark(s) on the cam(s) align with the marks on the inner timing belt cover.
➔ Here, we have marked them with typists' correction fluid (arrows).

❑ **STEP A2B:** Check also that the TDC mark on the lower pulley aligns with the pointer or mark on the inner timing belt cover, as applicable.

3-A2A

❑ **STEP A3:** Remove the following:
➔ The lower outer timing belt cover - see *Job 2*.
➔ The crankshaft pulley (see *Job 8*).
➔ First, turn it so that it aligns with the Top Dead Centre (TDC) marker (arrowed).

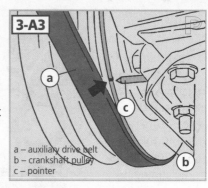

3-A3

a – auxiliary drive belt
b – crankshaft pulley
c – pointer

TOP TIP!

❑ **STEP A4:** • The centre crankshaft pulley nut will be difficult to turn.
• With the starter motor removed, you can have a helper prevent the flywheel ring gear from turning with a large screwdriver.
• Another way is to hold the pulley with a chain wrench, as shown, as you undo the nut.

3-A4

SOHC ENGINES

3-A5

❑ **STEP A5:** Before removing the timing belt, you will need to lock the engine to prevent it from turning:
➔ Check once again that all the TDC markers are aligned.
➔ Use a camshaft locking tool on the end of the camshaft. Remove the distributor (petrol) or vacuum pump (diesel). This is the Sykes-Pickavant locking tool prior to being bolted into place.
➔ Lock the crankshaft if possible. On engines with a clutch cover plate in the bellhousing (see *Chapter 7, Part B, Job 2*), remove the plate and fit a locking tool to the flywheel ring gear.

STEP A6: To remove the timing belt:

→ Loosen the three bolts holding the water pump to the block (a)...

→ ...and push on the belt or turn the water pump so that tension is removed from the belt.

3-A6

STEP A7: If the pump doesn't want to turn, use a spanner on the hexagon on the back of the water pump (position arrowed).

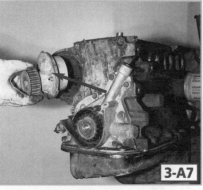

3-A7

DOHC ENGINES

STEP A8: Undo the camshaft sensor securing bolts, and move the sensor out of harm's way.

→ Undo the tensioner bolt. If the engine you are working on is a pre-1993 1.8 or 2.0-litre version, the tensioner will slide automatically, releasing the tension – see FACT FILE: TENSIONERS.

→ Place an Allen key in the hole on the tensioner, as shown.

→ Turn the unit anti-clockwise to relieve the tension on the belt.

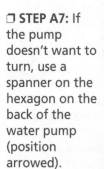

3-A8

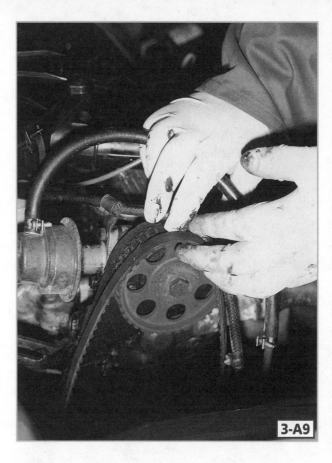

3-A9

STEP A9: Slide the timing belt off the pulley(s).

TOP TIP!

• To ensure that the replacement belt will be fitted with the valves (and injection pump, on diesel engines) in the correct position - essential to avoid serious engine damage - turn the engine until all the timing marks are aligned.

• The crankshaft and camshaft must not be turned independently of each other, or the valves may collide with the pistons – see the IMPORTANT NOTE near the beginning of this Job.

• Refitting is best done from the crank sprocket upwards. Fit the belt round the camshaft pulleys, then finally the water pump sprocket or the idler pulley on the inlet side of the engine.

SOHC ENGINES

STEP A10: To tension the timing belt:

→ Make sure the three water pump bolts are loose – see *Step A6*.

→ Turn the water pump using a wrench on the hexagon on the water pump shaft – see *Step A6*.

→ Tighten the three bolts enough to grip the pump while you check the tension.

→ Tighten the bolts to the correct torque when the tension is accurately set.

DOHC ENGINES

TENSIONERS

• The tensioner is found in slightly different positions, depending on which models of engine you are working on.

• Pre-1993 1.8 and 2.0-litre variants have the tensioner between the inlet camshaft pulley and the water pump (see illustration *10-8*).

• On other DOHC engines, however, the tensioner sits low on the end of the engine, between the water pump and the crankshaft pulley (see illustration *3-A11*).

• On the V6, the belt tensioner is attached to the rear cylinder head.

• This doesn't materially affect the way the tensioner tightens the belt, but in the latter case, it means that two idler pullies are used.

• There are also procedural differences for tightening the tensioners, which are set out in the next two steps.

❑ **STEP A11:** The tensioner (arrowed) found between the crankshaft pulley and the water pump has a marker.

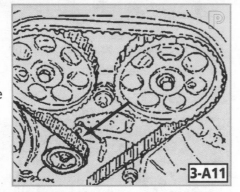

3-A11

➔ To correctly tension the belt, turn the tensioner securing bolt back clockwise with an Allen key, until this marker is in the middle of the 'V' in its back plate.

❑ **STEP A12: PRE-1993 1.8 AND 2.0-LITRE DOHC ENGINES:** These engines require a different tensioning procedure which is simpler than for other engines.

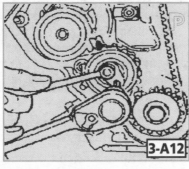

3-A12

➔ Loosen the tensioner mounting bolt, to allow the tension to be released.

➔ Remove the bolt, and loosely fit a new one.

➔ Push the pulley forwards and upwards until the belt is as tight as possible, then tighten the bolt enough to hold it.

➔ Check whether the belt is at the specified

tightness – see *Job 2, Step B3* – and slacken or tighten a little to get it right.

➔ Finally, fully tighten the mounting bolt.

V6 ENGINE

IMPORTANT NOTE:

➔ Vauxhall point out that if the toothed belt on the V6 engine is fitted 'out' by as little as one tooth, the pistons can collide with the valves causing severe engine damage.

➔ Vauxhall recommend the use of special tool KM-800, which comprises several parts and ensures that the engine is at TDC and the camshafts locked in position as the cambelt belt is replaced.

➔ We believe that it is possible to carry out the work described here without the use of special tools KM-800 but it will be necessary to work with great accuracy.

➔ If you would prefer not to take risks, you may wish to purchase the appropriate special tool from a Vauxhall agent, or have them carry out the work.

ADJUSTMENT

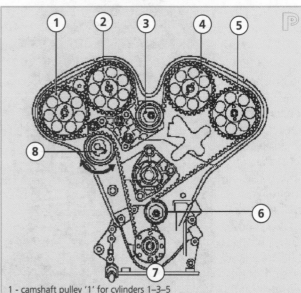

1 - camshaft pulley '1' for cylinders 1–3–5
2 - camshaft pulley '2' for cylinders 2-4-6
3 - toothed belt guide roller for adjustment of camshaft pulley for cylinders 1-3-5
4 - camshaft pulley '3' for cylinders 1–3–5
5 - camshaft pulley '4' for cylinders 2-4-6
6 - toothed belt guide roller for adjustment of camshaft pulleys for cylinders 2-4-6
7 - toothed belt drive gear
8 - toothed belt tension roller

3-A13

❑ **STEP A13:** The V6 engine timing gear. Refer to this illustration throughout this section.

STEP A14: Remove the outer camshaft belt cover – see *Job 2, Steps 14 to 18*. Referring to *Job 2, Step 15*, slacken the belt tension roller then turn the eccentric counter clockwise to its stop.

STEP A15: Tighten the tension roller slightly.

STEP A16: Using the tension roller, turn the engine in its normal direction of rotation through two complete crankshaft revolutions.

STEP A17: Slacken the tension roller, then turn the eccentric clockwise until the marks align.

STEP A18: Tighten the tension roller.

RENEWAL

IMPORTANT NOTE:
→ This job MUST be carried out with the engine cold.
→ Positioning marks are provided on the replacement belt and MUST be correctly aligned.

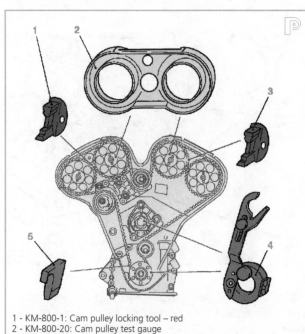

1 - KM-800-1: Cam pulley locking tool – red
2 - KM-800-20: Cam pulley test gauge
3 - KM-800-2: Cam pulley locking tool – green
4 - KM-800-10: Tool for locking crankshaft at TDC
5 - KM-800-30: Wedge to prevent cam belt slipping – yellow

3-A19

STEP A19: To renew the toothed drive belt it is necessary to arrange all four camshaft pulleys so that the marks align with those on the toothed belt drive rear cover – and to keep them in that position while the crankshaft is brought to TDC and the belt fitted.
→ Vauxhall special tool set KM-800, includes two devices (1, 3) which keep the camshafts locked in position as the toothed drive belt is replaced. This job describes the process WITH the special tools – if you wish to attempt the job without the tools then these Steps will provide a basic guide.
→ It is possible to fit a toothed drive belt without the Vauxhall special tools, but not recommended.

STEP A20: Remove the outer camshaft belt cover – see *Job 2, Steps 14 to 18*.

STEP A21: Remove the side engine compartment cover.

STEP A22: Remove the torsional vibration damper.

STEP A23: Turn the crankshaft to just before TDC, fit special tool KM-800-10 to the toothed belt drive gear, then turn the engine until it contacts the coolant pump flange – indicating that the engine is at TDC.
→ There is a movable lever on the tool – press this against the coolant pump flange flat and tighten the knurled bolt to lock it there.
→ This locks the crankshaft at TDC.

STEP A24: The marks on the toothed belt cover (on an old belt they might not be visible), the marks on the camshaft pulleys and the toothed drive belt rear cover should all now align, showing that all camshafts are at their correct position when No. 1 cylinder is at TDC position.

STEP A25: Fix each pair of camshaft pulleys in position by inserting KM-800-2 (green handle) between pulleys 3 and 4, and inserting KM-800-1 (red handle) between pulleys 1 and 2.
→ If the special locking tools cannot be inserted, slacken the relevant toothed belt roller and turn the toothed belt roller eccentric.

STEP A26: Release the toothed belt rollers, guide pulleys and remove the toothed belt.

IMPORTANT NOTE: If the toothed belt is being removed to allow other work, such as cylinder head removal to be carried out, it is highly recommended that the toothed belt drive is renewed at the same time.

STEP A27: The toothed drive belt has several marks – these are important – note the following.
→ The direction of belt travel (clockwise) is marked by arrows.
→ The double dash must align with the groove on the oil pump and the notch on the toothed belt drive gear.
→ There are also marks to align with the camshaft pulleys.

STEP A28: There are two ways to fitting the toothed belt, depending on the base plate fitted.
→ Follow the instructions according to the base plate fitted.

TOOTHED BELT TENSION ROLLER WITHOUT FRONT FLANGED WHEEL AND WITH CODE LETTER 'D' ON THE BASE PLATE

STEP A29: Install the toothed belt via the belt guide pulley and cam pulleys 3 and 4, ensuring that

the driving end (in between the pulley and cam pulley 4) is taut.
→ The positioning marks on the toothed belt MUST align with the notches on the cam pulleys and rear cover.

❑ **STEP A30:** Install the toothed belt above the tension rollers and cam pulley 1 and 2.

TOOTHED BELT TENSION ROLLER WITH FRONT FLANGED WHEEL AND WITH CODE LETTER 'E' ON THE BASE PLATE

❑ **STEP A31:** Install the toothed belt above the tension roller, camshaft pulleys 1 and 2, toothed belt guide pulley, cam pulleys 3 and 4 then toothed belt guide pulley - ensuring that the driving end (in between the pulley and cam pulley 4) is taut.
→ The positioning marks on the toothed belt MUST align with the notches on the cam pulleys and rear cover.

IMPORTANT NOTE:
→ Check that the double dash marks on the toothed belt align with the groove on the oil pump and on the toothed belt drive gear.
→ Wedge KM-800-30 between the oil pump and toothed belt drive gear to prevent the belt from moving.

BOTH TYPES OF BASE PLATE - ADJUSTMENT

❑ **STEP A32:** Turn the toothed belt guide roller eccentric anti-clockwise until its tip is at the 2 o'clock position, then fasten the bolt.

❑ **STEP A33:** Turn the eccentric of the toothed belt guide pulley anti-clockwise until its cone end is at the 11 o'clock position, then tighten its bolt.

❑ **STEP A34:**
Release the toothed belt tension roller and over extend it in an anti-clockwise direction until it makes slight contact with the 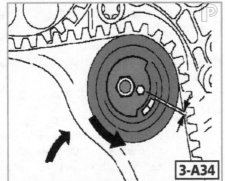 stop, then allow it back until a guide measurement of 1 mm is obtained.
→ Tighten the fastening nut.

❑ **STEP A35:** Check that all of the toothed belt marks align correctly, then remove the special tools KM-800-10.

→ GENTLY turn the crankshaft two turns in normal engine direction – stop if any resistance is felt (turn the crankshaft back and re-check the toothed belt mark alignment!) – to just before TDC.

❑ **STEP A36:** Fit tool KM-800-10, bring the crankshaft to TDC and lock the tool as previously described.

❑ **STEP A37:** Place test gauge KM-800-20 on camshaft pulleys 3 and 4.
→ Turn the toothed belt guide pulley anti-clockwise until the marks of the camshaft pulleys align with those of the test gauge, then fasten the toothed belt guide roller nut.
→ Remove the test gauge.

❑ **STEP A38:** Fit the test gauge to camshaft pulleys 1 and 2.
→ Turn the toothed belt guide roller anti-clockwise until the marks of the camshaft pulleys align with those of the test gauge, then fasten the toothed belt guide roller nut.
→ Remove the test gauge.

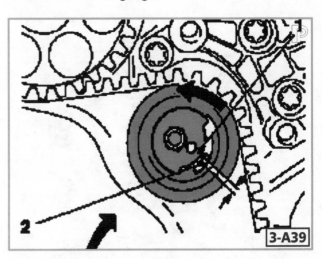

❑ **STEP A39:** Release the toothed belt tension roller and over extend anti-clockwise until making contact with the stop and allow it to move back until mark I is roughly 3 to 4 mm above mark 2.
→ Tighten the toothed belt tension roller.
→ Remove all tools.

❑ **STEP A40:** Repeat *Steps A35 to A38*.

IMPORTANT NOTE: The dash marks on the toothed belt will not align after one or more engine revolutions – they are there for fitting purposes only.

Section B: Timing belt replacement and adjustment - DURING ENGINE REBUILD.

The process is similar to the one followed when renewing the cam belt when the engine is in the vehicle, but you will have to establish the sprocket positions from scratch.

ENGINE REBUILDING

V6 ENGINE
It is strongly advised that a rebuild is only attempted if you have access to Vauxhall special tools KN-800-10. You could alternatively assemble the engine minus camshafts and toothed belt, and have a specialist complete the rebuild.

DIESEL ENGINES
When rebuilding an engine, and after the pump and camshaft alignment marks have been lost, it is not possible for the diesel pump to be set without specialist equipment. You can get as far as to fit the camshaft/timing belt or chain, but DO NOT attempt to run the engine without having a specialist set the injection pump timing for you, or serious engine damage can occur.

ALL ENGINES
When fitting a cam belt as part of an engine build, it is MOST IMPORTANT that the camshaft is fitted and the camshaft pulley set to its alignment mark while none of the pistons are at TDC. None of the engines covered by this manual are 'safe' engines, so the valves can collide with the pistons if they are at TDC.
➔ Once the valve positions are set, the crankshaft position can be aligned.
➔ Don't re-use an already used belt. It will be more prone to breakage once removed - replace it!

ALL ENGINES EXCEPT V6

❑ **STEP B1:** Turn the camshaft so that the camshaft sprocket aligns with the timing mark shown for your engine in *Section A*.

IMPORTANT NOTE:
➔ If you have followed the earlier IMPORTANT NOTE the pistons will all be part-way down the bores.
➔ If you are not sure, turn the camshaft with very great care, so that if a valve should come into contact with a piston, you can stop before any damage is done.

❑ **STEP B2:** Fit the sprocket to the crankshaft so that the alignment mark is in the position shown.

❑ **STEP B3:** Fit and tension the belt as described in *Section A: Timing belt replacement and adjustment - WITH ENGINE IN VEHICLE*.

❑ **STEP B4: PETROL ENGINES:** Check the ignition timing when the work is complete.

❑ **STEP B5: DISTRIBUTOR IGNITION:** Check that the mark on the distributor rotor arm (a) is pointing to the No. 1 cylinder mark

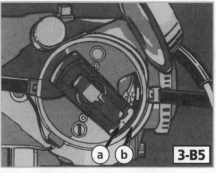

3-B5

on the distributor body (**b**). If not, turn the distributor or remove and refit it, so that the rotor arm is properly aligned. To do this, align the rotor arm with the arrow etched on the upper curve of the distributor body.

❑ **STEP B6: DISTRIBUTORLESS IGNITION SYSTEMS (DIS):** Make sure that the camshaft position sensor, if present, is bolted securely in place before you replace the timing belt cover.

V6 ENGINE

❑ **STEP B7:** In principle, this could be achieved without Vauxhall's special tools - though we recommend you do not try this – if an alternative method could be found to lock all four camshafts at the positions when number one cylinder is at TDC.
➔ Were this possible, then the crankshaft would have to be turned 60 degrees *BEFORE* TDC, the camshafts fitted and locked in the correct positions, then the crankshaft turned to TDC and the toothed belt fitted as previously described.

JOB 4: CYLINDER HEAD - *removal.*

IMPORTANT NOTE: Refer to the general notes on cylinder head removal in **Chapter 6, Part A**.

➔ SOHC versions house the camshaft in a casting that is entirely separate from the head itself, but bolted to the block using the head bolts - the head becomes effectively the filling in a metal 'sandwich' once the engine is re-assembled.
➔ DOHC engines and engines with timing chains, on the other hand, house their camshafts in the head casting itself. In this case the head can be removed with the camshaft *in situ*.

Section A: SOHC engines.

☐ **STEP A1:** Referring to *Chapter 8, Part B, Job 1*, drain the cooling system.

FUEL INJECTION MODELS

☐ **STEP A2:** If the model you are working on has fuel injection, depressurise the system, referring to *Chapter 9, Part A, Job 12*.

☐ **STEP A3:** If you are working on a SOHC engine, refer to *Job 2* and remove the timing belt outer cover(s).

☐ **STEP A4:** Refer to *Job 3* and remove the timing belt. Then remove the camshaft pulley, referring to *Job 5*. Finally, remove the timing belt inner cover.

☐ **STEP A5:** Disconnect the front exhaust pipe from the exhaust manifold.

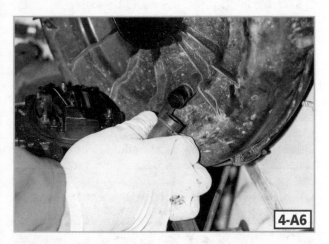

4-A6

☐ **STEP A6: CARBURETOR VERSIONS:** If the vehicle you are working on has carburetor induction:
➜ Undo the bolts fixing the air filter assembly, and move it out of the way.
➜ Remove the breather pipe as you do so. If necessary, remove it completely.
➜ Disconnect the fuel lines and accelerator and choke cables, as applicable, from the carburetor.

☐ **STEP A7: FUEL INJECTED VERSIONS:** If the vehicle you are working on has fuel injection:
➜ Remove the induction trunking from the throttle body.
➜ Disconnect the accelerator cable from the throttle body.
➜ Remove all sensor connections, such as the air mass meter.

☐ **STEP A8:** Disconnect the vacuum hoses from the inlet manifold. If necessary, remove the crankcase

breather hose from the cam housing or, depending on which model you are working on, the rocker cover.

☐ **STEP A9:** Undo the rocker cover securing bolts and remove the cover. Carefully remove its gasket.

4-A9

☐ **STEP A10:** Remove the coolant hoses from the coolant pump housing.

☐ **STEP A11:** Unbolt the distributor assembly retention plate, and remove it with the HT leads.
➜ If the vehicle you are working on has distributorless ignition, disconnect the coil lead and remove the HT leads.
➜ Alternatively, as applicable, remove the leads along with the Direct Ignition System unit - see *Chapter 9, Part A*.

4-A11

☐ **STEP A12:** Now double-check that you have disconnected, drained or removed the following items (refer to the relevant sections of this manual):
➜ Drain off the coolant into a suitable container and dispose of responsibly.
➜ Remove the air filter and housing and all associated air and vacuum lines, including the air pump fittings and EGR valve and lines, if applicable.
➜ **PETROL ENGINES:** Disconnect and remove the HT leads from the spark plugs and also all other connections to the distributor.
➜ **DIESEL ENGINES:** Disconnect and remove the glow plugs and the injectors - but ONLY after reading the relevant *Safety First!* notes.
➜ Disconnect the fittings to the turbocharger (when fitted) but leave the turbo connected to the exhaust manifold and remove as a unit.
➜ Disconnect the fuel line at the engine-end (and plug the line), disconnect the throttle cable, and choke cable if fitted.

→ Remove the carburetor or fuel injection system. See **Chapter 9, Ignition, Fuel, Exhaust**.

→ Remove the coolant hoses to the radiator, heater and choke.

→ Label and detach all remaining electrical connections.

→ Detach the exhaust pipe from the manifold.

☐ STEP A13: Remove the timing belt - see **Job 2**. You could remove the pulleys and inner belt cover, as shown here, or wait until later...

4-A13

TOP TIP!

• You may wish to remove the inlet manifold at this stage, because it is a large component and may get in the way.

☐ STEP A14: • If you 'crack' the exhaust manifold nuts at this stage, but leave the manifold in place for now, you will find it easier to lift the head off the block, and easy to undo any stubborn fixing nuts later.

☐ STEP A15: Remove the camshaft cover and the gasket.

☐ STEP A16: The cylinder head bolts pass right through the camshaft housing and hold both components in place.

→ Make sure that the engine is firmly held, such as in an engine stand, if out of the vehicle.

→ Use an extension on the socket, because of the large amount of force needed.

4-A16

→ Loosen the cylinder head bolts by half a turn at a time in the order shown (see **Step B14**), until all are free, then remove them.

☐ STEP A17: ...and remove the housing with camshaft and sprocket.

4-A17

☐ STEP A18: If the cylinder head has not sealed itself to the gasket, the head can be lifted straight off.

4-A18

TOP TIP!

• Do not use a wedge to break the seal that often occurs between cylinder head and block.

• Extra leverage can be gained by lifting and rocking carefully on the manifolds.

☐ STEP A19: Remove all traces of the cylinder head gasket from both block and head. Scrape off carbon deposits with a wooden or aluminium scraper, taking care not to gouge the aluminium surface of the head. Stuff rags into all openings in the cylinder head to stop debris dropping into them.

☐ STEP A20: Clean excessive carbon deposits from the crowns of the pistons without damaging the surface of the aluminium.

TOP TIP!

• If the engine is worn and you don't intend overhauling it at this stage, we strongly advise that you leave the piston crowns alone. The carbon on them can actually help preserve compression pressures when the engine is near the point of nearing overhaul.

• An aerosol gasket remover spray will help to remove pieces of stuck-on gasket.

☐ STEP A21: The cylinder head can be checked for distortion by use of a straight edge and feeler gauge. At the same time, check for excessive corrosion.

- Always have the cylinder head refaced before refitting to greatly reduce the risk of the cylinder head blowing after being refitted.
- This is not usually applicable to diesel engines because there is already a high compression with little clearance. If distorted, replace the cylinder head.

Section B: DOHC engines.

❏ **STEP B1:** Referring to *Chapter 8, Part B, Job 1*, drain the cooling system. Unbolt the coolant expansion tank and remove it, to give you more space to work in.

❏ **STEP B2:** Depressurise the fuel system, referring to *Chapter 9, Part A, Job 12*.

❏ **STEP B3:** Referring to *Job 3*, remove the timing belt. Then remove the camshaft pullies, referring to *Job 7*. Along with these, unbolt the timing belt tensioner and the idler pullies. Finally, remove the timing belt inner cover.

❏ **STEP B4:**
Remove the plastic cover from the top of the rocker cover. If the model you are working on has a cast alloy inlet manifold, unbolt and remove its upper half. Fill the lower inlet tracts with rags, to stop unnecessary dirt getting down them, and remove the rocker cover.

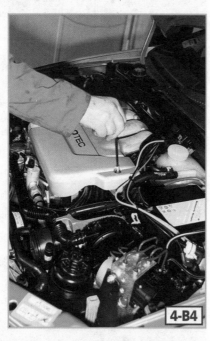
4-B4

❏ **STEP B5:** Disconnect the fuel return hose from the fuel pressure regulator unit. Fuel will spill out, so have a rag ready to soak it up, and a suitable rubber cap to seal the union (this will prevent any dirt getting into the fuel system).

❏ **STEP B6:** Disconnect the fuel hose from the fuel injector rail. Again, some fuel may escape, so have a rag ready to soak it up, and a suitable rubber cap to seal the union.

❏ **STEP B7:** Disconnect any vacuum hoses from the head assembly and disconnect all the hoses of the emissions control equipment, if the model you are working on has it.

❏ **STEP B8:**
Disconnect the accelerator cable (a) from the throttle or throttle body.
➜ Also remove the wiring connectors for the idle control valve and throttle position sensor.
➜ Then remove the throttle or throttle body altogether.

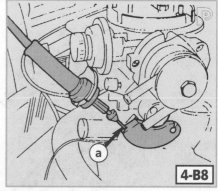
4-B8

❏ **STEP B9:** Disconnect the HT leads from the plugs, using the orange plastic tool stored on one of the leads.
➜ Referring to *Chapter 9, Part A*, unbolt the Direct Ignition System unit from the rocker housing and remove it.

❏ **STEP B10:**
Disconnect the exhaust front pipe from the exhaust manifold and, if applicable, the Lambda sensor wiring connector (a).
➜ Leave the sensor (1) in the manifold.

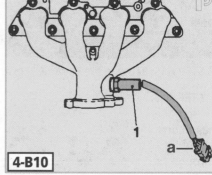
4-B10

❏ **STEP B11:** Disconnect the coolant hose from the thermostat housing. Then disconnect the coolant hose from the inlet manifold. This isn't easy to get to in every case.

❏ **STEP B12:** Undo the alternator mounting bolts, then move the alternator away from the engine. Some models have electrical earth/ground points on the alternator brackets, which should be disconnected.

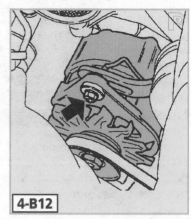
4-B12

☐ **STEP B13:** Disconnect all the wiring connectors left on the head-and-inlet manifold assembly. These include:
→ The fuel injector plug.
→ The crank position sensor.
→ The cam position sensor.
→ The inlet temperature sensor.
→ In each case, disconnect the wiring runs from all their clips on the head assembly.

TOP TIP!

• If you get confused as to which connector is which, simply put a tag of masking tape on each, and write their names on them.
• This can save a lot of head-scratching when it comes to reassembly!

IMPORTANT NOTE: Check *Section A, Steps A12 and A13* to ensure that everything is disconnected.

☐ **STEP B14:** Referring to *Part A, Job 3*, remove the cylinder head bolts, observing the order of loosening shown in the diagram and loosening them in gradual increments.

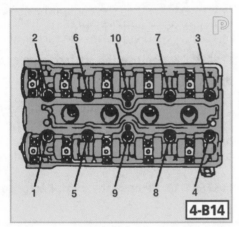

☐ **STEP B15:** Remove the cylinder head. If it is reluctant to be freed, refer to *Part A, Job 6*.

☐ **STEP B16:** Remove the head gasket, and make sure its mating surface on the block is completely free of traces of gasket, or corrosion. See *Steps A18 to A21*.

Section C: Engines with timing chains

☐ **STEP C1:** Drain the cooling system - see *Chapter 8, Part B, Job 1*.

☐ **STEP C2:** Referring to *Job 1, Section A*:
→ Disconnect the battery.
→ Remove the exhaust front pipe.
→ Remove the air filter box and auxiliary drive belt.
→ Remove the camshaft cover and turn the crankshaft until number one cylinder is at TDC on the compression stroke.
→ Undo the right hand engine mounting and raise the right hand side of the engine.
→ Remove the auxiliary drivebelt tensioning assembly.
→ Remove the upper chain tensioner cap and withdraw the plunger.
→ Remove the Allen head bolts and injection pump sprocket cover.
→ In the absence of special tools to lock the camshaft and injection pump. sprockets, make accurate alignment marks between pump flange and camshaft, and the chain and sprockets.
→ Remove the upper chain guide.
→ Remove the camshaft sprocket retaining bolt.
→ Remove the fuel pump injection sprocket.
→ Lift the camshaft sprocket complete with chain up out of the upper chain cover.

☐ **STEP C3:** Disconnect and remove the glow plugs - but ONLY after reading the relevant *Safety First!* notes. See *Chapter 9, Part B, Ignition, Fuel, Exhaust*.
→ Disconnect the fuel line at the engine-end (and plug the line), disconnect the throttle cable if fitted.
→ Remove the fuel injection system. See *Chapter 9, Ignition, Fuel, Exhaust*.
→ Remove the coolant hoses from the cylinder head.
→ Label and detach all remaining electrical connections.
→ Detach the exhaust pipe from the manifold.
→ Remove the brake vacuum pump. See *Chapter 12, Job 21*.

TOP TIP!

• You may wish to remove the inlet manifold at this stage, because it is a large component and may get in the way.

☐ **STEP C4:** Clean then remove the fuel feed and return hose unions from the fuel injector. See *Chapter 9, Part B, Ignition, Fuel, Exhaust*.
→ Release the hoses from their unions and position them clear of the cylinder head.

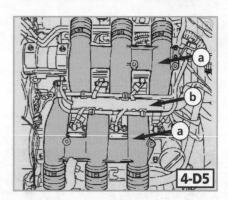

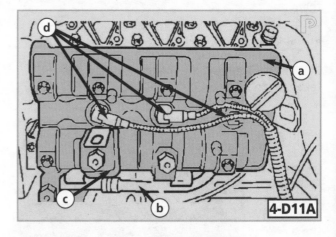

a – charge air pipe
b – fuel injection lines
c – timing chain bolts

4-C5

☐ **STEP C5:** Remove the top timing chain cover bolts.
→ Three bolts hold the cover to the head.
→ One bolt holds the cover to the block.
→ Disconnect the fittings to the turbocharger but leave the turbo connected to the exhaust manifold and remove as a unit.

☐ **STEP C6:** Slacken the cylinder head bolts, starting at the left hand front and working in a spiral pattern. See *Chapter 6, Part A, Job 3*.
→ Slacken each bolt by no more than half a turn and repeat the spiral pattern until all bolts can be turned by hand.
→ Lift out the cylinder head bolts and their washers.

☐ **STEP C7:** Remove the cylinder head. See *Steps A18 to A21*.

Section D: V6 engine.

IMPORTANT NOTE: Removing the cylinder heads from a V6 entails removing the camshaft toothed drive belt, a job we cannot recommend you attempt unless you have access to Vauxhall special tools KN-800-10. See *Job 3* for details.

☐ **STEP D1:** Remove the air filter.

☐ **STEP D2:** Remove the plenum chamber.
→ Disconnect the air intake hose from the throttle body.
→ Disconnect the throttle and cruise control cable ends.
→ Disconnect the plug from the throttle potentiometer.
→ Disconnect the engine vent and vacuum hoses from the throttle body.
→ Disconnect the exhaust gas recirculation pipe and intake manifold bracket.

☐ **STEP D3:** Remove the engine damper arm (if fitted).

☐ **STEP D4:** Remove the camshaft belt – see *Job 3*.

☐ **STEP D5:**
Remove the air intake bridges (a).
→ Disconnect the wiring harness plug from the coolant temperature sensor and sender.
→ Disconnect the wiring harness plug from the injection valves cable bundle (b) and the wiring harness plugs from the injectors.
→ The air intake bridge is held by six nuts.
→ Remove the air intake flange.

☐ **STEP D6:** Remove the camshaft pulleys.

☐ **STEP D7:** Remove the coolant bridge from between the cylinder heads.

☐ **STEP D8:** Remove the coolant pipe and the dipstick guide tube.

☐ **STEP D9:** Remove the secondary air pipe from the exhaust manifold of cylinders 1-3-5, the secondary air pipe cylinders 2-4-6 from the hose connection, and the air duct hose from the secondary air valve.

☐ **STEP D10:** Remove the air deflector covers.

4-D11A

☐ **STEP D11A: HEAD 2-4-6 CYLINDERS:**
→ Detach power steering pump pressure line (b) from intake manifold bracket (c).
→ Remove the spark plug HT leads (d).
→ Remove the camshaft covers (a).

STEP D11B: HEAD 1-3-5 CYLINDERS: Disconnect the cable channel (1) from its bracket and lay it on the rear.

→ Detach the cable channel bracket (2) from the inlet manifold and detach the inlet manifold bracket (3) from the cylinder head.

→ Disconnect the coolant lines (4) from the bracket and engine vent housing.

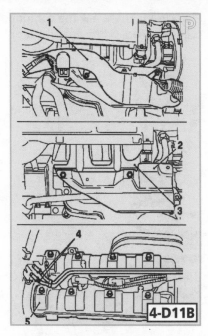
4-D11B

4-D12

STEP D12: Remove the exhaust camshafts.

STEP D13: Disconnect all wiring.

STEP D14: Remove cylinder head 1-3-5 first.

→ Work on the cylinder head bolts in the order shown here.

→ Undo each cylinder head bolt in turn by a quarter of a turn (ninety degrees), then undo each bolt half a turn still working in the correct order.

→ Finally, remove each bolt, in the correct order.

4-D14

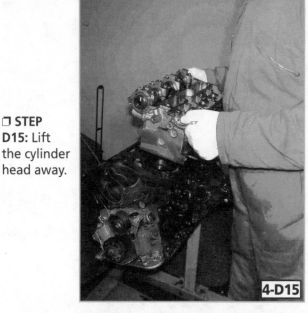

STEP D15: Lift the cylinder head away.

4-D15

STEP D16: Cylinder head refit instructions are given in *Job 6, Step 12*.

→ Use colourless wetting agent to camshaft cover sealing surfaces.

JOB 5: CYLINDER HEAD - *overhaul*.

IMPORTANT NOTE: While it isn't difficult to overhaul a cylinder head, many people find it more cost-effective to use a service exchange reconditioned unit instead, as an overhaul is a fairly time-consuming business. However, if you do overhaul the head, read *Part A, General Procedures* thoroughly beforehand.

FACT FILE
DIESEL and V6 ENGINES.

• Diesel engines with eight valves (8V) can be treated as single overhead camshaft (SOHC) engines for the purposes of overhaul, except that no diesel engine cylinder head can be skimmed.

• Diesel engines with sixteen valves (16V) are direct injection. The injectors are perpendicular and situated in between the four valves, fed by 'injection nozzle traverses'. Overhauling cylinder heads fitted to these engines should generally be entrusted to a Vauxhall specialist as it involves specialist tools.

• The V6 engine cylinder heads are overhauled as DOHC cylinder heads.

☐ **STEP A1:** These are the components of an SOHC cylinder head, with separate camshaft housing.

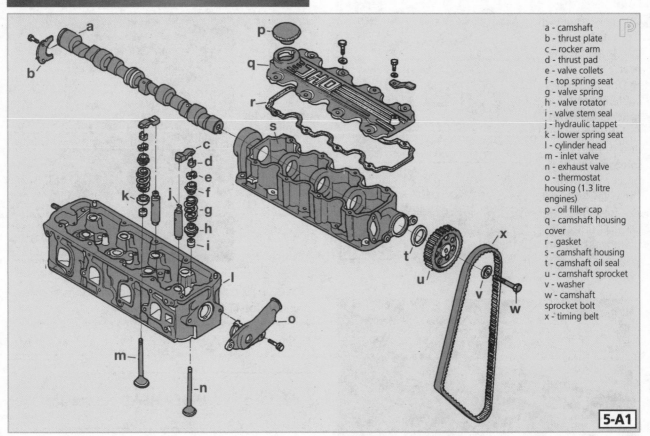

a - camshaft
b - thrust plate
c – rocker arm
d - thrust pad
e - valve collets
f - top spring seat
g - valve spring
h - valve rotator
i - valve stem seal
j - hydraulic tappet
k - lower spring seat
l - cylinder head
m - inlet valve
n - exhaust valve
o - thermostat housing (1.3 litre engines)
p - oil filler cap
q - camshaft housing cover
r - gasket
s - camshaft housing
t - camshaft oil seal
u - camshaft sprocket
v - washer
w - camshaft sprocket bolt
x - timing belt

5-A1

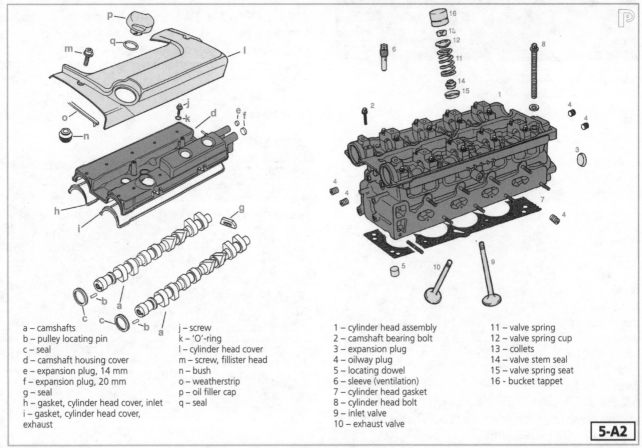

a – camshafts
b – pulley locating pin
c – seal
d – camshaft housing cover
e – expansion plug, 14 mm
f – expansion plug, 20 mm
g – seal
h – gasket, cylinder head cover, inlet
i – gasket, cylinder head cover, exhaust

j – screw
k – 'O'-ring
l – cylinder head cover
m – screw, fillister head
n – bush
o – weatherstrip
p – oil filler cap
q – seal

1 – cylinder head assembly
2 – camshaft bearing bolt
3 – expansion plug
4 – oilway plug
5 – locating dowel
6 – sleeve (ventilation)
7 – cylinder head gasket
8 – cylinder head bolt
9 – inlet valve
10 – exhaust valve

11 – valve spring
12 – valve spring cup
13 – collets
14 – valve stem seal
15 – valve spring seat
16 - bucket tappet

5-A2

☐ **STEP A2:** These are the components of a DOHC cylinder head.

STEP A3: Remove the spark plugs and discard them. You should replace them as part of the overhaul.

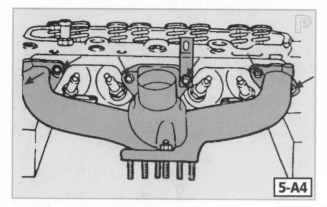
5-A4

STEP A4: If you left them on the head, unbolt the inlet and exhaust manifolds.
→ The exhaust manifold may have badly corroded nuts.
→ Try cleaning the stud threads with a wire brush and soaking the area with penetrating oil, but don't be surprised if the nut extracts the entire stud with it.
→ If this happens, don't worry, as the stud is threaded.
→ Simply use a new one when you refit the manifold.

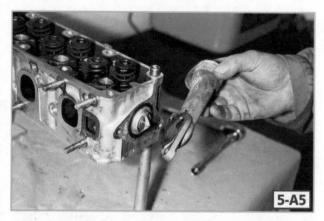
5-A5

STEP A5: Remove the thermostat housing, along with the thermostat.
→ Discard the thermostat, and replace it on refitting.
→ See *Chapter 8, Part B, Job 5*.

TOP TIP!

• As each valve, spring, its collets and seats and rocker arms should be replaced exactly where they came from, it is a good idea to devise a tray where they can lie, in order and undisturbed.
• Use a piece of card, with holes punched in it for each valve.
• Alternatively, bag each cylinder's components up and clearly label the bags.

STEP A6: Remove the camshafts - see *Job 6*.

STEP A7: If you are working on an SOHC engine, remove the cam followers and thrust pads, clean them and put them aside in order.

5-A7

→ Pull the hydraulic cam followers from the head.
→ This could be difficult, mainly because they are a close fit, and oil around them causes a vacuum as you try and pull upwards.

FACT FILE

CAM FOLLOWERS – ALL TYPES

• Although the cam followers can be overhauled, they are not prohibitively expensive to buy new.
• We recommend that, as part of a head overhaul, they are renewed.
• If you decide not to do so, keep them upright, in a plastic tub with clean engine oil in it and in the right order. Fit a piece of the card with holes pierced in it through which you can push the followers to hold them.

DOHC ENGINES

TOP TIP!

STEP A8:
• Remove the valve lifters. If they show signs of wear, or the engine made 'tappet' noises, replace them as a set.

5-A8

• Use a valve grinding tool to pull out each lifter.

16V DIESEL ENGINES

IMPORTANT NOTE: Overhauling the 16V diesel cylinder head should ideally be entrusted to a Vauxhall or engine overhaul specialist unless you have special tool KM-328-B, needed to remove the injectors.

FACT FILE

16 VALVES DIESEL ENGINES.

• These engines have sixteen valves operated by a single cam via a 'valve bridge'. The valve bridge is effectively a steel plate that rests on the valve stem tops; when the camshaft lobe faces downward, it pushes the plate down and with it both valve stems.

• The cylinder head assembly is complicated by the direct injection in which the injectors are mounted vertically and situated at the very top of the combustion chambers in between the four valve seats.

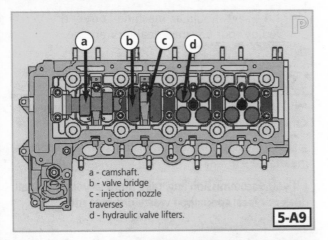

a - camshaft.
b - valve bridge
c - injection nozzle traverses
d - hydraulic valve lifters.

5-A9

□ **STEP A9:** Remove the injection nozzle traverses – see *Chapter 9, Part B*.

□ **STEP A10:** Remove the valve bridges and the hydraulic valve lifters, keeping them in sequence for correct reassembly.

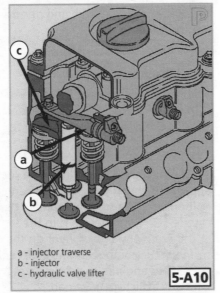

a - injector traverse
b - injector
c - hydraulic valve lifter

5-A10

□ **STEP A11:** Remove the seals from the injection nozzles.

□ **STEP A12:** Using special tool KM-328-B, remove the injectors or adapt a workshop slide hammer, as shown.
→ Remove the injector seals.

5-A12

ALL ENGINES

□ **STEP A14:** Remove the valve collets using a valve spring compressor.
→ Fit it carefully to the spring top and compress the spring.
→ With your fingers or (very carefully) a pair of long-nosed pliers, remove the two halves of the split collet.

5-A14

□ **STEP A15:** Pull out the spring cap, then the spring, and finally the spring seat. Clean these components, and place them in order, referring to the last *Top Tip*!

5-A16

□ **STEP A16:** Using long-nosed pliers or valve seat pliers, remove the valve stem oil seals.
→ IMPORTANT NOTE: The valve seals must be renewed when you re-assemble the head.

□ **STEP A17:** Alternatively, the old seals can be levered off.

5-A17

STEP A18: Try to move the valves from side to side. If there is any play, valve guide wear is indicated. New valve guides will be needed. See *Section B*.

STEP A19: Turn the head on its side, and remove the valves – see *Part A, General Procedures*. Store them in order with their springs seats, collets and springs. Lay them out on a piece of card, and keep it somewhere where there is no risk of them getting jumbled up. If there is this risk, employ a cardboard egg or fruit tray, which has compartments where components can sit.

STEP A20: Clean the head of oil, carbon deposits, remnants of gasket and any sealing compound. To do this, support it in a vice, and use a wire brush drill attachment.

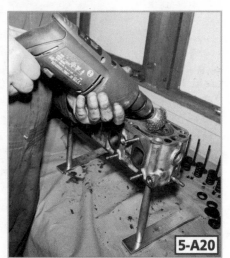
5-A20

→ Don't be over-zealous when cleaning the head.
→ Alloy heads are easily damaged.
→ If the wire brush sits in one place for too long it can score the head.

STEP A21: Once the head is clean, inspect it thoroughly for flatness.
→ Place a straight edge along the bottom to check for warping.
→ Use a feeler gauge to find out how far from true it is.
→ If this value is more than the figure given in *Chapter 1, Facts and Figures*, the head will have to be 'skimmed' flat at your Vauxhall dealer or an engine specialist.

5-A21

TOP TIP!

STEP A22: • If you are working on a DOHC engine, check that the small holes in the valve lifter mounts aren't clogged.
• If they are, gently poke them clear with a wire, and, if possible, blow compressed air through them.

STEP A23: Having wiped them clean, inspect the valves thoroughly.
→ First, spin them in your fingers to see if they are bent.
→ A bent valve must be replaced.
→ Check with a micrometer at several points along the valve stem (3 or 4 points). If there is any difference, the stem is worn and the valve should be replaced.
→ Look at the 'seat', where it closes against the head. Small pitting can be ground out on refitting, but major burning or pitting damage means the valve should be renewed.
→ If the valve seat is damaged, also look at the corresponding seat in the head. Severe damage means the seat should be machined out and renewed by your Vauxhall dealer or engine specialist.

SAFETY FIRST!

• DOHC engines have exhaust valves with an inner core of sodium.

• These must not be thrown away with other scrap metal.

• If you decommission any, take them to your Vauxhall dealer or local specialised waste disposal site.

STEP A24: Using new parts where necessary, refit the valves into the head, and grind them in - see *Part A, Job 10*.

STEP A25: Check the length of the valve springs for cracks or breakage.
→ Even if just one of the springs is shorter than the others, the whole set should be replaced.
→ In fact, it is highly advisable to renew them during engine overhaul.

STEP A26: Look carefully at the rocker arms and, if applicable, thrust pads, of SOHC engines.
→ It only takes a small amount of cam wear, in the case of SOHC engines, to ruin the latter.
→ If there is any sign of wear on the contact surfaces, the component must be replaced.
→ In SOHC engines, the rocker arm must be renewed with any afflicted thrust pad.

STEP A27: Lubricate the valve stems with clean engine oil and refit them, then fit new stem oil seals.

5-A27

☐ **STEP A28:** Refit the spring seat and place the spring and spring cap in position.

➔ With the spring compressor, carefully compress the spring.

➔ Refit the collets, having put a small blob of grease on each half first - this should hold them in place as you compress them by slowly releasing the spring compressor.

➔ Make sure they are pushed fully into their location groove in the valve stem.

☐ **STEP A29: 16V DIESEL ENGINE ONLY:** Refit the injectors with new seals.

☐ **STEP A30: 16V DIESEL ONLY:** Fit the injection nozzle traverses with new seals.

☐ **STEP A31:** Refit the valve lifters or followers, if applicable. Before fitting new items, soak them in a bath of clean engine oil to fill them.

☐ **STEP A32: 16V DIESEL ONLY:** Check the valve bridges for wear and refit them in their original locations.

☐ **STEP A33:** Refit the camshaft(s) and camshaft housing, as applicable - see *Jobs 7 and 8*.

☐ **STEP A34:** As applicable, refit the thrust pads and rocker arms.

Section B: Valve guide replacement.

If you accept the risk of damaging and ruining perfectly new valve guides, it is sometimes possible to replace them without the use of specialist equipment.

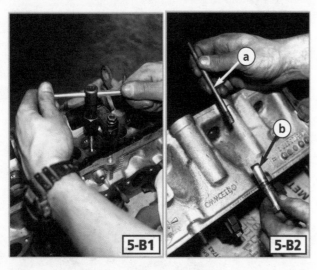

5-B1   5-B2

☐ **STEP B1:** Use a thread tap to make a thread inside each old guide.

☐ **STEP B2:** Screw a bolt (**a**) into the guide (**b**) and drift the guide (via the end of the bolt) out of the head. This drift is an old threaded stud – anything that slides inside the guide will do.

☐ **STEP B3:** Place another (smaller) bolt in the new guide.

➔ Drift down onto the bolt head, protecting the guide.

➔ The tool shown is a purpose-made drift.

➔ If you have a press, use it in preference to hammering!

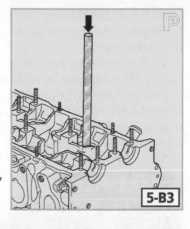

5-B3

☐ **STEP B4:** Make sure all guides are driven in to the correct depth – compare new with old.

5-B4

☐ **STEP B5:** Ream out the new guides with the correct size of hand reamer, using plenty of cutting fluid.

5-B5

JOB 6: CYLINDER HEAD – *refitting*.

Have the head skimmed (NOT diesel engines) by an engine overhaul specialist. Refitting a head that has not been skimmed is likely to lead to a blowing head gasket. The compression ratio of a diesel engine will be raised too high if it is skimmed.

IMPORTANT NOTE: Ensure that the cylinder head is fitted with none of the pistons at Top Dead Centre, in case yours is of the type where the valves may contact the pistons if rotated separately:

➔ On some engines, the valves will collide with the pistons if the camshaft/s is/are turned while the engine is set to Top Dead Centre (TDC).

➔ To avoid risk of damage, the camshaft/s should NEVER be replaced (or turned if already in place) while the pistons are at TDC.

➔ Vauxhall recommend that the crankshaft is turned

back (the *opposite direction* to normal rotation) by 60 degrees, moving the pistons safely a little way down the bore.

→ With the cylinder head fitted to the engine, the camshaft/s must now be turned to their alignment marks for when No. 1 piston is at TDC.

→ Before the timing belt is refitted, the crankshaft must be moved forward (in the *direction of* normal rotation) to the TDC position again.

FACT FILE

GASKETS AND BOLTS

• A cylinder head gasket may only be compressed once. After tightening, it loses its resilience and will not seal again properly. So, if you need to remove the head a second time, even if you have just fitted a new gasket and have torqued the head down, you will need to use yet another new gasket.
• Use all new gaskets for all applicable items, such as valve cover (early type), manifolds and so on.
• Cylinder head bolts may also only be used once. Once-used bolts should not be re-used because the torque settings will be incorrect and a second stretch will weaken them.
• When these stretch-type of bolts are used, there is no need to re-tighten the cylinder head bolts after a given period of running the engine as there would be with the earlier type of re-usable bolts.

❒ **STEP 1:** Check the cylinder head for cracks, especially around and between valve seats, and for distortion, using a straight edge and feeler gauges. If any cracking or distortion are found and you are not sure whether or not the head can be re-used, have the head checked by your main dealer or engine specialist.

6-2

❒ **STEP 2:** Check that all the cylinder head mating surfaces are thoroughly clean.

→ Ensure that the bolt holes in the block are free of debris and that the threads are free running.

→ Check that the pistons are cleaned and that there is no loose debris in the cylinders before offering up the new head gasket.

❒ **STEP 3:**
On one side of the cylinder head gasket is the part number, or the word "OBEN" or "TOP" (arrowed). This must be placed

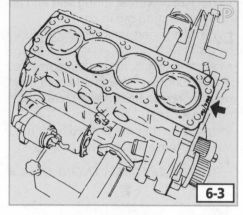
6-3

upwards, facing the cylinder head.

DIESEL ENGINES

❒ **STEP 4:**
Piston protrusion above the block at TDC must be measured. In place of the special tools shown here, you could use a perfectly straight steel

6-4

bar placed over the piston crown, and use a pair of feeler gauge sets to measure the parallel clearances between the block face and each side of the bar.

→ Set No. 1 piston at TDC and measure protrusion.

→ Measure maximum protrusion of each other piston.

→ Take the highest of the four values.

→ If the values differ greatly, discuss the supply of a better matched set of pistons with your supplier and/or check connecting rods.

❒ **STEP 5:**
Select and fit a diesel engine gasket as follows:
→ The number of notches (arrowed) or holes (punched in the front right-hand

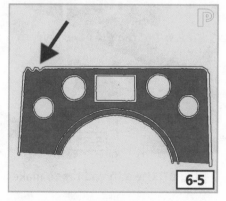
6-5

corner of the gasket) indicates gasket thickness. See FACT FILE on next page.

→ Fit the gasket with the part number uppermost.

GASKET - DIESEL ENGINE PISTON PROTRUSION (mm)

Number of Notches/Holes	Engine Designation Code	Piston Protrusion (mm)
One	15D, 15DT, 15DTR	0.58 to 0.64
Two		0.65 to 0.70
Three		0.71 to 0.78
None	16D, 16DA	Up to 0.75
One		0.76 to 0.85
Two		0.86 and above
None	17DT, 17DR, 17DTL	Up to 0.80
One		0.81 to 0.90
Two		0.91 and above
None	TC 4 EE 1	0.58 to 0.64
One		0.65 to 0.71
Two		0.72 to 0.78

IMPORTANT NOTE: Check with your supplier that the gasket you are buying is the correct one!

□ **STEP 6: SOHC ENGINES:** Fit the camshaft housing and gasket to the top of the cylinder head. See IMPORTANT NOTE near the start of this Job.

6-6

□ **STEP 7: ALL ENGINES:** Insert the (renewed) head bolts, but do not tighten them yet.

ALL ENGINE TYPES

□ **STEP 9:** Tighten the head bolts to their correct settings.

□ **STEP 10:** Tighten in the order shown, starting with the LOWEST NUMBER and working upwards.

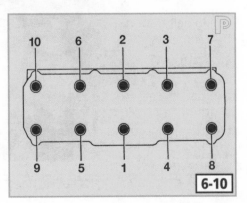

6-10

CYLINDER HEAD BOLT TIGHTENING METHODS

Tighten each bolt in the correct order, through four or five stages (depending on model). All the bolts should be tightened at each stage, before moving on to the next stage. These stages are set out in the table below:

Engine Designation Code	Stages
13S, 14NV, 16SV, C16NZ, X16SZ, C16SZR	1 - Tighten bolt to 25 Nm 2 - Turn a further 60 degrees 3 - Turn a further 60 degrees 4 - Turn a further 30 degrees 5 - Warm engine to normal operating temperature, then turn a further 30 degrees
16S, 16SH, 18E, C18NZ, 20NE, 20SHE, C20NE, E18NVR	1 - Tighten bolt to 25 Nm 2 - Turn a further 60 degrees 3.-.Turn a further 60 degrees 4 - Turn a further 60 degrees 5.- Warm engine to normal operating temperature, then turn a further 30 degrees.
C16XEL, X16XEL, X18XE, C20XE, 20XEJ	1 - Tighten bolt to 25 Nm 2 - Turn a further 65 degrees 3 - Turn a further 65 degrees 4 - Turn a further 65 degrees 5 - Warm engine to normal operating temperature, then turn a further 40 degrees.
C16NZ2, 18SV, 18NVR, C20LET, 20NEJ	1 - Tighten bolt to 25 Nm 2 - Turn a further 90 degrees 3 - Turn a further 90 degrees 4 - Turn a further 90 degrees
X18E1, X25XEV	1 - Tighten bolt to 25 Nm 2 - Turn a further 90 degrees 3 - Turn a further 90 degrees 4 - Turn a further 90 degrees 5 - Turn a further 45 degrees
X20XEV, X25XE, C20SEL, C22SEL, C25XE	1 - Tighten bolt to 25 Nm 2 - Turn a further 90 degrees 3 - Turn a further 90 degrees 4 - Turn a further 90 degrees 5 - Turn a further 15 degrees
16D, 16DA	1 - Tighten bolt to 25 Nm 2 - Turn a further 90 degrees 3 - Turn a further 90 degrees 4 - Turn a further 45 degrees 5 - Warm engine to normal operating temperature, then turn a further 30 degrees 6 - After 650 miles turn a further 45 degrees
17D, 17DR, 17DTL	1 - Tighten bolt to 25 Nm 2 - Turn a further 90 degrees 3 - Turn a further 90 degrees 4 - Turn a further 45 degrees 5 - Warm engine to normal operating temperature, then turn a further 30 degrees. 6 - Turn a further 15 degrees.
TC 4 EE 1	1 - Tighten bolt to 40 Nm 2 - Turn a further 70 degrees 3 - Turn a further 70 degrees.
X17DT	1 - Tighten bolts to 40 Nm 2 - Turn a further 60 to 75 degrees 3 - Turn a further 60 to 70 degrees
X20DTL, X20DTH	1 - Tighten bolts to 25 Nm + 65 degrees. 2 - Turn a further 65 degrees. 3 - Turn a further 65 degrees. 4 - Turn a further 65 degrees. 5 - Turn a further 15 degrees.

Y17DTR, X22DTR, Y22DTR, Y26SE – Cylinder head torque settings not available.

☐ **STEP 11:**
Refit the following:
➔ Using new gaskets place the exhaust and inlet manifolds onto their studs and tighten down.
➔ The exhaust down pipe can now be reconnected.

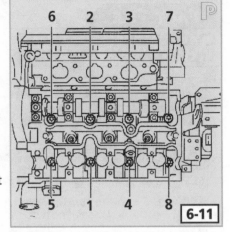

6-11

➔ Replace the carburetor or injection components.
➔ Fit a new timing belt. See *Job 3*.
➔ Reconnect all the remaining hoses, cables and wires to the cylinder head.

V6 ENGINE

IMPORTANT NOTE:
• **WE DO NOT RECOMMEND AT THAT YOU FIT THE CYLINDER HEADS WITH CAMSHAFTS UNLESS YOU HAVE ACCESS TO VAUXHALL SPECIAL TOOL KN-800-10. See *Job 3, V6 ENGINE*.**
• **ENSURE CRANKSHAFT IS AT 60 DEGREES BTDC BEFORE REFITTING THE CYLINDER HEADS.**

☐ **STEP 12:** Refit cylinder head 2-4-6 first.
➔ Tighten cylinder head bolts in the pattern shown here.
➔ Tighten each bolt in turn, first to 25 Nm (18 lbs. ft).
➔ Tighten each bolt in turn a further 90 degrees, working in the same pattern.
➔ Go around all bolts three more times in the same pattern, tightening each one in turn: Plus ninety degrees; Plus ninety degrees; Plus fifteen degrees.

☐ **STEP 13:** DO NOT attempt to fit the camshafts before reading *Job 3, V6 ENGINES*.

JOB 7:	CAMSHAFT(S) – *remove, check, replace.*

The job is substantially different between the three different types:
Section A: SOHC belt driven;
Section B: DOHC belt-driven;
Section C: SOHC chain driven
Section D: V6 engine camshafts are treated in principle as DOHC engine camshafts, but there are important differences, read *Job 3, V6 ENGINES*, BEFORE starting work.

Section A: SOHC engines except 16V diesel.

IMPORTANT NOTE: To remove/refit the camshaft of an SOHC engine, *in situ*, a special tool is required.
➔ The tool depresses the valve springs to allow the camshaft out of its housing.
➔ In this Job, we show this tool in use.
➔ If, however, you don't have access to the tool, you will have to remove the housing before stripping the cam out of it – see *Job 8*.

☐ **STEP A1:** Remove the air filter assembly, then unbolt the rocker cover and remove it.

☐ **STEP A2:** Remove the outer timing belt cover, referring to *Job 1*.

☐ **STEP A3:** Turn the crank pulley bolt until the engine is at Top Dead Centre, then turn it back 90 degrees.

☐ **STEP A4:** Unbolt the fixing plate (arrowed) and remove the distributor - or the Distributorless Ignition System (DIS) unit, depending on age of vehicle - from the transmission end of the head.

7-A4

➔ Refer to *Chapter 9, Part A*, and disconnect the HT leads from the plugs and coil as you do so.
➔ The camshaft thrust plate is now accessible.

☐ **STEP A5:** Unbolt and remove the fuel pump, and lay it carefully to one side.

☐ **STEP A6:** Remove the timing belt from the cam pulley. See *Job 3*.

☐ **STEP A7:** Unbolt the cam pulley, preventing the cam from turning with an open-ended spanner on its flat section.

7-A5

STEP A8: Fit the valve spring compression tool set as shown here, and follow the manufacturer's instructions. There are several variations – these are the Sykes-Pickavant compression tools.

7-A8

STEP A9: Remove the two thrust plate bolts and pull the plate itself from the housing. Mark it with typists'

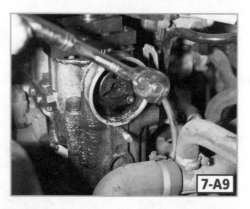

7-A9

correction fluid, so that you replace it with the same side facing outwards.

7-A10

STEP A10: Pull the camshaft carefully from the camshaft housing, via the distributor end of the engine.

STEP A11: Clean the camshaft housing, and carefully check the bearings. If they look worn or obviously pitted, the housing should be replaced in its entirety – see the next Job.

7-A12

STEP A12: Clean the camshaft and give it a thorough check.
→ Make sure there are no signs of scoring on the bearing contact surfaces – a sure sign of bearing wear.
→ Also check the cam lobes (arrowed).
→ These engines are prone to sudden camshaft failure, so if any lobe is noticeably scored or worn, as in the example arrowed here, the cam should be replaced.
→ IMPORTANT NOTE: If the cam does need replacing, its rocker arms should also be renewed along with the thrust pads.

STEP A13: Fit a new camshaft front oil seal:
→ With a small flat-bladed screwdriver, lever the camshaft oil seal from the camshaft housing.
→ Fit a replacement, using a suitably sized tube or socket to make sure it is driven gently and evenly into place.

STEP A14: Replace the 'O' ring seal for the other end of the camshaft – this sits in a groove on the distributor or DIS unit housing.

STEP A15: Oil the bearings and the oil seal with clean engine oil, then carefully refit the camshaft.

STEP A16: Refit the thrust plate.
→ Once it is in place, use a feeler gauge (position of curved arrow) to check the end-float, in the cam.
→ There's

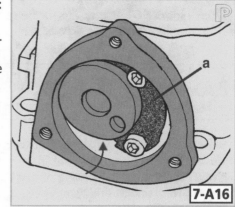

7-A16

not much room to insert a feeler gauge!
→ If the end-float is more than the figure given in **Chapter 1, Facts and Figures**, the endplate (**a**) is worn and must be replaced.

☐ **STEP A17:** Refit the distributor or DIS unit, and refit the HT leads.

☐ **STEP A18:** Refit the fuel pump, then refit the camshaft pulley.

☐ **STEP A19:** Refit the timing belt, and adjust it - see *Job 3*.

☐ **STEP A20:** Once the timing belt is adjusted, release the valve spring compression tool.

☐ **STEP A21:** Replace the rocker cover and other components removed, referring to *Step B1*.
→ Make sure the engine oil level is correct.
→ If you have fitted a new camshaft, run the engine for one minute at 2000 rpm, drop the revs to 1500 rpm, raise them to 3000 rpm for a further minute, then back to 2000 rpm for a final minute.
→ That should be sufficient to fill the cam followers with oil and quieten them down.

Section B: DOHC engines.

TOP TIP!

• It isn't essential, but it is a good idea to replace camshafts in pairs.
• If one has worn badly, it is an indication that the other may well be at the same stage in life.

7-B1

☐ **STEP B1:** Remove the plastic cover from the top of the rocker cover and remove the spark plugs.
→ If the model you are working on has a cast alloy inlet manifold, unbolt and remove its upper half, before removing the rocker cover itself.
→ Fill the lower inlet tracts with rags, to stop dirt from getting down them and remove the rocker cover.

☐ **STEP B2:** Unbolt the Distributor-less Ignition System (DIS) module from the transmission-end of the head.

7-B2

☐ **STEP B3:** Remove the timing belt - see *Job 3*.
→ Unbolt the cam pulley, holding the cam in position with an open-ended spanner on its flat section.

☐ **STEP B4:** Make sure the bearing caps carry etched numbers. In most cases they do.
→ However, if they don't, clean the cap with a rag and etch a number into each one, or use a system of blobs of typists' correction fluid.
→ It's usual to start numbering with '1' at the front of the engine.

☐ **STEP B5:** Unlike the SOHC camshaft, there is no lock plate on the DOHC unit. Use a feeler gauge to check the lateral play, or 'end-float' in the cam. If it exceeds the value given in *Chapter 1, Facts and Figures*, the camshaft must be replaced.

TOP TIP!

☐ **STEP B6:** These are the bearing cap numbers:
• Leave caps 2 and 4; 7 and 9 done-up while you completely remove all the other caps.
• These four caps will hold the camshaft down against the valve springs.
• Undo, half-a-turn at a time, the bolts for bearing caps 6, 8, 10.

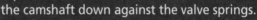

7-B6

☐ **STEP B7:** Once spring pressure is released, studs can be fully unscrewed. Remove the caps fully then remove the camshafts, lifting them away.

7-B7

❏ **STEP B8:** Wipe the camshaft clean and inspect it thoroughly. If the bearing contact surfaces or the cam lobes are worn significantly, the cam should be replaced, along with the valve lifters, in the case of lobe wear.

TOP TIP!

• The camshaft runs straight in the head – there are no shells.

❏ **STEP B9:** Carefully inspect the bearing surface of the bearing caps and in the head.
➜ If they are significantly scored or worn, the head should be replaced, referring to - see *Job 6*.
➜ If the head is renewed, it should come with a matched set of bearing caps.
➜ It is usually more cost effective to buy a service exchange head, ready assembled.
➜ It is possible to have these heads machined and bearings fitted by a machine shop.

❏ **STEP B10:** With a small flat-bladed screwdriver, prise the camshaft oil seal from its camshaft mounting. Fit a replacement, using a suitably sized tube or socket to make sure it is driven gently and evenly into place.

❏ **STEP B11:** Prepare the contact surfaces before refitting the cam.
➜ The valve lifter (tappet) faces and cam lobes should be dressed liberally with molybdenum grease, while the bearing surfaces should be oiled.
➜ On the mating faces of the timing belt end bearing caps, and their correspondents on the head, put a smear of sealing compound.

❏ **STEP B12:** Lay the cam in position in the head.
➜ You need to make sure it is lying in Top Dead Centre (TDC) position.
➜ To determine this, fit the camshaft pulley finger tight, and rotate the camshaft accordingly.
➜ Then refit the bearing caps, but loosely, consulting the marks on the caps to make sure that they go in the right places.

❏ **STEP B13:** First, tighten numbers 2 and 4 (or 7 and 9) bearing cap bolts – see illustration *7-B6*.

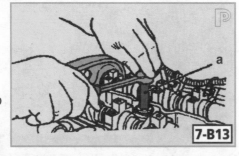

7-B13

Tighten progressively in stages of half-a-turn at a time, using a long socket (a).

❏ **STEP B14:** Tighten each camshaft's bolts to the torque specified in *Chapter 1, Facts and Figures*, in the pattern indicated by the arrow in this diagram.

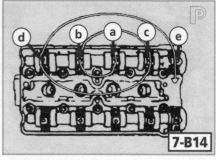

7-B14

❏ **STEP B15:** Tighten the camshaft pulley bolt to its specified torque.

❏ **STEP B16:** Refit the timing belt and adjust it - see *Job 3*.

❏ **STEP B17:** Follow *Step A21*.

Section C: 16V diesel engines.

IMPORTANT NOTE: 16V DIESEL ENGINES.
• When the camshaft is renewed it is essential that the valves bridges are also renewed.

❏ **STEP C1:** Referring to *Job 1, Section A*.
➜ Disconnect the battery.
➜ Remove the exhaust front pipe.
➜ Remove the air filter box and auxiliary drive belt.

7-C1

➜ Remove the camshaft cover and turn the crankshaft until No. 1 cylinder is at TDC on the compression stroke.
➜ Raise the right hand side of the engine.
➜ Remove the auxiliary drivebelt tensioning assembly.
➜ Remove the upper chain tensioner cap and withdraw the plunger.
➜ Remove the Allen head bolts and injection pump sprocket cover.
➜ In the absence of special tools to lock the camshaft and injection pump sprockets, make accurate alignment marks between pump flange and camshaft, and the chain and sprockets.
➜ Remove the upper chain guide.
➜ Remove the camshaft sprocket retaining bolt.
➜ Remove the fuel pump injection sprocket.
➜ Lift the camshaft sprocket complete with chain up out of the upper chain cover.

7-C2

☐ **STEP C2:** Make sure the bearing caps carry etched numbers. If not, number them.

☐ **STEP C3:** Starting from the outside and working in a spiral pattern to the inside (See *Step 7-B14*), slacken the camshaft bearing cap bolts one turn each, then repeat the pattern until the valve spring pressure on the camshaft has subsided.
→ Then remove the bolts and lift the caps away.
→ Lift the camshaft.

☐ **STEP C4:** Wipe the camshaft clean and inspect it thoroughly. If the bearing contact surfaces or the cam lobes are worn significantly, the cam should be replaced, along with the valve bridges, in the case of lobe wear.

☐ **STEP C5:** Carefully inspect the bearing surface of the bearing caps and in the head.
→ If they are significantly scored or worn, the head should be replaced. See *Job 6*.

☐ **STEP C6:** Check that the crankshaft is at TDC then, if they've been removed, refit
→ The hydraulic valve lifters.
→ The injector traverses.
→ The valve bridges.

☐ **STEP C7:** Lay the camshaft in position with the number one cylinder lobes facing up. See *Chapter 6, Part A, Job 1*.
→ Prepare the contact surfaces before refitting the cam.
→ The valve lifter (tappet) faces and cam lobes should be dressed liberally with molybdenum grease, while the bearing surfaces should be oiled.
→ On the mating faces of the timing chain end bearing caps, and their correspondents on the head, put a smear of sealing compound.

☐ **STEP C8:** Refit the bearing caps loosely, consulting the marks on the caps to make sure that they go in the right places.

☐ **STEP C9:** Tighten numbers 2 and 4 (or 7 and 9) bearing cap bolts, progressively in stages of half-a-turn at a time, using a long socket

☐ **STEP C10:** Tighten each camshaft's bolts to the torque specified in *Chapter 1, Facts and Figures* as shown in illustration *7-B14*.

☐ **STEP C11:** Refit the camshaft sprocket with chain. See *Job 1, Section A*.

☐ **STEP C12:** Follow *Step A21*.

Section D: V6 engines.

IMPORTANT NOTE:
• The crankshaft must be turned back 60 degrees before TDC before any work on the camshafts or toothed belt is carried out.
• Read *Job 3*, V6 engine, before working on the camshafts.

☐ **STEP D1:** Carry out the steps detailed in *Job 4, Steps D1 to D11*.

☐ **STEP D2:** Remove the camshaft pulleys.

☐ **STEP D3:** Make sure the bearing caps carry etched numbers. In most cases they do.
→ However, if they don't, clean the cap with a rag and etch a number into each one, or use a system of blobs of typists' correction fluid.
→ It's usual to start numbering with '1' at the front of the engine.

☐ **STEP D4:** Use a feeler gauge to check the lateral play, or 'end-float' in the cam. If it exceeds the value given in *Chapter 1, Facts and Figures*, the camshaft must be replaced.

TOP TIP!

☐ **STEP D5:**
• Undo, half-a-turn at a time, the bearing cap bolts in the spiral order shown in *Job 6, Step 11*.

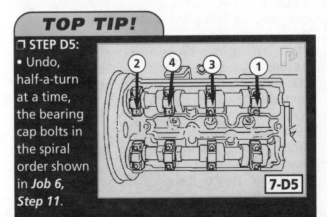

7-D5

☐ **STEP D6:** • Once spring pressure is released, studs can be fully unscrewed. Remove the caps fully then remove the camshafts, lifting them away.

STEP D7: Wipe the camshaft clean and inspect it thoroughly. If the bearing contact surfaces or the cam lobes are worn significantly, the cam should be replaced, along with the valve lifters, in the case of lobe wear.

STEP D8: Carefully inspect the bearing surface of the bearing caps and in the head.
➜ The camshaft runs straight in the head – there are no shells.
➜ If bearing surfaces are significantly scored or worn, the head should be replaced - see *Job 6*.
➜ If the head is renewed, it will be supplied with a matched set of bearing caps.
➜ It is usually more cost effective to buy a service exchange head, ready assembled.
➜ It is possible to have these heads machined and bearings fitted or resurfaced by a machine shop.

STEP D9: With a small flat-bladed screwdriver, lever the camshaft oil seal from its camshaft mounting. Fit a replacement, using a suitably sized tube or socket to make sure it is driven carefully and evenly into place.

STEP D10: Prepare the contact surfaces before refitting the cam.
➜ The valve lifter (tappet) faces and cam lobes should be dressed liberally with molybdenum grease, while the bearing surfaces should be oiled.
➜ Put a smear of sealing compound on the mating faces of the timing belt end bearing caps, and their correspondents on the head.

STEP D11: Set the crankshaft at 60 degrees BTDC.
➜ Lay the cam in position in the head.
➜ You need to make sure that the camshaft is turned to the Top Dead Centre (TDC) position – see *Part A, Job 2*.
➜ To determine this, fit the camshaft pulley finger tight, and rotate the camshaft accordingly.
➜ Then refit the bearing caps, but loosely, consulting the marks on the caps to make sure that they go in the right places.

STEP D12: Tighten the bearing cap bolts, progressively in stages of half-a-turn at a time, using a long socket in the opposite of the direction during removal.

STEP D13: Tighten each camshaft's bolt to the torque specified in *Chapter 1, Facts and Figures*.

STEP D14: Tighten the camshaft pulley bolts to their specified torque.

STEP D15: Refit the timing belt and adjust it – see *Job 3, V6 ENGINES*.
➜ Unless you have access to Vauxhall special tools KM-800-10 we recommend this is carried out professionally.

STEP D16: Replace the rocker covers and other components removed, referring to *Step D1*.
➜ Make sure the engine oil level is correct.
➜ If you have fitted a new camshaft, run the engine for one minute at 2000 rpm, drop the revs to 1500 rpm, raise them to 3000 rpm for a further minute, then back to 2000 rpm for a final minute.
➜ That should be sufficient to fill the cam followers with oil and quieten them down.

JOB 8: CAMSHAFT HOUSING (SOHC ENGINES) – *remove, replace.*

This Job applies only to vehicles with SOHC engines

STEP 1: Drain the cooling system. See *Chapter 8, Part B, Job 1*.

STEP 2: If the model you are working on has fuel injection, depressurise the system - see *Chapter 9, Part A, Job 12*.

STEP 3: Remove the timing belt outer cover(s) – see *Job 2*.

STEP 4: Now remove the following:
➜ The timing belt – see *Job 2*.
➜ Then remove the camshaft pulley, - see *Job 7*.
➜ Finally, remove the timing belt inner cover – see *Job 2*.

8-4

STEP 5: Undo the rocker cover securing bolts and remove the cover. Carefully remove its gasket.

8-5

STEP 6: Disconnect the vacuum hoses from the inlet manifold. If necessary, remove the crankcase breather hose from the cam housing or, depending on which model you are working on, the rocker cover.

8-6

☐ **STEP 7:** Disconnect any other electrical connectors, wiring clips or pipes that you find on the housing.

☐ **STEP 8:** Referring to *Part A, Job 3*, remove the cylinder head bolts, observing the order of loosening shown there and loosening them in gradual increments.
→ Make sure you label them so that they are refitted correctly.

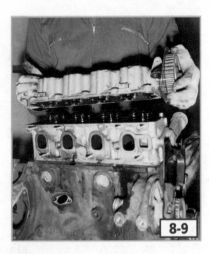

☐ **STEP 9:** Lift the camshaft housing from the head.
→ If it is reluctant to come off the head, tap it carefully with a soft-faced mallet.
→ The engine shown here has been removed from the vehicle, but the principal is exactly the same.

☐ **STEP 10:** Remove and refit the camshaft, referring to the last Job. However, you will not need to compress the valve gear, of course.

☐ **STEP 11:** Replace the housing as the reverse of removal.
→ Be sure to lift off the cylinder head and renew the head gasket - see *Job 5*.
→ Whenever a head gasket is disturbed, it must be renewed, or it will blow.
→ Check the cylinder head face for distortion – see *Part A*.
→ Also replace the housing gasket, fitting it with a suitable sealing compound.

IMPORTANT NOTE: Remember that new cylinder head bolts must be used each time they are replaced. They stretch to provide the correct torque and must not be re-used.

JOB 9: **ENGINE/TRANSMISSION –** *remove, refit.*

☐ **STEP 1:** Before dismantling, disconnect the battery leads and remove the battery. *See Chapter 10, Electrical, Dash, Instruments, Fact File: Disconnect the Battery.*

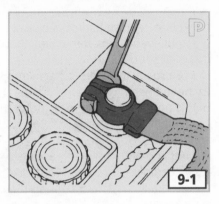

☐ **STEP 2:** Referring to *Chapter 8, Part B, Job 1*, drain the cooling system.

☐ **STEP 3:** Disconnect all the coolant hoses from the engine, including the connections to the heater matrix on the bulkhead.

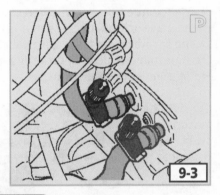

☐ **STEP 4:** If the model you are working on has fuel injection, depressurise the system - see *Chapter 9, Part A, Job 12*.

☐ **STEP 5:** Disconnect all the fuel hoses from the engine.
→ Catch spilt fuel in a suitable container, then seal up

the hose ends to stop dirt getting in them.
→ Appropriately sized pieces of steel rod held into the hoses with the original hose clips may be the best thing to use.

☐ **STEP 6:** Drain the oil from the sump into a suitable container.

☐ **STEP 7:** Drain the transmission fluid from the gearbox - see *Chapter 7, Part B, Job 2*.

☐ **STEP 8:** Remove the radiator fan and radiator assemblies - see *Chapter 8, Part B, Jobs 3 and 4*.

☐ **STEP 9:** Remove the air cleaner assembly, or induction tubing and air box assembly, as appropriate.

☐ **STEP 10:** Disconnect the following:
→ The throttle, choke and clutch cables from the engine, as applicable.
→ The speedometer drive cable connection from the transmission (see illustration *9-18, item a*).
→ Tie them out of the way.

☐ **STEP 11:** Disconnect all vacuum pipes, including those in the fuel system and those from the brake servo.

☐ **STEP 12:** Remove the drive belts.
→ Unbolt the air conditioning and power steering pumps (power steering pump adjuster/lower mounting shown) from their mountings on the engine.
→ Tie them securely out of the way in or near the engine bay.

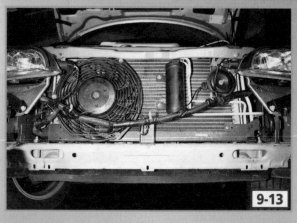

SAFETY FIRST!

☐ **STEP 13:** It will probably not be possible to avoid having the air conditioning system depressurised (if fitted).

• The system is filled with a gas which is pressurised and potentially harmful.

• There are usually only rigid pipes to the pump so it can't be moved without opening hoses.

• Have a main dealer or air-con. specialist empty the system, then check and refill it when the work is complete.

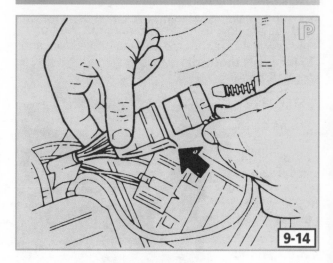

☐ **STEP 14:** Disconnect all wiring connectors from the engine.
→ As you go round, it may help to label each connector, making things easier when it comes to re-assembly.
→ Look out for tabs (arrowed) to lift or press to enable the two halves of connectors to be separated.

TOP TIP!

• Do a 'sweep' round the engine/gearbox, disconnecting any wiring or tubing which may still be attached to the engine.

REMOVE UP OR DOWN?

☐ **STEP 15:**
Decide
which way
you are
going to
remove the
engine/
transmission
unit.

9-15

• The V6
HAS to be
removed upwards.

• We show an engine being lifted upwards, out of
the engine bay.

• However, non-V6 engines with ancillaries such as
power steering and air conditioning systems may
well prove better lowered downwards.

• You will need to be able to raise the vehicle high
enough.

☐ **STEP 16:** Remove the bonnet - see *Chapter 13,
Part B, Job 2*.

☐ **STEP 17:** You will now need to work underneath
the front of the vehicle.

➜ If you are now using a hoist, raise the front of
the vehicle. Make sure the parking brake is on, and
a rear wheel chocked.

☐ **STEP 18:**
Disconnect
the
gearchange
linkage (**b**)
and
speedometer
cable (**a**)
from the
transmission
– see *Chapter
7, Part B, Job
8*.

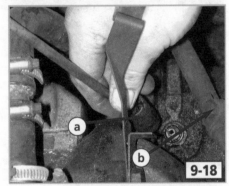

9-18

☐ **STEP 19:** Unbolt the
exhaust from the
exhaust manifold.

➜ The manifold nuts
are difficult to reach –
a socket, extension
bar and knuckle joint
make life easier.

➜ If you are lowering
the engine, you will
need to remove at
least the front section.

9-19

➜ It is often quicker to remove the entire exhaust -
see *Chapter 9, Part C, Job 3*.

☐ **STEP 20:** Remove the earth/ground strap from the
transmission unit, and disconnect the wiring
connectors from the alternator and starter motor.

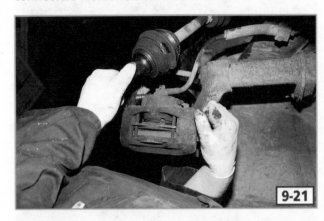

9-21

☐ **STEP 21:** Remove the driveshaft inboard ends from
the differential casing - see *Chapter 7, Part B, Job 11*.

TOP TIP!

• Disconnect the drive shafts from the hubs.
• Do not allow the driveshafts to droop under their
own weight, and tie them safely out of the way
with strong cable ties or wire.

☐ **STEP 22:** At this point, support the weight of the
engine/transmission with an engine hoist - see *Part
A, Job 11*.

➜ Alternatively, support the assembly from
underneath using a trolley jack with a block of wood
placed on top of the trolley jack head.

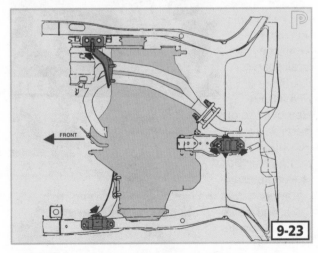

9-23

☐ **STEP 23:** Remove all the bolts holding the rear
engine/transmission mounting in place, and remove
the mounting.

➜ Repeat the exercise with the other two
engine/transmission mountings.

➜ Have an assistant help support the
engine/transmission as you carefully raise or lower it
out of the engine bay.

STEP 24: Raise the engine/ transmission slightly, then check that nothing still connects the engine to the vehicle before lifting the assembly.

9-24

➜ Have an assistant help steady the assembly as it is lifted.
➜ It is a good idea to place protection on all the bodywork surrounding the engine bay in case the assembly grazes it as it is lifted.

STEP 25: On some vehicles it will be necessary to manoeuvre the assembly to clear engine bay panels and components.

9-25

STEP 26: There are several ways you can safely move an engine around:
➜ A small car trailer is relatively stable and a good height – but can be tricky to manhandle an engine out of.
➜ Place a piece

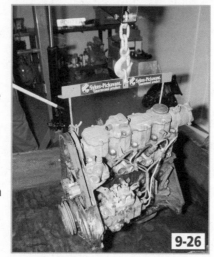
9-26

of timber across the legs of the Clarke engine hoist, lower the engine on to this platform and use the hoist as a trolley.

SAFETY FIRST!

• As the assembly is lowered to the ground where you intend to work on it, chock it securely upright.

• Otherwise, it could easily topple over and injure someone - and possibly crack a casting at the same time!

STEP 27: By far the safest – and most convenient – way of working on an engine is to fit it to an engine stand, such as the Clarke model seen in action here.

9-27

9-28

STEP 28: Separate the engine and transmission by unbolting the bellhousing blanking plate and removing the engine to gearbox bolts (arrowed).
➜ IMPORTANT NOTE: If the unit you are working on has automatic transmission, remove the torque converter first - see *Chapter 7, Part B, Job 1, Section B*.
➜ With somebody to help support the engine, pull the gearbox cleanly off the engine.
➜ Don't let it droop, which will place strain on the input shaft and could cause a future oil seal leak.

STEP 29: Replace the engine/gearbox assembly as the reverse of removal.
➜ It is usually best to attach the transmission to the engine and fit the two as an assembly. See – *Chapter 7, Transmission, Clutch*.

IMPORTANT NOTE: When refitting the engine, make sure all mounting bolts are tightened to their specified torque settings.
➜ Refer at each stage of refitting to the section of this manual relevant to the particular system you are working on.

JOB 10: ENGINE BLOCK COMPONENTS – *dismantle*.

With the engine separated from the gearbox, it can be dismantled.
→ The best way to do this is to support it on a specialised engine stand, which allows you to turn the engine whichever way you need in safety – see *Job 9, Step 25*.

❑ **STEP 1:** These are the crankshaft-related components.
→ Remove the flywheel (**g**). You may have to do this to attach the engine to an engine stand, depending on the size of the engine.

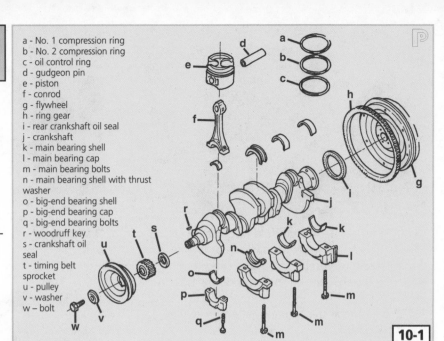

a - No. 1 compression ring
b - No. 2 compression ring
c - oil control ring
d - gudgeon pin
e - piston
f - conrod
g - flywheel
h - ring gear
i - rear crankshaft oil seal
j - crankshaft
k - main bearing shell
l - main bearing cap
m - main bearing bolts
n - main bearing shell with thrust washer
o - big-end bearing shell
p - big-end bearing cap
q - big-end bearing bolts
r - woodruff key
s - crankshaft oil seal
t - timing belt sprocket
u - pulley
v - washer
w – bolt

10-1

10-2

❑ **STEP 2:** Remove the alternator and its brackets from the block assembly.

❑ **STEP 3:** Unbolt and remove the starter motor, if you haven't already done so.

10-3

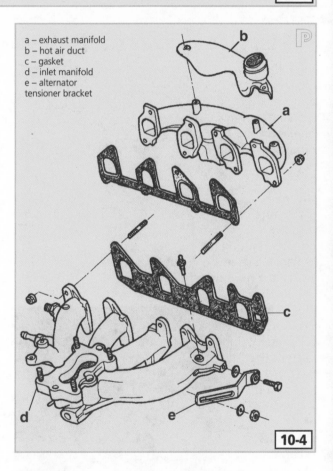

a – exhaust manifold
b – hot air duct
c – gasket
d – inlet manifold
e – alternator tensioner bracket

10-4

❑ **STEP 4:** Remove both the inlet and exhaust manifolds.
→ If you are working on a turbo-diesel engine, the turbo and its ancillaries should be removed - see *Chapter 9, Part B, Job 6*.

10-5

10-9

STEP 5: The exhaust manifold nuts may well be very corroded, and so stuck that one or two of their studs unscrew themselves from the block.
→ If this happens replace the offending studs.

STEP 6: Remove the coolant gallery from the rear side of the engine.

STEP 7: Remove the oil filter. If the model you are working on has an oil cooler.
→ Disconnect the feed pipes from the adapter housing (**a**).
→ Remove its securing nut (**b**).
→ Pull the housing off the block.

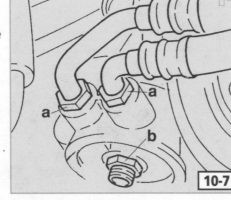

10-7

STEP 8: These annotation letters refer to illustration **3-A1**. Remove:
→ Crankshaft pulley (**a**).
→ Outer timing belt cover and timing belt.
→ Timing belt tensioner (**c**).
→ Idler pulley (**d**).
→ Camshaft pulleys (**e**).

10-8

STEP 9: Take out the inner timing belt cover bolts...

STEP 10: ...and remove the inner timing belt cover – see **Job 2**. You may have to unbolt the camshaft position sensor first.

10-10

STEP 11: Remove the water pump - see **Chapter 8, Part B, Job 8**.
→ If you are working on a DOHC engine, unbolt the timing belt tensioner and the idler pulley.

10-11

10-12

STEP 12: Remove the oil pump mounting bolts...

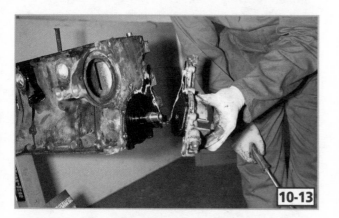

10-13

STEP 13: ...and remove the oil pump from the block.

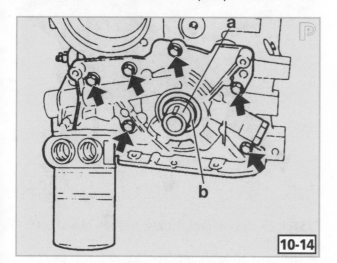

10-14

STEP 14: Before removing the pump, remove the Woodruff key from the crank nose (**a**) so that the oil seal (**b**) is not damaged as the pump is removed.
→ On some engines, you will have to unbolt a crankshaft speed sensor and its bracket from the pump housing first, and disconnect the wiring connector from the oil pressure warning light switch which is part of the assembly.
→ Discard the pump assembly. While it can be overhauled, it is most cost effective to buy a new unit complete.
→ Remove the pump - bolts arrowed - and gasket.

STEP 15: Remove the dipstick, by removing the bolt (arrowed) holding its tube housing to the block.
→ Remove the thermostat housing and thermostat - see *Chapter 8, Part B, Job 6*.

10-15

10-16

STEP 16: If you are working on a diesel engine, remove the fuel injection pump – see *Chapter 9, Part B, Job 4*.

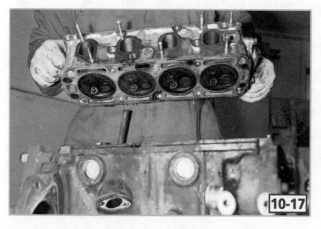

10-17

STEP 17: Remove the cylinder head assembly – see *Job 4*.

STEP 18: Strip the block of any other ancillary components that remain, such as fuel hoses or brackets. If you think they might be difficult to identify on rebuild, label them as you clean them.

> **TOP TIP!**
> • Write down notes about any components you think you may have trouble refitting.
> • A little time spent at this stage will save hours of head-scratching later!

STEP 19:
Remove the sump. Make sure you have a drip tray under the engine as you do so!

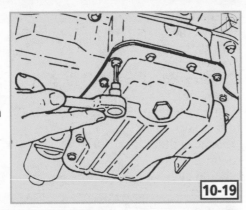
10-19

10-20

☐ **STEP 20:** Remove the oil pump pick-up tube. Make sure it's clear.

BALANCING SHAFTS – X20XEV 1998-ON AND C22SEL ENGINES

FACT FILE

BALANCING SHAFTS

- The X20XEV engine from 1998 and all C22SEL engines have two differential balancing shafts situated in a casting within the upper sump.
- The balancing shafts are gear driven from the crankshaft.

STEP 21:
Remove the lower oil pan sump (**a**) and balancer shaft housing (**b**).

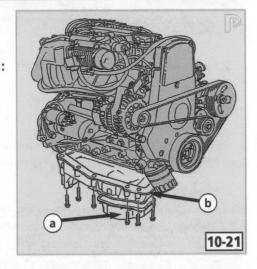
10-21

STEP 22:
These are the balancer shafts viewed from the direction (large arrow) shown in *Step 23*.

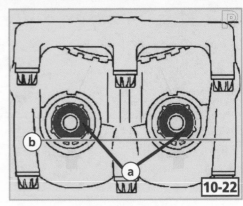
10-22

➔ Turn the crankshaft until the flats visible on the right hand ends of the balancer shafts (**a**) are facing downward and aligned with each other – imaginary line (**b**).

➔The engine will now be at TDC.

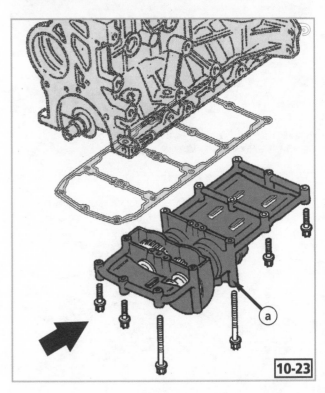
10-23

☐ **STEP 23:** Unbolt and remove the balancer shaft housing (**a**).

➔ The bolts are differing lengths, and must be replaced in the same positions.

➔ The spacer piece can also be removed.

☐ **STEP 24:** If any work is carried out on the crankshaft or bearings, the balancing shaft tooth backlash must be adjusted. To adjust backlash:

➔ Fit the correct differential/cylinder block spacer; a special tool is needed to measure backlash, and in the absence of this the unit should be taken to a Vauxhall specialist to have the backlash measured and to supply the appropriate spacer.

ALL ENGINES

STEP 25: Check that the big end caps are marked with the number of the cylinder they belong to (arrowed).

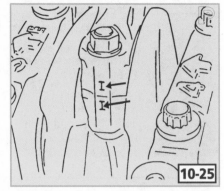

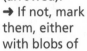

➡ If not, mark them, either with blobs of typists' correction fluid, or light centre-punch marks.

➡ It is essential that not only are they replaced on the correct piston rod, but that they are replaced the right way round.

➡ Paint a line of correction fluid (or paint), or score a light line across the joint between the two.

➡ Referring to *Part A, Job 12*, remove the big end cap joint bolts and remove each cap with its bearing shell. Keep the two together.

STEP 26: Turn the engine on one side, and feel at the top of the cylinder bore for a ridge caused by carbon deposits.

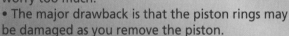

• This illustration shows a bearing scraper being used to remove the ridge – but you'll be lucky!

• However, if it is not significant, don't worry too much.

• The major drawback is that the piston rings may be damaged as you remove the piston.

• See *Part A* for information on honing the cylinder bores.

STEP 27: Remove the conrod bearings.

➡ Carefully tap the pistons out from the bores with a blunt, soft instrument such as the handle of a hammer.

➡ Again, make sure the bearing shells are stored carefully, and in order, with their pistons. See *Part A* for full information.

STEP 28: Strip the pistons of their rings (if you intend re-using the piston crowns) and free the piston from the conrod.

➡ To do this, the gudgeon pin (arrowed) has to be removed.

➡ On all the engines covered by this manual except the C20XE DOHC variant, the gudgeon pins are an interference fit.

➡ The piston assembly must be taken to your Vauxhall dealer or engine specialist, who will use heating and press equipment to remove the pins.

STEP 29: If you are working on a C20XE, DOHC engine, the pistons employ floating gudgeon pins, held in the piston crowns with circlips.

➡ Lever one of the clips out with circlip pliers and push the pin out.

➡ Remove the other circlip.

➡ These must be renewed, as must the pin AND conrod if the pin appears worn.

TOP TIP!

• Before refitting the gudgeon pin, coat it lightly with clean engine oil, and make sure it rotates freely in the piston and conrod small end.

STEP 30:
Referring to *Part A, Job 12*:
→ Make sure the caps are numbered (**a**) and marked for direction.
→ Remove the main bearing cap bolts (**b**).

STEP 31: Use a breaker bar – lots of torque needed on main bearing bolts!

STEP 32: Lift out the caps, along with their bearing shells. Store them in order, as with the big end/piston assemblies.

STEP 33: Lift out the crankshaft, then remove its upper bearing shells.

FACT FILE

BLOCK FLUSHING

• Thoroughly flush out and degrease the block, including all the oil and water ways.
• It is advisable to extract the oilway plugs and core plugs and refit these.
• The oilway plugs are screwed back into place after having their threads coated in thread sealant.
→ The core plugs are coated with sealant and tapped into place using a socket as a drift.

JOB 11: ENGINE – *reassemble*.

IMPORTANT NOTE: Before rebuilding the engine block assembly, carefully read *Part A, Jobs 16 and 17*!
→ These Steps provide information that is specific to the engines covered by this manual.

STEP 1: Lubricate a new rear oil seal, and slide it over the rear end of the crankshaft.

STEP 2:
To refit the pistons:
→ Fit a piston ring compressor to the piston and slide the piston into its bore until the compressor touches the block.
→ Check that the big end journal you are aiming to connect with is at the bottom of its downward sweep, then carefully begin to tap the piston into the bore, using the wooden handle of a hammer.
→ As you are doing so, locate the lubricated bearing shell on the conrod, and guide it onto the journal.
→ Check that the piston is located the right way round by looking for an arrow marked on the piston crown top.
→ If there is one, it should point to the flywheel end of the engine.
→ Repeat the fitment and checking exercise with the other three pistons.

STEP 3: Note that grooves in the main bearings (arrowed) should be filled with silicone sealer before fitting.

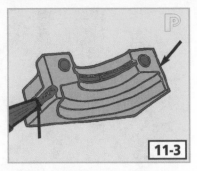

11-3

20XEV 1998-ON AND C22SEL ENGINES

STEP 4: Refit the balancing shaft differential assembly – see *Job 10, Step 21-on*.
→ The crankshaft MUST be at TDC when the assembly is fitted, and both flats visible on the shaft ends must face downward and be aligned.

ALL ENGINES

STEP 5: Make sure that the sump is properly sealed:
→ The sump and baffle assembly should be refitted with a thin, unbroken line of sealant on the mating face of the block (arrowed).
→ The sealant must pass INSIDE each sump bolt hole.

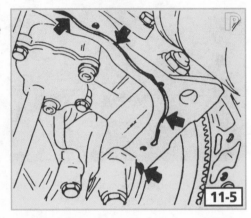

11-5

STEP 6: Refill the engine with oil and coolant, referring to the relevant sections of this manual. It is a good idea to turn the engine over for a minute with the spark plugs or glow plugs removed and the supply lead to the coil or DIS module disconnected, so the oil system can refill itself and thoroughly lubricate the engine. Run the engine in for at least 500 miles, without exceeding 3000 rpm.

JOB 12: FLYWHEEL/DRIVEPLATE – *remove, check, refit*.

STEP 1: If the transmission is still attached to the engine, remove it, as described in *Chapter 7, Part B*.

STEP 2: Remove the clutch (manual transmission) or torque converter (auto. transmission) assembly as described in *Chapter 7, Part B, Job 3*.

TOP TIP!

STEP 3: Using a punch or typists' correction fluid, make an alignment mark (arrowed)

12-3

between the crankshaft end and the flywheel/driveplate (arrowed).
• The bolts actually have an offset pattern that sets the position, but this helps you to refit the heavy flywheel in the right place, first time.

STEP 4: Hold a lever into the starter ring teeth to prevent the flywheel turning, and remove its mounting bolts.

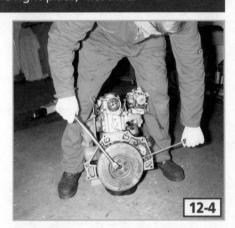

12-4

STEP 5: Remove the flywheel. Be careful, as it is fairly heavy.

STEP 6: Inspect the flywheel for cracks or excessive wear/scoring on the friction face. Light wear can be machined out by an engineering shop, but serious damage condemns the flywheel to replacement.

STEP 7: Inspect the starter ring teeth:
→ If they are badly burred, worn or broken, the ring should be replaced.
→ You can remove the old one yourself, by carefully sawing or chiseling through it.
→ When you break it, it will spring apart and fall off the flywheel.
→ Installing the new ring, however, is a job for an engineering specialist or your Vauxhall dealer, who will heat it to a specific heat to expand it, fit it in place, and allow it to contract to a tight friction fit.

STEP 8: Replace the flywheel, aligning the mounting marks you made before removal.
→ Use suitable thread lock on each bolt.

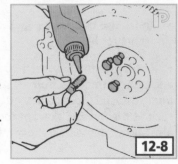

12-8

➜ Tighten the mounting bolts to the torque settings specified in *Chapter 1, Facts and Figures*.

JOB 13: ENGINE/TRANSMISSION MOUNTINGS *– remove, check, refit.*

The models covered by this manual mount the engine/transmission assembly to the body with three rubber-bushed mountings and, on some models, a bushed 'damping bar' attached to the front cross member. Over time, and if contaminated by road-borne pollution and engine oil, the rubbers can deteriorate. This allows the engine too much freedom of movement, translating into vibration, which can damage components and makes for a very uncomfortable ride!

☐ **STEP 1:** The part of an engine mounting that suffers wear is the rubber bush. Sometimes, a visual inspection will show up deterioration as cracking or tear marks. If this isn't immediately obvious, move the joint around with a large screwdriver, or ask someone to rock the engine/transmission while you inspect the bushes.

SAFETY FIRST!

☐ **STEP 2:** The mountings can be replaced individually, leaving the others in the vehicle. However, the engine/gearbox assembly must be supported so well that it cannot move at all, to avoid

overstraining the other mountings.

• Careful use of a trolley jack and a piece of wood can achieve this...

• ...but by far the best method is to support the engine/transmission from its left-hand side engine hook, attached to a dedicated engine support bar, such as the Sykes-Pickavant tool seen in use here. It also allows much freer movement under the vehicle.

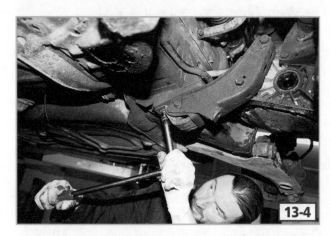

☐ **STEP 3:** Remove the bolts holding the mounting assembly to the engine/transmission.

☐ **STEP 4:** Remove the bolts holding the mounting assembly to the body.

☐ **STEP 5:** Remove the nuts holding the damping bar in place (if applicable). Remove the bar.

☐ **STEP 6:** Remove the rubber bush from the mounting assembly. In most cases, this is bolted into place, but in some right-hand side engine mountings the rubber lives in the body mounting bracket, which simply pulls free of the mounting.

STEP 7: Clean the mounting, the mounting bolt threads and, if necessary, the mounting threads on the bodywork (cut these clean with a suitable tap).

STEP 8: Fit the new bushes.

STEP 9: Coat the threads of the mounting bolts with locking solution, and refit the mounting assembly. Tighten to the specified torque setting.

TOP TIP!

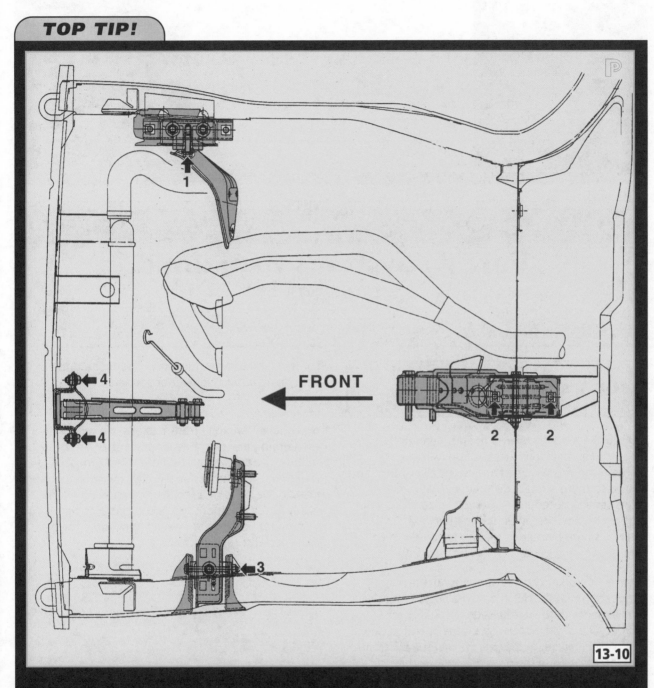

FRONT

13-10

STEP 10: If you have removed the engine/transmission for major maintenance, you may wish to take the opportunity to inspect and, if necessary, replace the mounting bushes.

• On certain models the right-hand mounting (**1**) cannot be replaced with the engine in the vehicle.

• If you are replacing all the mountings, refit them in the order shown in the diagram:

1 – right-hand engine to body

2 - rear transmission-to-floorpan/subframe

3 – left-hand side transmission to body

4 - front damping bar, if present

STEP 11: Remove the engine support.

CHAPTER 7: TRANSMISSION, CLUTCH

*Please read **Chapter 2 Safety First** before carrying out any work on your car.*

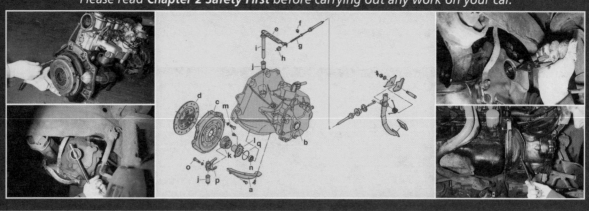

	Page No.			Page No.
Part A: Systems Explained	7-1		**Part B:** Repair Procedures	7-2

Part A: Systems Explained

CONTENTS

Page No.

JOB 1: SYSTEMS EXPLAINED.

Among the models covered by this manual, the range of transmissions fitted is fairly simple. All manual transmission units are essentially of the same family, so there are few differences that materially affect repair procedures. Externally, size is the main difference between models. The fifth gear mechanism on five-speed models is housed in an extension to the generic transmission casing.

In terms of working on the transmission, the major difference is that the clutch on some SOHC models can be changed with the transmission unit still in the vehicle. All manual models use a cable operated clutch, and a lever linkage for the gearchange. The differential is an integral part of the transmission assembly, and shares its lubrication with the gear cluster.

☐ **STEP 1:** These are the gear change, linkage and (in the centre) speedo. drive components associated with the 'generic' manual transmission.

Vauxhall's automatic transmissions, too, are closely related to each other. Vehicles to 1988 were fitted with a three speed unit, later models with a four speed unit. However, due to the complexity of these units, the number of jobs we recommend as suitable for DIY mechanics is limited to direct replacement of the transmission unit and adjusting/changing of the

control cables. Any other work should be done by your local Vauxhall dealer or specialist.

In view of the ready availability of well-priced, overhauled Vauxhall transmissions, we recommend exchanging any transmission if it suffers internal failure. The cost – both in money and time – of rebuilding a transmission simply cannot be justified next to the price of a service exchange unit from a specialist. In addition, repair components can sometimes be difficult or even impossible for the non-specialist repairer to buy.

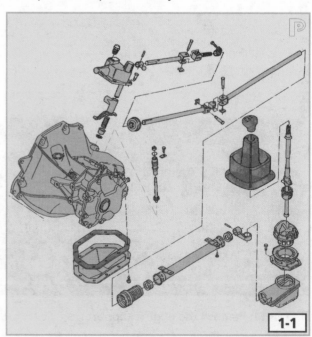

1-1

Part B: Repair Procedures

CONTENTS

Page No.

Before dismantling, disconnect the battery negative (-) earth/ground terminal. See *Chapter 10, Electrical, Dash, Instruments, Fact File: Disconnecting the Battery* BEFORE doing so!

JOB 1: TRANSMISSION (ENGINE IN VEHICLE) - *remove, refit.*

On many of the models covered by this manual, it is decidedly easier to remove the engine and transmission complete, before splitting the transmission unit from the engine. However, it is possible to do the job with the engine left in the bay. The procedures for this job are different for manual and automatic transmission so we have dealt with it in two separate sections.

Section A: Models with manual transmission.

TOP TIP!

• Where possible, remove the clutch before removing the transmission – it gives more room for manoeuvring the 'box out of the engine bay.

❏ **STEP A1: VECTRA:** Remove the battery and mounting plate.

❏ **STEP A2:** Disconnect the clutch cable or hydraulic hose.

➜ **CAVALIER:** Disconnect the clutch cable from the transmission, and tie it out of the way with a cable tie – see *Job 3*.

➜ **VECTRA:** Clamp the clutch fluid hose and remove the end fitting from the transmission - see *Job 4*.

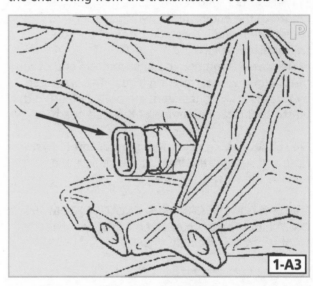

1-A3

❏ **STEP A3:** Pull the wiring connector from the reversing light switch (arrowed). This is found on the front of the transmission casing.

1-A4

❏ **STEP A4:** Undo the speedometer cable connector from the transmission (arrowed), and move the cable out of the way.

STEP A5: DOHC ENGINES: On these models, the engine bay is likely to be too crowded to enable access to the transmission bolts.

→ Drain the cooling system before removing the expansion tank and lower cooling hose - see *Chapter 8, Part B*.

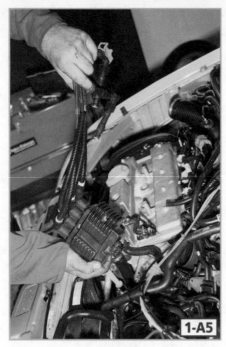

1-A5

→ Remove the solid cooling tube that runs the length of the engine block, which at the transmission end is secured by one of the top transmission bolts.

→ Remove the ignition module from the end of the head, along with its ignition leads. (This is secured by three bolts, and the leads are pulled up using the plastic tool attached to one of the lead heads).

STEP A6: Taking note of *Chapter 2, Safety First!* raise and support the front of the vehicle and remove the wheels.

STEP A7: ALL VECTRA AND CAVALIER FROM 1998 ONLY: Referring to *Chapter 11, Job 16*, remove the front subframe assembly.

TOP TIP!

STEP A8: Support the engine, using a suitable bar located in the front wing runners, such as this Sykes-Pickavant tool shown here.

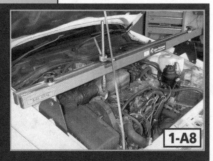

1-A8

• If you don't have access to this tool, use an engine hoist.
• The aim is to steady the engine from its right-hand side lifting hook.
• However, the Sykes-Pickavant bar is much the best, as it supports the engine as you jack the vehicle up to remove the transmission.
• Otherwise, you will need someone to act as 'hoist man' when raising or lowering the vehicle.

STEP A9: Disconnect the anti-roll bar from the left-hand side suspension lower control arm.
→ Unbolt the suspension control arm support – see *Chapter 11, Steering, Suspension*.

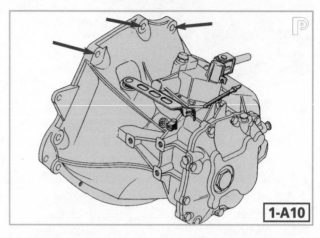

1-A10

STEP A10: Remove the upper three transmission-to-engine bolts (arrowed), the only ones accessible from above the vehicle.

STEP A11: CAVALIER ONLY: Referring to *Job 2*, disengage the transmission input (first motion) shaft.

STEP A12: Remove the exhaust pipe to transmission casing, where fitted. Drain the transmission fluid - see *Chapter 5*.

STEP A13: Remove the driveshaft inboard ends from the differential casing - see *Job 11*.
→ Using wire or strong cable ties, hang the driveshafts out of the way.
→ They must not be allowed to droop to the full flex of the CV joints, or these will be damaged.

TOP TIP!

• Even though you've drained the transmission fluid, more will probably leak out as you release the driveshafts. Have a suitable container ready to catch it

STEP A14: If the model you are working on has an engine earth/ground strap connected to the transmission casing, remove it.

STEP A15: Support the transmission using an engine hoist from above, or a trolley jack and a sturdy piece of wood from beneath. The former is much the better way!
→ On some models, the strap from an engine hoist will need to pass through the large aperture in the battery tray. If this is the case with the vehicle you are working on, disconnect the battery negative (-) earth/ground terminal. See *Chapter 10, Electrical, Dash, Instruments, Fact File: Disconnecting the Battery* BEFORE doing so!

STEP A16: Disconnect the gear linkage from the transmission, and its guide bracket on the rear engine mounting – see *Job 6*.

STEP A17: You will now need to remove:
→ The left rear engine mounting.
→ The left-side engine/transmission mounting.

STEP A18: On models with a pressed steel sump, remove the clutch housing cover plate.

1-A18

STEP A19: Make sure the hoist or trolley jack is supporting the weight of the transmission unit, and remove the lower transmission bolts.

STEP A20: Check that all cables and wires have been removed from the transmission casing. Pull the transmission unit away from the engine and lower it to the ground.

TOP TIP!

• If you find it extremely difficult to manoeuvre the transmission, more space can be created by lowering the engine a few centimetres.
• However, you will have to disconnect the exhaust from the manifold, and release it from its front mounting rubber, in order to do this.

STEP A21: Having ensured the clutch is correctly centred (see *Job 2*), refit the transmission as the reverse of removal.
→ Tighten all bolts to their specified torque settings.
→ Use locking compound on the engine/transmission mountings' bolts.

STEP A22: Refill with transmission oil - see *Chapter 5*.

STEP A23: If necessary, refill the cooling system - see *Chapter 8, Part B, Job 1*.

STEP A24: Adjust the clutch cable - see *Job 4*.

Section B: Models with auto. transmission.

STEP B1: Before dismantling, disconnect the battery negative (-) earth/ground terminal. See *Chapter 10, Electrical, Dash, Instruments, Fact File: Disconnecting the Battery* BEFORE doing so!
→ **VECTRA:** Remove the battery, along with it's mounting plate and the starter motor.

STEP B2: Disconnect the selector cable from its lever on the transmission, and tie it out of the way with a cable tie – see *Job 7*.

STEP B3: Unclip the vent pipe from the front of the transmission unit and drain the transmission fluid into a suitable container. The drain plug is found on the differential housing.

STEP B4: Unplug the wiring connections from the transmission unit.
→ If the vehicle you are working on has emission control equipment,

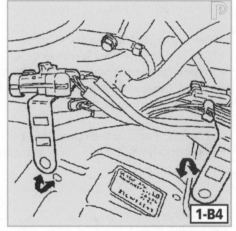
1-B4

the oxygen sensor wiring may also be clipped to the transmission.
→ If so, unclip and disconnect the plugs (see arrows).

STEP B5: On some models, the engine bay is likely to be too crowded to enable access to the transmission bolts. You can improve access in the following ways:
→ Drain the cooling system before removing the expansion tank and lower cooling hose – see *Chapter 8, Part B*.
→ Remove the solid cooling tube that runs the length of the engine block, which at the transmission end is secured by one of the top transmission bolts.
→ Remove the ignition module from the end of the head, along with its ignition leads. (This is secured by three bolts, and the leads are pulled up using the plastic tool attached to one of the lead heads).

STEP B6: Taking note of *Chapter 2, Safety First!* jack up the front of the vehicle and remove the wheels.

□ **STEP B7:** Support the engine, using a suitable bar located in the front wing runners. If you don't have access to this tool, use an engine hoist. The aim is to steady the engine from its right-hand side lifting hook. However, the bar is best, as it supports the engine as you jack the vehicle up to remove the transmission. Otherwise, you will need someone to act as 'hoist man' when raising or lowering the vehicle.

□ **STEP B8: ALL VEHICLES FROM 1988:** Referring to *Chapter 11, Job 16*, remove the front subframe assembly.

□ **STEP B9: CAVALIERS BEFORE 1988:** Referring to *Chapter 11, Steering, Suspension*:
→ Disconnect the anti-roll bar from the left side suspension lower control arm.
→ Unbolt the suspension control arm support.
→ Disconnect the rear left transmission to body mounting.
→ Disconnect the left lower suspension arm support.

□ **STEP B10: CAVALIERS BEFORE 1988:**
→ If fitted, remove the kickdown cable – see *Job 8*.
→ Remove the starter motor – see *Chapter 10, Electrical, Dash, Instruments*.
→ Remove the fluid dipstick.

□ **STEP B11:** Remove the upper three engine-to-transmission bolts, the only ones accessible from above the vehicle.

□ **STEP B12:** If the model you are working on has them, clamp the transmission fluid cooler pipes and disconnect them from the transmission. Fluid will leak out, so have a suitable container ready to catch it.

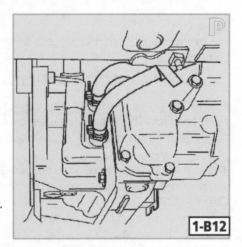
1-B12

□ **STEP B13:** Remove the driveshaft inboard ends from the differential casing - see *Job 9, Step 3*.
→ Using wire or strong cable ties, hang the driveshafts out of the way.
→ They must not be allowed to droop to the full flex of the CV joints, or these will be damaged.

TOP TIP!

• Even though you've drained the transmission fluid, more will probably leak out as you release the driveshafts. Have a suitable container ready to catch it.

□ **STEP B14:** If the model you are working on has an engine earth/ground strap connected to the transmission casing, remove it. Again, fluid could well leak from the casing as you do this, so have a suitable container ready to catch it.

□ **STEP B15:** Support the transmission using an engine hoist, or a trolley jack and a sturdy piece of wood. The former is much the best way.

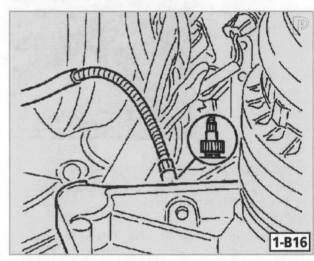
1-B16

□ **STEP B16:** Disconnect the speedometer cable from the transmission unit.

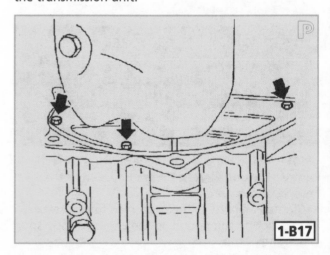
1-B17

□ **STEP B17:** Remove the torque converter housing cover and undo the three bolts (arrowed) holding the converter to the drive plate.
→ You will have to turn the engine over to bring each one into view.
→ Hold the converter steady with a broad-bladed screwdriver wedged in the teeth of the drive plate as you undo each bolt.

❏ **STEP B18:** Remove the rear engine mounting.

❏ **STEP B19:** Remove the left-hand side engine/transmission mounting.

❏ **STEP B20:** Remove the lower transmission -to-engine bolts (arrowed).

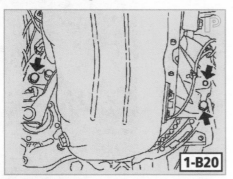

1-B20

TOP TIP!

❏ **STEP B21:** If you find it extremely difficult to manoeuvre the transmission, more space can be created by lowering the engine a few centimetres. However, you will have to disconnect the exhaust from the manifold, and release it from its front mounting rubber, in order to do this.

❏ **STEP B22:** Make sure the hoist or trolley jack is supporting the weight of the transmission, then pull it away from the engine and lower it to the ground.

IMPORTANT NOTE: The torque converter can fall away as you remove the transmission. This could damage it, and cause a large spillage of fluid. Make sure it is held as you lower the transmission, and carefully removed as soon as possible.

❏ **STEP B23:** Refit the transmission as the reverse of removal. Tighten all bolts to their specified torque settings, and use locking compound on the engine/transmission mountings' bolts.

IMPORTANT NOTE: Renew the torque converter mounting bolts.

❏ **STEP B24:** Refill with transmission oil - see *Chapter 5.*

❏ **STEP B25:** If necessary, refill the cooling system - see *Chapter 8, Part B, Job 1.*

JOB 2:	CLUTCH AND CLUTCH RELEASE BEARING - *remove, refit.*

FACT FILE

TRANSMISSION REMOVAL

STEP A-1A: On Cavalier models to 1992, the clutch can be replaced without removing the transmission unit.

2-A1A

• During 1992 a heavier flywheel was introduced to all models covered by this manual. To remove the clutch from models with the newer flywheel, the transmission must be removed see *Section B.*

Section A: Cavaliers to 1992.

❏ **STEP A1-B:** Taking note of *Chapter 2, Safety First!* jack up the front of the vehicle and remove the left-hand side wheel.
➜ If you are not using a vehicle hoist, support the vehicle on axle stands.
➜ These are the clutch, release mechanism, cable and pedal arrangements.

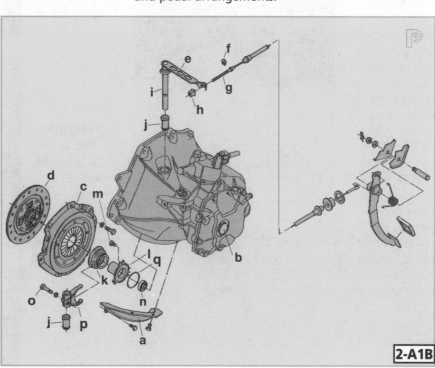

2-A1B

➜ The letters relating to this drawing are referred to in the following Steps.
➜ Drain the transmission fluid.

2-A2

❒ **STEP A2:** Remove the clutch housing cover plate (**a**).

❒ **STEP A3:** If the model you are working on has an engine earth/ground strap connected to the transmission casing, remove it.

❒ **STEP A4:** See *Job 1, Steps A16 to A18*:
➜ Disengage the transmission input shaft from the flywheel as follows.

❒ **STEP A5:** If the model you are working on has a plug (*2-A1B, item b*) at the end of the input shaft (on the end plate of the transmission unit), unscrew and remove it.
➜ If not, remove the end plate, and store its gasket safely.

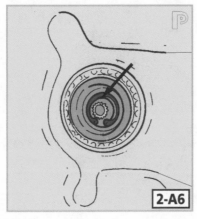

2-A5

❒ **STEP A6:** Behind the plug, you will see a circlip (arrowed).
➜ With circlip pliers, remove the circlip securing the input shaft.

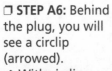

2-A6

❒ **STEP A7:** Remove the input shaft Torx screw.

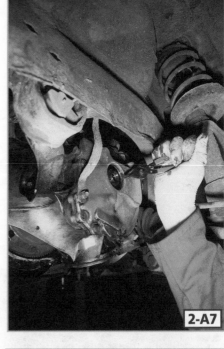

2-A7

❒ **STEP A8:** Pull the shaft away from the engine to disengage it from the flywheel.
➜ There are two ways of doing this:
➜ **EITHER:** A specialist tool such as the Sykes-Pickavant puller shown here…

2-A8

❒ **STEP A9:** …**OR:** A suitable lever (such as a long screwdriver) acting on a bolt screwed into the shaft.

2-A9

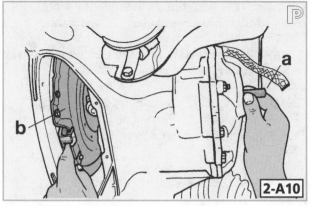

2-A10

STEP A10: Pull the input shaft (**a**) until it is clear of the clutch (**b**).

STEP A11:
Ask someone to depress the clutch pedal - or hold the pedal down with a piece of wood.
➜ You can now clamp the clutch plate to the cover plate using three suitable clamps, such as the Sykes-Pickavant ones seen here (see inset).

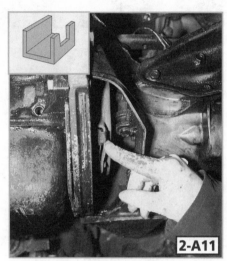

2-A11

➜ Turn the engine (using the crank pulley bolt) to bring each of the three clamping points into view, one after the other.

TOP TIP!

• Turning the engine over using the crank pulley bolt, is made much easier if you remove the spark plugs.

STEP A12:
Turning the engine to bring them into view, undo the six bolts holding the clutch cover plate to the flywheel (**a**) – each side

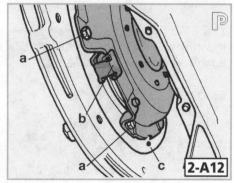

2-A12

of where the clamps (**b**) will have been located.
➜ If the engine turns as you are undoing a bolt, hold the flywheel still by wedging a broad-bladed screwdriver into the teeth of the flywheel's ring gear.

STEP A13: Pull the clutch assembly down, out of the clutch housing.

FACT FILE

FLYWHEEL WEAR

• Check the surface of the flywheel, where it mates with the clutch, for scoring or significant micro-cracking - caused by the excessive heat generated by clutch slip.
• If there is any doubt about the condition of the flywheel face, replace it.
• A scored face will quickly ruin a new clutch plate, and a cracked flywheel is at risk of disintegration.

STEP A14: Hold the clutch assembly in a vice with soft jaw protection strips, and compress it.
➜ Remove the clamps.
➜ Undo the vice.
➜ The pressure plate (illustration *2-A1B, item c*) will be separated from the friction plate (**d**).

STEP A15:
Disconnect the clutch cable from the control arm on the transmission – see illustration *2-A1B*. To do this:
➜ Move the lever (**e**) back a little…

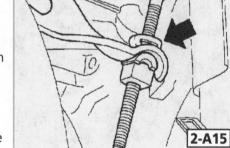

2-A15

➜ …unscrew the plastic clip (arrowed and *2-A1, item f*) from the threaded section…
➜ …then slide the threaded section (**g**) from the lever without disturbing the adjuster (**h**).

STEP A16: The illustration letters here refer to illustration *2-A1B*.
➜ Working in the clutch housing, undo the bolt (**o**) holding the release fork (**p**) to the control arm shaft (**i**).
➜ Pull the shaft upwards, which will free the release fork and bearing (**k**).
➜ If the bushes (**j**) are worn, press out and replace both.

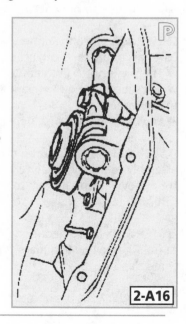

2-A16

☐ STEP A17: IMPORTANT NOTE: Carry out this Step only if the release bearing guide (**l**) is worn, or there is an oil leak from behind it.
→ Detach the release bearing guide (**l**) by taking out the three bolts (**m**).
→ Drift out the oil seal (**n**) and replace. Wipe a smear of grease on the lips of the new seal.
→ Replace the O-ring (**q**) – fit dry.

☐ STEP A18: Lightly grease the surface of the release bearing and the nylon pivot arm bushes with molybdenum grease, and refit the bearing as the reverse of removal.

☐ STEP A19: If the replacement clutch assembly does not come fitted with compression clamps, fit them, compressing the clutch in a soft-jawed vice.

☐ STEP A20: Smear the splines of the hub on the friction disc lightly with high melting point grease. Offer the clutch assembly up into position, and ask someone to hold it there.

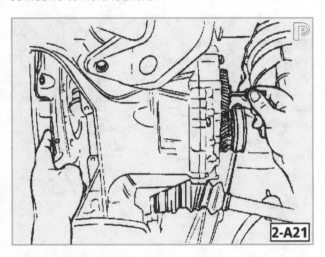

2-A21

☐ STEP A21: Push the input shaft into position, through the friction disc hub. This could well prove difficult. However, DO NOT hammer the shaft, as this will damage components inside the transmission.

☐ STEP A22: Once the input shaft is in place, turn the cover assembly to its correct position.
→ There is a notch in the outer edge of the plate, which should line up with an 'X' or dot etched into the flywheel.
→ See illustration **2-A12, item c**.

☐ STEP A23: Refit the six bolts holding the cover assembly to the flywheel, turning the engine over as for removal. Tighten them to their specified torque settings.

☐ STEP A24: Ask someone to press the clutch pedal, or hold it down with a piece of wood. Remove the three clamps.

☐ STEP A25: Refit the bolt to the end of the input

shaft, and refit its circlip. Refit the input shaft end cover, or transmission end plate, as applicable.

☐ STEP A26: If necessary, refit the earth/ground strap to the transmission end cover.

☐ STEP A27: Refit the clutch housing cover plate.

☐ STEP A28: Refit the wheel, and lower the vehicle off its jacks.

☐ STEP A29: Refit the clutch cable, and adjust it. See **Job 3**.

Section B: Cavaliers from 1992.

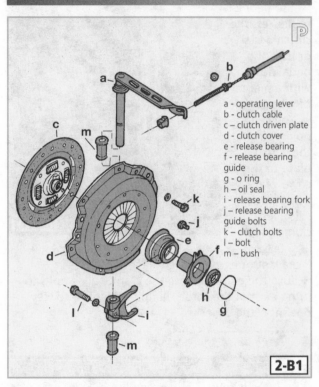

a - operating lever
b - clutch cable
c – clutch driven plate
d - clutch cover
e - release bearing
f - release bearing guide
g - o ring
h – oil seal
i - release bearing fork
j – release bearing guide bolts
k – clutch bolts
l – bolt
m – bush

2-B1

☐ STEP B1: Refer also to **Section A** where relevant.
→ This drawing shows the clutch components.
→ Remove the transmission, as described in **Job 1**.

☐ STEP B2: Undo the six clutch cover bolts (illustration **2-B1, item k**), working progressively round them to evenly ease the spring pressure in the clutch assembly (**d**). Remove the bolts.

2-B2

STEP B3: Ease the cover out of the recess in the flywheel and remove it. The driven plate (**c**) will fall free at this point, so catch it.

TOP TIP!

STEP B4: Check the inside of the clutch bellhousing for contamination by oil.
• This indicates a leak from either the crankshaft rear seal or the transmission input shaft seal.
• A faulty seal must be replaced immediately, or it will ruin the new clutch plate with further oil contamination.
• Oil can cause judder and slip.
• Replace the oil seal (**h**) from the end of the release bearing guide (**f**) – grease seal lips.
• Replace the O-ring dry.

FACT FILE
FLYWHEEL WEAR

STEP B5: Check the surface of the flywheel, where it mates with the clutch, for scoring or significant micro-cracking - caused by the excessive heat generated by clutch slip.
• If there is any doubt about the condition of the flywheel face, replace it.
• A scored face will quickly ruin a new clutch plate, and a cracked flywheel is at risk of disintegration.

STEP B6: Inside the bellhousing, undo the bolt (**l**) holding the release fork (**i**) to the control arm shaft (**a**).
➜ Pull the shaft upwards, which will free the release fork and bearing.
➜ Check the release fork pivot and its bushes (**m**) for wear.
➜ The bushes should be replaced if they are worn.

STEP B7: Detach the release bearing.

STEP B8: Lightly grease the new release bearing and the nylon fork pivot bushes with molybdenum grease, and refit them.

STEP B9: Clean the flywheel and new clutch friction surfaces with a degreasing agent, to remove any traces of grease or protective film.

STEP B10: Offer the driven plate to the flywheel.
➜ The side with its hub projecting further out should face away from the engine.

STEP B11: Place the clutch cover in its location recess in the flywheel. The flywheel has an 'X' etched into its face or a dot, which lines up with a notch in the rim of the cover plate.

STEP B12: Fit the six clutch cover bolts, tightening them progressively until the driven plate is very lightly gripped, but will still move laterally.

STEP B13: Using an alignment tool (**2**) pushed into the bush in the end of the crankshaft, make sure the clutch driven plate is perfectly centralised in relationship to the output shaft.

2-B13

IMPORTANT NOTE: Check to see if the bush is okay.
➜ If disintegrated, pull it out.
➜ Use a thread tap screwed in to grip it with.
➜ Fit a new bush.

STEP B14: Some clutch assemblies come supplied with a plastic centralising tool, but not all of them. If the clutch you are fitting does not have one, a tool set such as the Sykes-Pickavant one shown here will do the job.

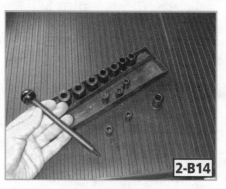

2-B14

STEP B15: Progressively working round the plate, tighten the six cover bolts in increments, until they are at their recommended torque settings.

STEP B16: Smear a little 'copper' grease on the release bearing guide and transmission input shaft.

STEP B17: Refit the transmission. See *Job 1*.

Section C: Vectras.

The Vectra clutch is operated by an hydraulic system, comprising a master cylinder and a release cylinder, which is concentric with the transmission first motion (input) shaft.
➜ The release cylinder/release bearing assembly is a sealed unit. If the release bearing is worn or the assembly leaks, the whole assembly must be renewed.

❒ **STEP C1:** Follow *Steps B1 to B5*.

❒ **STEP C2:** Clean the release cylinder, undo the hydraulic pipe union nut and disconnect the pipe.

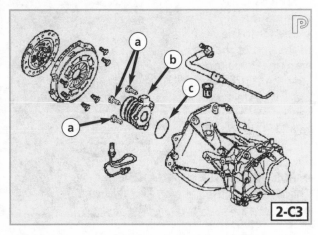

2-C3

❒ **STEP C3:** Undo the three bolts (**a**) and remove the release cylinder assembly (**b**).
➜ Remove the release cylinder/transmission casing sealing ring (**c**), which MUST be renewed on reassembly.
➜ Take care not to allow any dirt to enter the transmission.

❒ **STEP C4:** Follow *Steps B9 to B17*.

JOB 3:	CLUTCH CABLE - *remove, refit, adjust*.

TOP TIP!

❒ **STEP 1:** Measure the amount of thread that sticks out of the adjuster.
➜ When you fit the new cable, it will need to be correctly adjusted.
➜ This

3-1

measurement will give you a good starting point.

❒ **STEP 2:** Disconnect the clutch cable from the control arm on the transmission. To do this, move the lever back a little, unscrew the plastic clip from the threaded section (arrowed), then slide the threaded section from the lever.

3-2

❒ **STEP 3:** Release the cable from the retainer cast into the top of the transmission casing.

❒ **STEP 4:** Disconnect the clutch cable from the clutch pedal. Pull the top of the return spring (arrowed) out of the pedal.

3-4

3-5

❒ **STEP 5:** Pull the cable (arrowed) forwards through the bulkhead, into the engine bay.

❒ **STEP 6:** Make a note of the route the cable takes through the engine bay, then remove the cable.

❒ **STEP 7:** Refit it, as the reverse of removal and adjust.

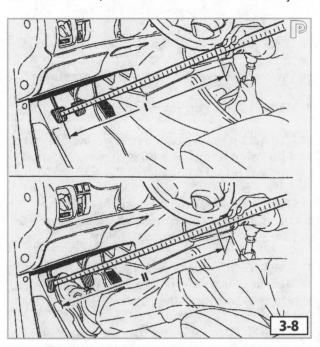

3-8

❒ **STEP 8:** To adjust the clutch, measure the distance between the clutch pedal and the base of the steering wheel (**I**). Then measure the distance with the clutch pushed fully to the floor (**II**).

☐ **STEP 9:** The indicator for adjustment is the difference between these two measurements.
→ Screw the plastic stop (arrowed) along the threaded shaft of the cable's

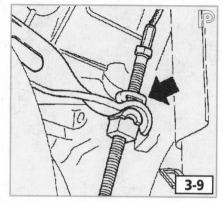

transmission end until the correct distance is achieved.
→ When you have finished, the clutch pedal should sit slightly higher than the brake pedal.

JOB 4: CLUTCH HYDRAULIC SYSTEM.

The Vectra clutch is operated by an hydraulic system comprising a master cylinder and a release cylinder. Details of renewing the release cylinder are in *Job2, Section C*.

Section A: Bleeding the system.

This operation is very similar to bleeding brakes, with the exception that the clutch pedal should be pumped several times then held down before the bleed valve (on the transmission casing) is slackened to allow air to escape.

☐ **STEP A1:** Check the fluid level in the reservoir and top up if necessary.

☐ **STEP A2:** Attach a bleed pipe (other end immersed in brake fluid) or brake bleeding kit to the bleed valve.

☐ **STEP A3:** An assistant should depress the clutch pedal several times and then hold it down to pressurise the system.

☐ **STEP A4:** Open the bleed valve and allow air/fluid to escape until the fluid has no air bubbles.
→ Check the clutch pedal 'feel'.
→ Check the fluid level in the reservoir and top up if necessary.

Section B: Master cylinder removal, replacement.

The master cylinder cannot be repaired; if faulty, it must be renewed.

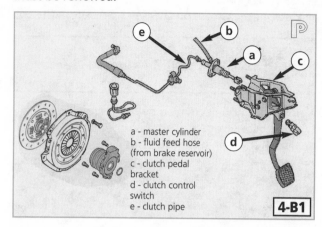

a - master cylinder
b - fluid feed hose (from brake reservoir)
c - clutch pedal bracket
d - clutch control switch
e - clutch pipe

☐ **STEP B1:** Remove the brake servo - see *Chapter 12, Job 11*.
→ On left hand drive models, also remove the relay box.

☐ **STEP B2:** Clean the master cylinder (*4-B1, item a*) and place cloth underneath to catch hydraulic fluid.
→ To minimise fluid leakage, remove the reservoir cap, place a film of plastic over the filler aperture and refit the cap.

☐ **STEP B3:** Remove the retaining clip and pull the hydraulic pipe (*4-B1, item e*) from the master cylinder. Immediately plug the pipe end and master cylinder to minimise fluid loss.
→ Remove the sealing ring from the union – this must be renewed.
→ Refit the pipe retaining clip to the master cylinder so it doesn't get lost!

☐ **STEP B4:** Disconnect the feed hose (*4-B1, item b*) from the master cylinder. Plug the pipe end and the master cylinder to keep dirt out.

☐ **STEP B5:** Working inside the vehicle, remove the clevis pin and its retaining clip (arrowed) from the master cylinder pushrod.
→ Lift the clip slightly with a screwdriver then pull it off.
→ On left hand drive vehicles, remove the trim panel to get to the pushrod.

☐ **STEP B6:** Remove the master cylinder securing nuts, and lift the master cylinder from the engine bay.
➜ The gasket MUST be renewed.

☐ **STEP B5:** Replacement is the opposite of removal.
➜ The master cylinder gasket and union sealing ring MUST be renewed.
➜ Place a little grease on the clevis pin before fitting.

JOB 5: GEAR LEVER – *remove, refit.*

Among the models covered by this manual, four types of gear lever are used. Accordingly, we shall deal with this Job in four sections.

Section A: Early Cavalier.

☐ **STEP A1:** Place the lever in the neutral position. Release the rubber boot.
➜ The triangle (arrowed) faces forwards when refitting the gaiter.

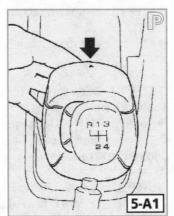
5-A1

☐ **STEP A2:** Depending on the model, it may be necessary to remove the centre console – see *Chapter 14, Job 10*.

☐ **STEP A3:** Remove the circlip (arrowed), push the lever to the left and withdraw it.

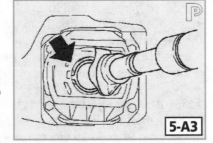
5-A3

☐ **STEP A4:** When refitting, grease the ball and socket.

Section B: Later Cavalier.

☐ **STEP B1:** If the vehicle you are working on has a centre console, remove it - see *Chapter 14, Job 10*.

☐ **STEP B2:** Move the lever to the neutral position, release the gaiter at the bottom, and slide it up the lever.

☐ **STEP B3:** Undo the four selector bolts.
➜ Lift the assembly out.

☐ **STEP B4:** There is a clip at the base of the lever (**1**). Lever it free with a flat-bladed screwdriver and remove it, along with the pin (**2**).
➜ The lever can now be separated from the housing.

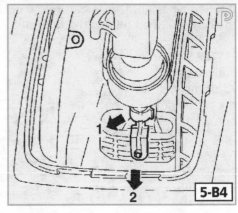
5-B4

☐ **STEP B5:** When refitting, grease the ball and socket.

Section C: Cavalier from 1988.

☐ **STEP C1:** Free the gaiter from the centre console.

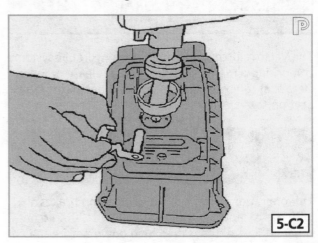
5-C2

☐ **STEP C2:** Remove the clip from the lever, pull the pivot pin out and remove the lever.

Section D: Vectra.

The gear lever fitted to the Vectra can be removed only with the gear linkage; Vectra gear lever removal is dealt with in *Job 6, Steps C1-C7*.

JOB 6: GEAR LINKAGE/ GEARCHANGE ROD ASSEMBLIES – *remove, refit.*

Among the models covered by this manual, three types of gear lever are used. Accordingly, we shall deal with this Job in three sections.

Section A: Cavalier to 1988.

The gearchange lever does not have to be removed on these models.

❐ **STEP A1:** These are the components of the gearchange linkage.
➔ All of the annotations letters in the following Steps refer to this illustration.

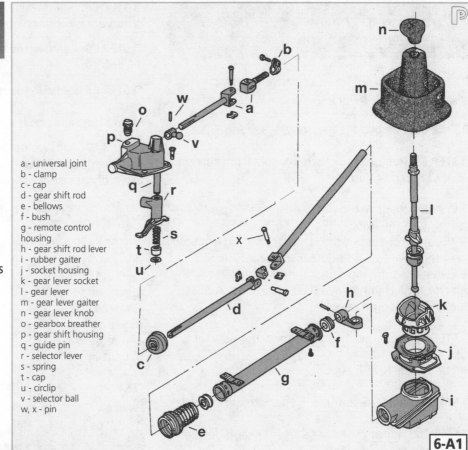

a - universal joint
b - clamp
c - cap
d - gear shift rod
e - bellows
f - bush
g - remote control housing
h - gear shift rod lever
i - rubber gaiter
j - socket housing
k - gear lever socket
l - gear lever
m - gear lever gaiter
n - gear lever knob
o - gearbox breather
p - gear shift housing
q - guide pin
r - selector lever
s - spring
t - cap
u - circlip
v - selector ball
w, x - pin

6-A1

6-A2

❐ **STEP A2:** Mark (arrowed) the clamp connecting the gear linkage to the gearchange rod with typists' correction fluid.
➔ Undo the pinch bolt at the universal joint (**a**) and disconnect the rod.

❐ **STEP A3:** Undo the bolts holding the gearchange rod's protective 'tube' (**g**) to the underside of the vehicle.
➔ Lever free the rubber cover over the gear lever's joint with the rod.

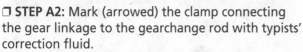

6-A3

❐ **STEP A4:** Disconnect the rod from the bottom of the gear lever.

❐ **STEP A5:** Withdraw the rod, along with the 'tube'.

❐ **STEP A6:** To remove the rod lever (**h**) from the rod, drift out the pin (**x**).

6-A5

❐ **STEP A7:** Replace the linkage components as the reverse of removal, using the correction fluid mark to set the clamp at its old position.

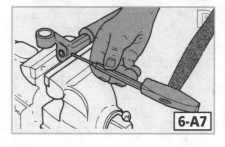

6-A7

Section B: Cavalier 1988-on.

☐ **STEP B1:** Undo but do not remove the gear selection/linkage clamp bolt.

☐ **STEP B2:** Remove the gear lever – see *Job 5*.

☐ **STEP B3:** Remove the centre console front section – see *Chapter 14, Job 10*.

☐ **STEP B4:** Remove the bolts securing the gear lever surround to the floor.
➜ The selector tube and gear lever housing can now be withdrawn toward the rear of the vehicle.
➜ When the tube is partially withdrawn, remove the selector tube clamp to avoid damaging the rubber boot in the upright panel.

☐ **STEP B5:** It is good policy to renew the selector tube bush.
➜ This can be levered out.

☐ **STEP B6:** Replacement is the opposite of removal.

☐ **STEP B7:** Adjust the gear selector linkage before tightening the tube clamp.

☐ **STEP B8:** Remove the adjuster hole plug, insert the shank of a 4.5 mm drill bit in the hole and turn the gear selector tube until the drill can be pushed into its adjustment hole.

☐ **STEP B9:** For pre-1993 vehicles, an assistant is needed to hold the gear lever in the neutral position, resting against the reverse stop.
➜ Tighten the clamp bolt.

☐ **STEP B10:** On 1993-on models, an assistant is not necessary, because the gear lever can also be locked by placing a second drill bit shank in the holes in the lever and base.

Section C: Vectra.

☐ **STEP C1:** Raise the front of the vehicle and remove the engine undertray if fitted.

☐ **STEP C2:** Make alignment marks on the selector rod and clamp, then slacken the clamp bolt but do not remove it.

☐ **STEP C3:** Move the gear lever into 4th gear position.

☐ **STEP C4:** Pull the selector rod from the clamp and remove the rubber gaiter.

☐ **STEP C5:** Remove the centre console – see *Chapter 14, Job 10*.

☐ **STEP C6:** Remove the gear lever surround switch panel.
➜ Pull the connectors apart.

☐ **STEP C7:** Remove the four bolts holding the gear lever assembly to the floor, and withdraw the assembly and selector rod.

SHIFT GUIDE REMOVAL

☐ **STEP C8:** Push the universal joint pin (arrowed) to compress the detent spring and press the pin out.

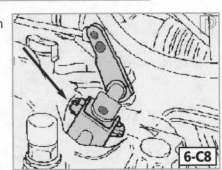

6-C8

☐ **STEP C9:** Remove the linkage bracket retaining clips (a).
➜ The bracket can now be lifted away.

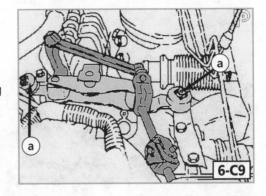

6-C9

☐ **STEP C10:** When refitting, apply grease to bearing surfaces and adjust the gear selector linkage as described in *Steps B7 to-B10*.

JOB 7:	AUTOMATIC GEAR SELECTOR CABLE – *replace, adjust*.

☐ **STEP 1:** Remove the centre console – several different types - see *Chapter 14, Job 10*.

☐ **STEP 2:** Remove the cable:
➜ Disconnect the selector cable inner from its clamp (**1**) on the lever assembly.
➜ Remove the cable outer from the retainer (**2**).

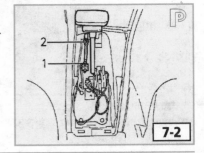

7-2

STEP 3: Disconnect the cable from the selector mechanism on the transmission.
➔ On Vectras, remove the battery and plate to gain access. See

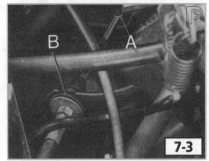

Chapter 10, Electrical, Dash, Instruments, Fact File: Disconnecting the Battery BEFORE doing so!
➔ On Cavaliers to 1988, this connection is a clamp.
➔ On later Cavaliers and all Vectras, it is a clip.

STEP 4: Pull the cable through into the engine bay, noting its route.

STEP 5: Refit the cable as the reverse of removal.

STEP 6: Move the selector lever to the 'P' position, if it is not already there, and fit the cable so that there is no slack in it, but is not unduly tight.

STEP 7: Ask someone to move the selector lever through all its position, checking at the transmission end that they all engage. On some models, the 'P' and 'N' positions are marked on the transmission to help with this.

STEP 8: Adjust the cable as necessary by moving its position in the clamp at the selector lever.

STEP 9: When the cable is correctly adjusted, refit the centre console.

JOB 8:	'KICKDOWN' THROTTLE CABLE – *replace, adjust*.

This Job applies to carburetor-engined vehicles only.

STEP 1: Remove the air cleaner assembly.

STEP 2: Undo the ball-joint connection to the carburetor lever by pulling its clip off and levering the joint apart with a flat-bladed screwdriver.

STEP 3: Unscrew the outer sleeve retaining bolt and disconnect the cable transmission end by pulling it upwards.

STEP 4: Remove the adjusting mechanism by squeezing in the lug.

STEP 5: Refit the cable as the reverse of removal.

STEP 6: Make sure there is slight slack in the cable.

STEP 7: Ask someone to slowly push the throttle pedal down until it just touches the kickdown switch.
➔ Make sure the throttle in the carburetor is wide open at this point.
➔ Adjust the kickdown and/or the throttle cable(s) to achieve this.

STEP 8: Remove the adjuster pin and ask your helper to push the throttle pedal fully down. The adjuster ratchet will move to the correct location.
➔ Replace the pin.

STEP 9: Refit the air cleaner assembly.

JOB 9:	DRIVESHAFT – *remove, replace*.

STEP 1: Release the outboard end of the driveshaft from the strut assembly – see *Chapter 11, Job 8*.

STEP 2: Support the outboard end of the driveshaft by hanging it on wire or a strong cable tie.

STEP 3: Using a suitable lever, remove the inboard end of the driveshaft from the differential housing.
➔ Transmission fluid will run out, so have a suitable container ready to catch it.

TOP TIP!
• Sometimes, disengaging the circlip in the driveshaft splined end can be difficult.
• A light hammer blow on the lever can help.

STEP 4: Replace the driveshaft as the reverse of removal.

STEP 5: Refill the transmission with transmission fluid - see *Job 1*.

JOB 10:	OUTER CV JOINT – *replacement*.

TOP TIP!
• If the gaiter to be replaced is an outboard one, proceed as follows.
• If you are only replacing an inboard one, remove the driveshaft - see *Job 9*.
• Support it in a vice, and do the work there.

STEP 1: Release the outboard end of the driveshaft from the strut assembly – see *Chapter 11, Job 8*.

STEP 2: Support the outboard end of the driveshaft by hanging it on wire or a strong cable tie.

❏ **STEP 3:** Remove the clips holding the gaiter in place.

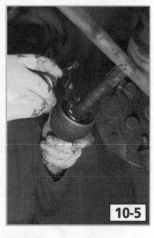

10-4   10-5

❏ **STEP 4:** Push the gaiter away from the CV joint, exposing it. Wipe away the excess grease so the innards of the joint are visible.

❏ **STEP 5:** With appropriate circlip pliers, such as the Sykes-Pickavant tool seen here, open out the circlip holding the CV joint onto the driveshaft.

❏ **STEP 6:** Pull the joint off the driveshaft. You may need to gently tap it with a soft-headed hammer to persuade it to move.

JOB 11: DRIVESHAFT GAITER – *remove, replace.*

❏ **STEP 1:** Remove, as follows:
→ The driveshaft - see *Job 9*.
→ The outer c.v. joint - see *Job 10*.

TOP TIP!

• If the gaiter to be replaced is an outboard one, proceed as follows.
• If you are only replacing an inboard one, remove the driveshaft – see *Job 9*.
• Support it in a vice, and do the work there.

❏ **STEP 2:** Pull the gaiter off the driveshaft. If you are also replacing the inboard gaiter, remove its clips and slide it along the driveshaft.

11-2

❏ **STEP 3:** Thoroughly clean the driveshaft of grease and dirt.

❏ **STEP 4:** Replace the gaiters and CV joint as the reverse of removal. Before securing the gaiters in place, make sure the CV joints are packed with high melting point grease.

IMPORTANT NOTE: A new circlip must be used to secure the CV joint.

❏ **STEP 5:** Refit the strut assembly and wheel - see *Chapter 11, Job 8*.

JOB 12: DRIVESHAFT OIL SEAL – *remove, replace.*

❏ **STEP 1:** Remove the inboard end of the driveshaft from the differential casing – see *Job 9*.

❏ **STEP 2:** Support the driveshaft by hanging it on wire or a strong cable tie.

❏ **STEP 3:** Remove the oil seal from the differential housing.
→ Although a specialist Vauxhall tool (**1**) is seen in use here, the job can be done simply with a broad-bladed screwdriver.

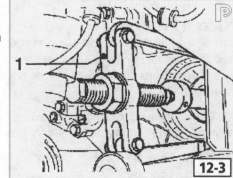

12-3

❏ **STEP 4:** Clean the oil seal mount and smear a thin film of transmission fluid on the sealing flange of the new seal.

❏ **STEP 5:** Using a soft-headed hammer on an appropriately sized socket or drift, gently drive the new seal into place (arrowed), until it sits flush with the differential casing.

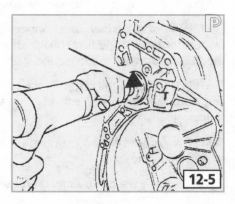

12-5

❏ **STEP 6:** Replace the inboard end of the driveshaft into the differential casing – see *Job 9*.

❏ **STEP 7:** Refit the wheel and lower the vehicle off its supports.

CHAPTER 8: COOLING SYSTEM

*Please read **Chapter 2 Safety First** before carrying out any work on your car.*

Part A: Systems Explained

CONTENTS

Page No.

JOB 1: SYSTEMS EXPLAINED.

IMPORTANT NOTE:
See **Chapter 5, Servicing** for information on refilling the cooling system with coolant.

→ It is a pressurised system, powered by a pump **(1)** driven from the timing belt.

→ This forces water round the cylinder block **(2)** and head **(3)** and, via hoses **(4)** through an aluminium radiator **(5)**, which cools it.

→ It works by transferring the water's heat to air passing through it as the vehicle is on the move.

→ If the vehicle is stationary, a thermostatic switch will switch on an electric fan to draw air through the radiator when the coolant reaches a certain temperature.

→ When the vehicle starts up, the engine is cold. In order to get the engine to working temperature as quickly as possible, a thermostatic valve **(6)** is closed, restricting the water to circulating within the engine block and head.

→ When the engine's working temperature is reached the thermostat opens, allowing the water to travel through the radiator and be cooled.

→ The heating circuit also diverts through the cabin to the heater (see arrows) radiator, or 'matrix'.

To heater

From heater

1-1

☐ **STEP 1:** All the models covered by this manual have, in principal, the same cooling system.

TOP TIP!

• Whenever you have drained and disconnected any part of the cooling system, always run the engine, after reassembly, to normal working temperature and check for leaks.

❏ **STEP 2:** This diagram shows how the heater dissipates heat from the water in the heater matrix (**1**) to air forced through it by the forced air flow (**2A**) and/or heater fan (**2B**).

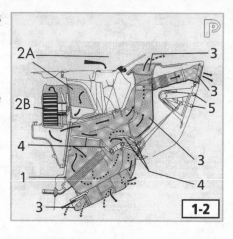

➜ The warmed air is then distributed into the cabin via ducts and vents (**3**), controlled by flaps (**4**) connected to the heater controls (**5**).

➜ Even vehicles with air conditioning use this system, along with an additional cooling circuit, which works with the heating function to maintain the air in the cabin at a temperature set by the driver.

SAFETY FIRST!

• The electric fan can come on if the engine is hot.

• Make sure loose clothing, hair, tools or your hands are never near the fan while checking the cooling system.

• Remember that on many vehicles, the fan can turn itself on even if the engine is stopped and the ignition switch turned to OFF.

Part B: Repair Procedures
CONTENTS

JOB 1: COOLANT - *change.*

SAFETY FIRST!

• Antifreeze is an irritant to your skin, and harmful to vehicle paintwork.

• Wear impervious gloves while dealing with it, and if it gets on any body panels, rinse it off with water.

• DO NOT open the cooling system while the engine is hot!

❏ **STEP 1:** Raise the front of the vehicle - see *Chapter 2, Safety First!* If you are not using a vehicle hoist, secure it on axle stands.

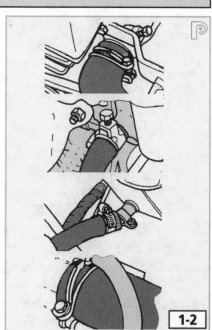

❏ **STEP 2:** These illustrations show the four types of securing clips that Vauxhall use.

STEP 3: Undo the clip (arrowed) securing the lower hose to the radiator and slide it down the hose, away from the joint.

➔ Place a bucket under the hose and work it off the joint - see *Job 2*.

STEP 4: If the hose is seized onto the radiator outlet, it can be cut away using a sharp craft knife.

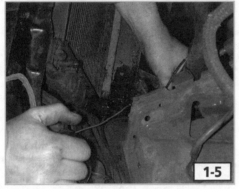

STEP 5: When all the coolant has run out, check there is no corrosion on the joint. If there is, clean it off with wire wool, before reattaching the hose and tightening its clip.

STEP 6: This is a good opportunity to test the condition of the hoses by squeezing them – see *Chapter 5, Servicing*. Replace any that have deteriorated – see *Job 2*.

STEP 7: Refill the system.
➔ Prepare a 50/50 mix of clean water and coolant additive (anti-freeze) ready for filling the system.
➔ Allow it to vent by removing the wiring connector from the coolant temperature sender and unscrewing the sender – see *Job 7*.
➔ If you are working on a DOHC engine, you can instead simply remove the bleed screw from the thermostat housing cover.
➔ Slowly pour new coolant into the expansion tank. The more gradually you do this, the less likely it is that air bubbles will become trapped in the circuit.
➔ When the water level has reached the 'Cold' or

'Kalt' marker line on the expansion tank, replace the temperature sender or bleed screw, as applicable.
➔ Start the engine. This allows the water pump to circulate the coolant, and expel any trapped air.
➔ The coolant level may drop swiftly as you do this. Pour in more coolant as it does so, to make sure no extra air is drawn into the engine.
➔ Continue to fill the system until the coolant is level with the 'Max' mark on the expansion tank, and bubbles have stopped appearing on the surface.
➔ Turn off the engine.

STEP 8: Refit the filler cap.

JOB 2: COOLANT HOSES - *change*.

SAFETY FIRST!

• **DO NOT drain coolant while the engine is hot!**

STEP 1: Raise the front of the vehicle off the ground. If you are not using a vehicle hoist, secure it on axle stands.

STEP 2: Drain the cooling system – see *Job 1*.

STEP 3: Undo the clips securing the hose you are replacing.
➔ If the hose is old, it may not come off. Hoses can 'weld' themselves to their stubs.
➔ Use too much force and you could cause damage! Instead, cut through the hose with a sharp knife and replace it.
➔ If the stub is a plastic one, take extra care not to cut through the stub as well!

STEP 4: Check there is no corrosion on the joint. If there is, clean it off with wire wool, before reattaching the hose and tightening its clip.

STEP 5: Refill the system – see *Job 1* – and refit the radiator cap.

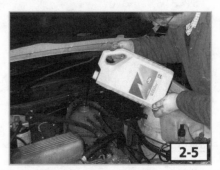

JOB 3: COOLING FAN - *remove, refit.*

SAFETY FIRST!

• If the engine is warm, the fan can start, whether or not the ignition is on.

• Before working on the fan, disconnect the earth/ground lead from the battery.

• On some models, this can affect the radio's memory, so instead remove the cooling fan's fuse from the fuse box – see *Chapter 15, Wiring Diagrams*.

☐ **STEP 1:** Disconnect the fan's electrical connectors.
→ On early models there will be just one connector.
→ On later models there may be several.

☐ **STEP 2:** Remove the bolts holding the fan and shroud assembly to the radiator.

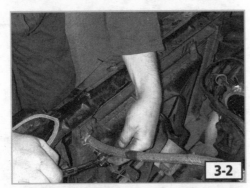

3-2

☐ **STEP 3:** Lift the assembly up and away from the radiator to remove it.

3-3

TOP TIP!

• If the fan assembly mounting bolts prove difficult to undo, remove the fan complete with the radiator – see *Job 4*.

☐ **STEP 4:** Remove the bolts holding the fan motor to the shroud and pull the fan free. It is a non-serviceable unit, so if there's anything wrong with it, it should be replaced.

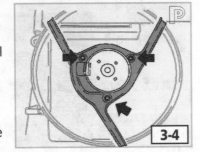

3-4

☐ **STEP 5:** If applicable, remove the clip securing the fan to the motor shaft as shown, and separate them.

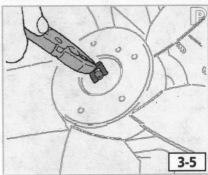

3-5

☐ **STEP 6:** Refitting is the reverse of removal.
→ Make sure the fan blade locates correctly on the motor shaft, and the shroud is properly fitted into its clip(s) on the radiator.

3-6

JOB 4: RADIATOR - *remove, clean, refit.*

SAFETY FIRST!

ALL MODELS

• DO NOT open the cooling system while the engine is hot!

MODELS WITH AIR CONDITIONING.

• DO NOT open the pressurised pipework on the air conditioning system.

• The air conditioning system contains pressurised gas and specialist equipment is needed to extract it. It is illegal to discharge the gas into the atmosphere.

• The gas used in air conditioning systems is extremely harmful to the environment and may cause injury to the person carrying out the work if released in an uncontrolled manner.

• If air conditioning units are in the way of carrying out other work, move units to one side with pipes still attached.

• If this is not possible, or there is a risk of damage to pipes or connections, have an air conditioning specialist de-pressurise the system so that you can dismantle it.

☐ **STEP 1:** Drain the cooling system - see *Job 1*.

☐ **STEP 2:** Disconnect the electrical connectors at the cooling fan.

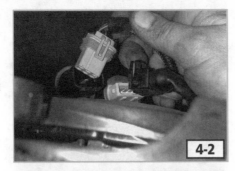

4-2

☐ **STEP 3:** If fitted, unbolt the air conditioning consender/fan assembly – see *Safety First!* at the start of this Job.

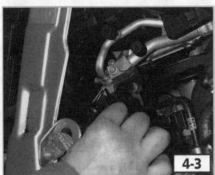

4-3

TOP TIP!

☐ **STEP 4:** If the vehicle you are working on has power steering, you can improve access to the radiator.

➜ Unbolt the power steering fluid reservoir, and secure it out of the way with a cable tie.

☐ **STEP 5:** Disconnect the cooling hoses from the radiator - see *Job 2*.

TOP TIP!

• If the engine bay is cramped, first disconnect radiator hoses from the ends furthest away from the radiator.

• You can then remove the hoses with the radiator assembly.

☐ **STEP 6:** Disconnect the coolant thermostatic switch(es) from their wiring connector(s).

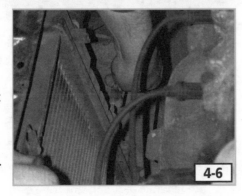

4-6

☐ **STEP 7:** If the vehicle you are working on has automatic transmission and/or power steering:

4-7

➜ Hoses may run alongside the radiator (depending on model) and may need to be removed or opened up.

➜ Clamp the hoses running to and from the transmission oil cooler next to the radiator.

➜ Disconnect the hoses, and seal the ends to stop dirt getting in.

☐ **STEP 8:** Undo the bolts holding the top retaining brackets to the radiator, and remove the brackets.

4-8

☐ **STEP 9:** On some models the radiator top fixing is a clip (arrowed).

➜ Undo this by squeezing with pliers, as shown here.

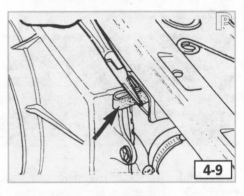

4-9

4-10

☐ **STEP 10:** Lift the radiator out of the engine bay. Note that its lower mountings are rubber bushes in the front cross member, into which locating pegs fit.

4-11

STEP 11: On models with air conditioning the workspace is cramped. Take care not to damage the radiator matricies.
➜ Remove the cooling fan assembly, referring to the last Job.

STEP 12: This is a good opportunity to flush the radiator. Use a garden hose to flush it, in the reverse direction to normal flow.

STEP 13: Clean dust and road debris from the radiator fins using a soft brush or an air hose, but never a high-pressure water jet. This can easily distort the fins.

TOP TIP!

• If the radiator is leaking, it is possible to repair it.
• However, replacements for Vauxhall vehicles are reasonably priced, so it will probably prove more cost-effective to renew it than to have a specialist repair it.

STEP 14: Check that the lower mounting rubbers are in good condition, and replace them if not. Refitting is the reverse of removal.

JOB 5:   THERMOSTATIC SWITCH(ES) - test, replace.

SAFETY FIRST!

• DO NOT open the cooling system while the engine is hot!

FACT FILE

VEHICLES WITH AIR CON.

• Models with air conditioning have two thermostatic switches on the radiator.
• The lower one switches on the cooling fan (and air conditioning cooling fan) when the coolant reaches 100 degrees C.
• Should the coolant temperature get to 105 degrees, the higher switch opens, and both fans begin to operate at a higher speed. However, they should both be tested in the same way.

5-1

STEP 1: Disconnect the wiring plug(s) from the terminals on the switch(es).

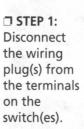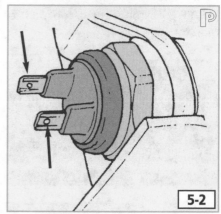

5-2

STEP 2: To test a switch, run the engine up to normal working temperature, but (in case the fan is not working) don't run the engine to the point where the coolant boils, or the temperature gauge says the engine is overheating. The fan should come on at this point.
➜ First, check to see if the fuse has blown.
➜ If not, pull the connector from the switch and bridge the gap between the two plug terminals. This should make the fan start.
➜ Alternatively, place a test light across the terminal (arrowed) on the switch – the test light should illuminate when the engine is hot. (DON'T overheat the engine while carrying out this test!)
➜ If it does not, there may be a wiring fault, and the loom should be checked for breakages or corroded connectors.

→ If it does, the switch may not be doing its job, and should be replaced.

→ Allow the engine to run until the temperature gauge indicates the engine is about to overheat, then turn off the engine and switch the ignition back on. Test the upper switch in this way, as it should be operative by this point.

❏ **STEP 3:** If a switch requires replacing, undo the expansion tank cap, then unscrew the thermostatic switch from the radiator. Have a container ready to catch the coolant that will run out.

❏ **STEP 4:** When coolant has drained to the level of the switch aperture:

→ Refit as the reverse of removal.

→ Use a new sealing washer.

→ Top up the coolant - see *Chapter 5, Servicing*.

JOB 6:	THERMOSTAT - *remove, test, replace*.

SAFETY FIRST!

• **DO NOT open the cooling system while the engine is hot!**

❏ **STEP 1:** Drain the cooling system – see *Job 1*.

❏ **STEP 2:** When necessary (certain models only) remove the outer timing belt cover - see *Chapter 6, Part C, Job 1*.

→ On the Vectra, it is necessary to remove the timing belt and idler, camshaft sprocket and rear timing belt cover – see *Chapter 6*.

❏ **STEP 3:** Disconnect the radiator top hose from the thermostat housing.

→ You can, alternatively, leave the top hose attached.

6-4

❏ **STEP 4:** Remove the bolts securing the thermostat housing from the engine.

6-5

❏ **STEP 5:** Pull the thermostat (arrowed) from its housing.

→ If it is stuck, waggle it from side to side to free it.

❏ **STEP 6:** To test the function of the thermostat:

→ Lower it, on a length of wire, into a pan of boiling water. It should open.

→ Remove it from the water, and as it cools it should close.

→ If this does not occur, replace the unit.

❏ **STEP 7:**
Refitting is the reverse of removal. However, the thermostat's rubber seal (when fitted) should always be renewed.

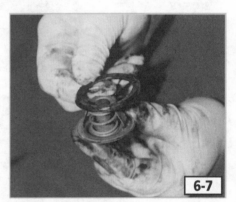

6-7

❏ **STEP 8:** Refill the system with coolant - see *Chapter 5, Servicing*.

JOB 7:	TEMPERATURE GAUGE SENDER - *remove, replace*.

❏ **STEP 1:** Drain the cooling system, as described in *Job 1*.

❏ **STEP 2:**
Disconnect the wiring connections from the sender unit.

7-2

❏ **STEP 3:** Unscrew the sender.

☐ **STEP 4:** Refit, as the reverse of removal.

JOB 8: **COOLANT PUMP** - *remove, replace*.

On all models, the coolant pump is a non-serviceable item, and must be replaced if it is worn.

SAFETY FIRST!

• **DO NOT** open the cooling system while the engine is hot!

☐ **STEP 1:** Drain the cooling system - see *Job 1*.

☐ **STEP 2:** Loosen the alternator mounting bolts and swing the alternator towards the engine, to slacken the auxiliary drive belt. Remove the belt.

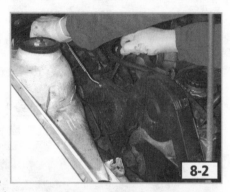
8-2

☐ **STEP 3:** Remove the outer timing belt cover - see *Chapter 6, Part C, Job 1*.

☐ **STEP 4:** Loosen the timing belt - see *Chapter 6, Part C, Job 5*.
➔ It may be helpful to fully remove the timing belt before removing the pump.

8-3

☐ **STEP 5A:** Remove the coolant pump upper mounting bolt.

8-5A

☐ **STEP 5B:** Remove the coolant pump lower mounting bolt.

8-5B

☐ **STEP 6:** The pump may well be stuck, and require a few light taps with a hammer and drift before it can be freed.

8-6

TOP TIP!

• This is an ideal opportunity to replace the timing belt, which is a relatively inexpensive part for Vauxhall vehicles - see *Chapter 6, Part C, Job 3*.

☐ **STEP 7:** Remove the coolant pump.

☐ **STEP 8:** Fitting a new coolant pump.
➔ Ensure that a new O-ring gasket (arrowed) is used, and smeared with a light film of white silicone grease before fitting.
➔ Tighten the pump mounting bolts to their specified torque settings.

8-7

➔ However, remember that on SOHC engines, this should not be done until the timing belt is adjusted correctly - see *Chapter 6, Part C, Job 3*.

8-8

❏ **STEP 9:** The lower water pump bolt is difficult to fit – use a long Allen key or a hexagonal bit in a socket extension – to clear the bodywork.

8-9

❏ **STEP 3:** Disconnect the following:
➜ The wiring connector(s) (**a**) from the pump(s).
➜ The fluid hoses (**b**) from the pump(s).
➜ If the

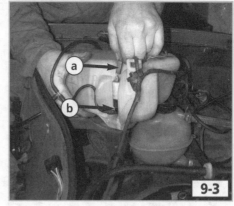

9-3

hoses are old, this could be difficult. Cut the pipe as close to the pump nozzle as possible, then slice the remainder carefully off the nozzle when you have removed the pump.

❏ **STEP 4:** Pull the pump away from and out of the reservoir.
➜ You will probably need to lever it carefully away with a screwdriver.

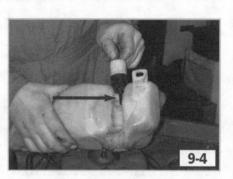

9-4

➜ Take care not to break the submersed part of the pump (arrowed).

❏ **STEP 5:** Refitting is the reverse of removal.
➜ Note that the pump mounting rubber seal can deteriorate and leak. If necessary, renew it.
➜ If the rubber sealing ring comes out with the pump, pull it from the nozzle and refit it to the reservoir before refitting the pump.

JOB 9:	WASHER PUMP - *remove, replace*.

9-1

❏ **STEP 1:** Depending on the model you are working on, it may be necessary to remove the reservoir first to remove the pump.

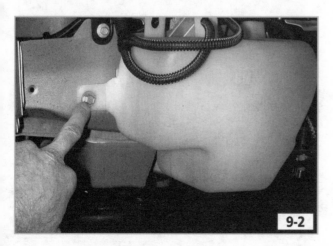

9-2

❏ **STEP 2:** On Vectras, the reservoir is situated behind the bumper – see *Chapter 13, Job 15.*

JOB 10:	WASHER JET - *remove, replace*.

❏ **STEP 1:** The washer jets can often be unclipped from their mounting holes – from above - by hand.

❏ **STEP 2:** If this proves difficult, reach the rear of the nozzle/s - lift the plastic engine bay panel, as applicable - and push

10-2

the nozzles through from behind.
→ The tailgate nozzle could require a small screwdriver to lever it out.
→ Be very careful not to damage the paintwork while doing so.

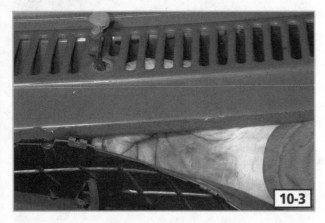

10-3

❏ **STEP 3:** Pull the hoses off the nozzle units.

TOP TIP!

• Be careful not to let the tailgate hose fall back into the tailgate.
• It will drip water down inside the unit, and may also prove hard to retrieve.
• Secure it outside the hole with tape.

❏ **STEP 4:** Refitting is the reverse of removal.

JOB 11:	HEADLIGHT WASHER PUMP - *remove, replace*.

❏ **STEP 1:** Some models have headlight washers fitted. The system uses the windscreen washer reservoir.

❏ **STEP 2:**
On the Vectra and later Cavaliers with headlight wash, remove the front bumper to get to the pump. Similarly, the wheel and inner arch trims also have to come off.

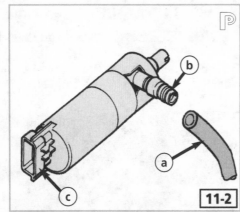

11-2

→ Undo the hose connections (**a**) to the pump (**b**). Be prepared for fluid to escape as you do so.

→ Undo the wiring connector (**c**) to the pump.

❏ **STEP 3:**
Pull the pump from the tank.
→ You will probably need to lever it carefully away with a screwdriver.
→ If the rubber sealing ring comes out with the pump, pull it from the nozzle and refit it to the reservoir before refitting the pump.

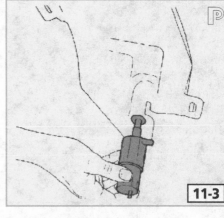

11-3

❏ **STEP 4:** Refill the reservoir and check for leaks.

JOB 12:	HEADLIGHT WASHER NOZZLES - *remove, replace*.

❏ **STEP 1:** Remove the front valance - see *Chapter 13, Bodywork*.

❏ **STEP 2:**
Disconnect the water supply pipes from the nozzles by lifting the locking tab (inset), and withdrawing the pipe.

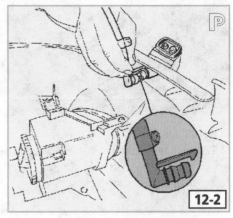

12-2

❏ **STEP 3:**
Undo the securing nuts (arrowed) holding the nozzle in place, from inside the valance, then pull the nozzles away from the valance.

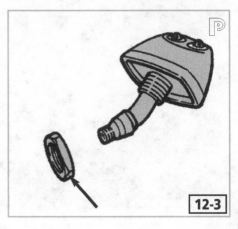

12-3

❏ **STEP 4:** Refit as the reverse of removal.

JOB 13: HEATER MATRIX - *remove, replace*.

☐ **STEP 1:** Remove the expansion tank cap to allow air to enter the system as coolant flows out.

☐ **STEP 2:** In order to gain access to the heater matrix hose connections on the bulkhead, it may be necessary to remove the battery. See *Chapter 10, Electrical, Dash, Instruments, Fact File: Disconnecting the Battery* BEFORE doing so!
➜ Remove any other components necessary to access the hoses, such as air intake trunking.

☐ **STEP 3A:** Clamp the hoses with hose clamps or a self-grip wrench.
➜ Disconnect them from the matrix – some are held with pipe clips...

☐ **STEP 3B:** ...while others have to be removed by holding one hexagon, to prevent it from turning, while

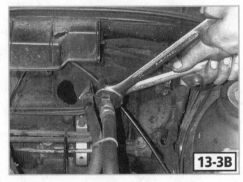

undoing the hose-end nut, as shown.
➜ Seal the ends and tie the hoses upwards with cable ties, to stop coolant escaping.

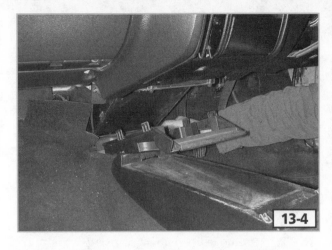

☐ **STEP 4:** Depending on which model you are working on, you may have to remove items of trim, such as the centre console - see *Chapter 14, Interior,*

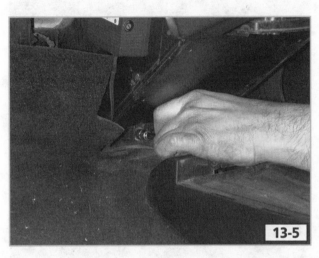

☐ **STEP 5:** Inside the vehicle, remove the matrix cover bolts.

☐ **STEP 6:** Remove the matrix cover...

13-7

☐ **STEP 7:** ...remove the matrix top fixings...

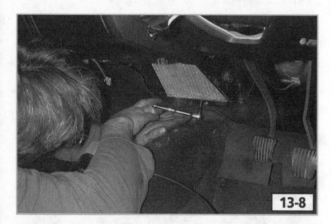

13-8

☐ **STEP 8:** ...and the lower bolts.

13-9

☐ **STEP 9:** The heater matrix can now be lifted out.

TOP TIP!

• It's a good idea to place a large absorbent cloth under the heater as you remove the matrix, to absorb any spilt coolant.
• Take care to protect upholstery, but if coolant splashes any trim, wipe it off immediately with a damp cloth.
• Remember that the matrix is full of coolant!

☐ **STEP 10:** Replace the grommets in the bulkhead.

13-11

☐ **STEP 11:** Refit the matrix as follows:
➜ Slide the hoses (**F**) at a slant from the left, through the openings in the bulkhead.
➜ Swing the heater core (**E**) under the air distributor housing and snap behind the air distributor housing.
➜ Remember to bleed the cooling system and top up the coolant level afterwards – see *Part A, Job 2*.

JOB 14: AIR CONDITIONING

14

It is MOST IMPORTANT that air conditioning is not worked on or opened without specialist knowledge or equipment. See *Chapter 10, Job 11*.

CHAPTER 9: IGNITION, FUEL, EXHAUST

*Please read **Chapter 2 Safety First** before carrying out any work on your car.*

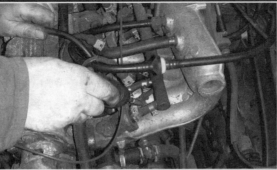
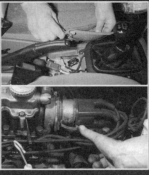

We have placed both **Ignition** and **Fuel** systems in this, the same chapter. In former days, the two areas could be treated separately, but that is no longer the case, and with today's electronic devices, ignition and fuel systems are integrated.

SAFETY FIRST!

• Take very great care when working on any vehicle equipped with electronic ignition.

• ELECTRONIC IGNITION SYSTEMS ALL INVOLVE VERY HIGH VOLTAGES! All manufacturers recommend that the high tension circuit (coil, distributor and HT wiring) is handled only by trained personnel if any more than routine dismantling is required.

• It is absolutely ESSENTIAL that people wearing medical pacemaker devices avoid the ignition system completely.

• Only connect or disconnect electrical leads on the ignition system - including HT leads and testing unit connections - when the ignition is switched off.

• If the engine has to be turned over on the starter but without starting, for example when checking compressions, disconnect all the cables from the distributor and remove the HT lead from the coil.

• Refer to *Chapter 2, Safety First!* and also to the *Safety First!* notes in this chapter.

• Disconnect the battery before carrying out any work on the ignition or fuel systems. However, if the car you are working on has a 'code memory' security feature on the stereo head unit, it will stop working after disconnection. Therefore, either make sure that you have the security code written down, or provide a backup power supply to the radio - see *Chapter 5, Servicing*.

Part A: Petrol/Gasoline Engines

SAFETY FIRST!

• When working on engines with electronic ignition systems, note the following, to avoid personal injury or damage to the ignition system.

• Make sure the ignition is switched off before removing any wires from the ignition system.

• If the engine is to be turned on the starter without starting:

EITHER - NON-ELECTRONIC IGNITION: Pull the centre HT lead from the distributor and connect it firmly to earth/ground.

OR - ELECTRONIC IGNITION: Disconnect the multi-pin plug from the distributor or ignition transformer (i.e. whichever has the HT spark plug leads connected to it).

CRANKING AN ENGINE WITH ELECTRONIC IGNITION: Severe damage can be caused to the ignition system if the engine is cranked with plugs/leads removed – and a severe electric shock can be experienced.

1. With very latest 'distributorless' systems, disconnect the multi-plug from the base of the 'black box' ECU unit.

2. With the two earlier systems, disconnect the multi-plug leading to the distributor/control unit from the ignition coil behind the battery.

PLUS - Where fuel injectors are fitted with electrical plugs (which is the case for all 16-valve engines covered by this manual, for example), disconnect the plug from each of the four injectors.

• Do not connect a capacitor/condenser to the coil terminals.

• Do not replace the rotor arm with a non-standard one, even to reduce radio interference.

• When suppressing interference, only resistors with 1 kohm and spark plug connectors with 5 kohms may be used on HT leads.

• If the car has to be towed with the ignition turned on, after the system has developed a fault, disconnect the wiring plug from the coil.

• The ignition should be switched off before washing the engine or the engine bay.

• A starting boost with a quick starter, or a supplementary battery, should be limited to no more than 30 seconds at 12.5 volts max.

• If electric arc, MIG or spot welding is carried out on the car, BOTH battery terminals must be disconnected.

JOB 1: SYSTEMS EXPLAINED.

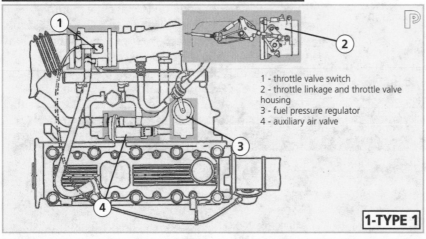

1 - throttle valve switch
2 - throttle linkage and throttle valve housing
3 - fuel pressure regulator
4 - auxiliary air valve

1-TYPE 1

❑ **TYPE 1:** Bosch LE Jetronic.

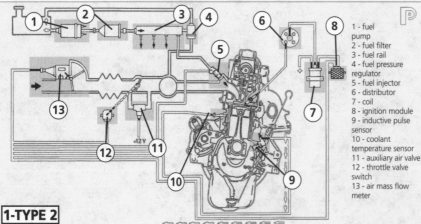

1 - fuel pump
2 - fuel filter
3 - fuel rail
4 - fuel pressure regulator
5 - fuel injector
6 - distributor
7 - coil
8 - ignition module
9 - inductive pulse sensor
10 - coolant temperature sensor
11 - auxiliary air valve
12 - throttle valve switch
13 - air mass flow meter

❑ **TYPE 2:** LE Jetronic.

1-TYPE 2

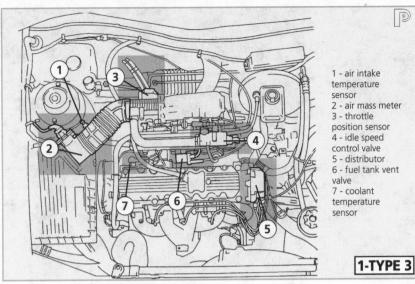

1 - air intake temperature sensor
2 - air mass meter
3 - throttle position sensor
4 - idle speed control valve
5 - distributor
6 - fuel tank vent valve
7 - coolant temperature sensor

❐ **TYPE 3:** Motronic M 1.5.2.

1-TYPE 3

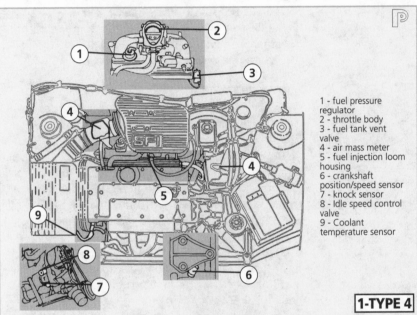

1 - fuel pressure regulator
2 - throttle body
3 - fuel tank vent valve
4 - air mass meter
5 - fuel injection loom housing
6 - crankshaft position/speed sensor
7 - knock sensor
8 - Idle speed control valve
9 - Coolant temperature sensor

❐ **TYPE 4:** Motronic M 2.5.

1-TYPE 4

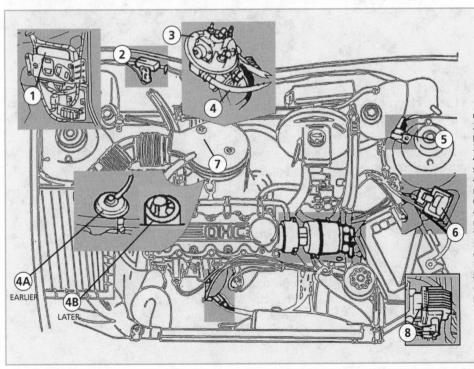

1 - electronic control unit
2 - manifold absolute pressure sensor
3 - fuel injector
4 - throttle position sensor
4A - exhaust gas recirculation valve (earlier type)
4B - exhaust gas recirculation valve (later type)
5 - octane rating plug
6 - ignition coil
7 - distributor
8 - DIS unit. Replaces 6 & 7 in later cars. Fits in place of distributor.

EARLIER
LATER

❐ **TYPE 5:** Multec single point.

1-TYPE 5

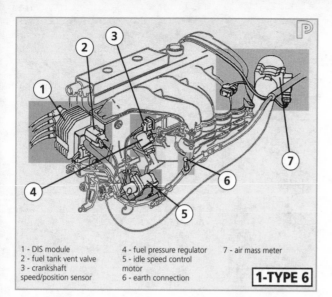

1 - DIS module
2 - fuel tank vent valve
3 - crankshaft speed/position sensor
4 - fuel pressure regulator
5 - idle speed control motor
6 - earth connection
7 - air mass meter

1-TYPE 6

❐ **TYPE 6:** Multec 'S'.

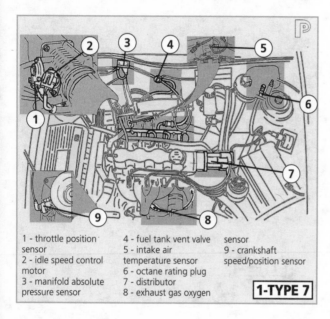

1 - throttle position sensor
2 - idle speed control motor
3 - manifold absolute pressure sensor
4 - fuel tank vent valve
5 - intake air temperature sensor
6 - octane rating plug
7 - distributor
8 - exhaust gas oxygen sensor
9 - crankshaft speed/position sensor

1-TYPE 7

❐ **TYPE 7:** Multec MP.

As a general rule, the earlier the model you are working on, the more easy the fuel and ignition systems are to service and repair. The electronics involved have become, over the time span covered by this manual, progressively more complex, to the point where servicing on the most modern cars is largely a matter of waiting for a component to fail completely and replacing it.

The upside of this is that the solid-state electronics employed in the latest cars is of a very high quality and highly stable. It also enables engines to run at the peak of their efficiency for most of their working lives. This saves the owner money, and prolongs the life of other systems.

Section A: Ignition.

FACT FILE

IGNITION TYPES

There are four basic types of system fitted to models covered by this manual. They represent a gradual advance in ignition technology, and were applied gradually to different models. Some engine variants wore several different ignition systems during their production life. When buying parts, however, it is essential to quote the engine designation code, etched onto the front of the engine block - as well as the year the car was made.

CONTACT BREAKER DISTRIBUTOR TYPE
Fitted to some 1.3, 1.6 and 1.8 litre engines in the Cavalier. This system features a single coil, mounted in the engine bay, and a distributor.

BREAKERLESS TYPE
Later, the contact breaker points system gave way to a 'breakerless' distributor without points.

'HALL SENSOR' TYPE
This electronic ignition system employs an ECU to communicate between the fuel and ignition systems, but still has a distributor, which has a 'Hall-effect' sensor to govern the spark output.

DISTRIBUTORLESS IGNITION SYSTEM (DIS) TYPE
Eventually, the distributor was replaced by a DIS module, which works on instructions from an ECU (Electronic Control Unit). This uses a camshaft position sensor to tell the ECU how fast the engine is turning, and at what point in the combustion cycle each piston is.

❐ **ILLUSTRATION A1:** DIS (*Distributorless Ignition System*) systems require no adjustment, and minimal maintenance. In typical, logical, Vauxhall style, the DIS module (**a**) bolts to the position on the end of the cylinder head (**b**) once occupied by the distributor.

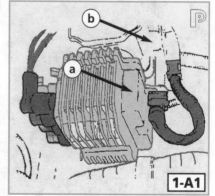

1-A1

Section B: Fuel intake/injection.

Here, the evolutionary process passes through:
→ Carburetors
→ Single-point injection systems, which fit onto inlet manifolds, more or less as a replacement for carburetors.
→ More sophisticated multi-point injection systems.

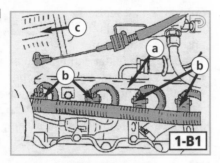

❑ **ILLUSTRATION B1:** Multi-point injection systems feature a fuel rail (**a**) in which fuel is held at high pressure. It is released directly into the cylinders through four injectors (**b**), under instructions from the DIS module (**c**).

Section C: Fuel supply.

All the models covered in this manual have a 'flow and return' type of fuel feed system. The tank is vented, a system which in later cars incorporates evaporative control systems to prevent the escape of fumes to the atmosphere. Earlier cars use a camshaft-driven, mechanical fuel pump, while later variants have an electric pump, immersed in the fuel tank itself.

Section D: Exhaust systems.

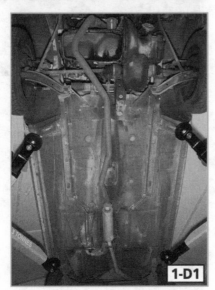

❑ **ILLUSTRATION D1:** Prior to the early '90s, all the systems covered here are straightforward affairs.

❑ **ILLUSTRATION D2:** The largest changes to exhaust systems came about when emission control equipment was fitted, mandatory in the UK from 1992.
→ A catalytic converter was incorporated into the exhaust. To make the emissions control equipment effective, the engine management ECU has to be informed of the exhaust gas temperature and composition, so a Lambda sensor is bolted into the exhaust, upstream of the 'cat'.
→ A further development of Vauxhall's emissions control system was the exhaust gas recycling (EGR) circuit. This recirculates a small amount of the exhaust gas, via a vacuum-controlled valve, into the inlet manifold. The result of this is that the level of nitrogen oxides in the final exhaust emission is lowered.
→ Some models also feature a secondary air injection system. This injects air into the exhaust manifold to burn unburnt gasses when the engine is not yet at full working temperature. As a result, the exhaust gas temperature rises at an artificially rapid rate, to get the catalytic convertor to working temperature as quickly as possible.

Section E: Engine management.

❑ **ILLUSTRATION E1:** To unite the induction, ignition and exhaust electronics in the newer models this manual covers, there is a computer-based electronic control unit, or ECU.
→ The ECU requires a whole range of data, so it can control the major ignition and fuel supply components with maximum efficiency.
→ Newer engines have a whole range of sensors, including those to measure throttle position, inlet air mass and inlet air temperature.

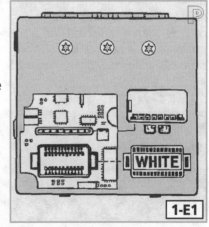

JOB 2: DIAGNOSTICS - *carry out checks.*

Section A: Background.

Vehicles are becoming increasingly dependent on electronics. When things are going well, this has great advantages for the driver, including:
→ Greater efficiency.
→ Better reliability.
→ Less maintenance.

Of course, almost everything has its downside, and vehicle electronics are no exception. Aspects often seen as 'problems' include:
→ The need for special tools - mainly electronic diagnostic equipment.
→ The need for knowledge and information not found in traditional manuals.
→ The fact that it's impossible to see faults with electronic systems.

However, the problems that vehicle electronics present can be overcome. Indeed, such problems can be turned on their heads. Then, instead of appearing as trouble, they can be seen as a set of extra advantages for the mechanic, such as:
→ Rapid fault diagnosis.
→ Less guesswork.
→ Fewer 'grey' areas caused by wear - electronics tend to either work or not work. They fail completely, rather than wearing out in the way mechanical components do.

Turning the minuses into pluses is a matter of knowing where to obtain information and how to use it. It is not possible to fault-find many areas affected by electronics on an old-fashioned, common-sense basis. But fortunately, the information is not too difficult to put to use - and it's even easier to find, if you know where to look!

TOP TIP!

• It's a myth that only the manufacturer's own diagnostic gear, found at your Vauxhall dealer, can do the job! But you will need a plug-in diagnostics tester to interrogate electronic-based systems.
• There are several makes of tester on the market. We feature the pro-oriented Sykes-Pickavant ACR System Tester. It's typical of the best of the testers available.

❑ **STEP A1:** It is not possible - even in a manual of this size - to cover diagnostics in specific detail; there's no room and no need. We will describe how to use diagnostic test equipment and the equipment

itself will do the rest. At least, that's the theory! What diagnostic test equipment, such as the Sykes-Pickavant ACR System Tester, is good at is identifying specific fault areas:

2-A1

→ If a component such as the engine knock sensor, the hall sender unit or the Lambda sensor is failing to give a reading, the tester can tell you almost instantly.
→ If you want to make a component operate without the engine running, an advanced unit such as this can simulate a signal to it, so that you can listen and feel for a solenoid clicking, a flap opening and closing, or the buzz of an EGR valve working.

What it can't always tell you is the source of the fault. Before fitting an expensive replacement, you should run physical checks, to find out:
→ Is the wiring to or from the component at fault?
→ Are the component's wiring connectors corroded or broken?
→ Has a fuse blown?
→ Has someone else wired it up incorrectly?

To analyse the results, you will need to go a step further. You will still need a wiring diagram, to check circuits and connections. And you will need extra diagnostics information to know how to check out potential faults. Most of the information you will need is in this manual - but if it had all been printed here, the book would have been too thick to carry!

❑ **STEP A2:**
Detailed diagnostic information is available from a number of sources:
→ You can order the manufacturer's own data, although some of it may be

2-A2

restricted information and not widely available. Try your local Fiat dealer's parts department.
→ There are several companies who specialise in producing detailed data of this sort. The one shown here is produced by CAPS (Computer Aided Problem Solving). It covers, as do all of them, a very wide range of manufacturers and models. All of these data compilations are expensive, typically costing many times the price of this manual!

Where Vauxhall cars are concerned, there are two rules of thumb operative concerning diagnostic equipment applications:

→ Such equipment can only be used on models with electronic ignition and fuel injection systems - i.e, later cars.

→ When setting up diagnostic equipment it is often the engine model, denoted by the number stamped on the front of the block, that is your most important reference point - rather than the model of car. Vauxhall tends to use a single engine across several model ranges.

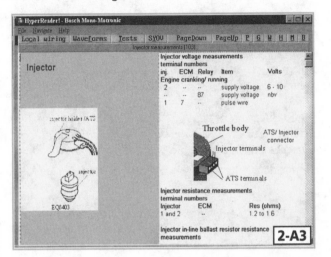

2-A3

STEP A3: CAPS lists an immense amount of data on each disk, including:

→ ECU pin settings.

→ Values expected for various test readings using, for instance, a standard test meter.

→ Waveforms for oscilloscope component testing.

→ Virtually everything you may need in order to carry out extensive diagnostic testing, in conjunction with a fault code reader.

STEP A4: A good quality multi-meter will be an essential tool for the diagnosis of more basic faults, such as continuity and correct connections. This Sykes-Pickavant multi-meter comes complete with full instructions for use and is both versatile and accurate. We have come across

2-A4

cheaper tools that are highly inaccurate, which makes them next to useless. How can you tell if there is a fault with the level of current (for instance) if the meter reads inaccurately?

STEP A5: We look later at a more comprehensive and complete diagnostic tool from Sykes-Pickavant, but this sensor simulator is an extremely useful piece of kit in its own right. Although it is not designed to actually drive solenoids and actuators (it will test both of these to see if they are receiving a control signal), it will test the following: Camshaft sensor; Crankshaft sensor; Mass Air Flow Meter; Manifold absolute pressure sensors; Oxygen sensor; Throttle position sensor.

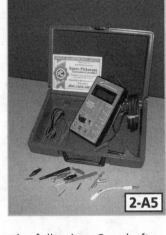

2-A5

STEP A6: An efficient Lambda sensor is essential for protecting the catalytic converter, allowing the vehicle to run efficiently and enabling the engine to obey emissions laws.

2-A6

The Sykes-Pickavant tester can be used to check the sensor while it is *in situ*.

FACT FILE

DIAGNOSTIC TESTING

BACKGROUND INFORMATION FROM SYKES-PICKAVANT
Electronic fuel injection and ignition systems on modern cars have a computer (called an Electronic Control Unit – ECU) - to provide the proper control of the fuel-air mixture. The ECU works by measuring many different characteristics of the car. Then, using this information, it calculates the correct quantity of fuel, the time of the fuel injection and the time for the ignition spark.

The ECU is a computer, like any other you could buy from a High Street shop, but instead of a keyboard it has SENSORS providing inputs, and instead of a visual display and peripherals as outputs, it has ACTUATORS. These are the computer's output; for example, the ECU controls the fuel injectors by opening and closing them for a few thousandths of a second. During this time fuel is sprayed into the inlet manifold (near the inlet valves) in the exact proportions required for efficient combustion. The ECU also controls the engine's idle speed, via the 'idle speed control valve'. *Continued...*

Continued...

Sensors are prone to failure, mainly because they are exposed to the harsh engine environment, where wiring harness connections can deteriorate over a few years. When this kind of fault occurs the ECU can be misled by the information being sent from the sensor. This results in the wrong quantity of fuel or timing for the sparks. Soon, the car can allow too much of the wrong exhaust gases to escape, can become difficult to drive or may even refuse to start altogether.

ECUs are able to detect faults in sensor signals by comparing the value with a range of values which are programmed by the person who designed, or 'mapped' the ECU. If a signal is outside the expected range then something is wrong. The ECU does not 'know' exactly what, but it can sense that the signal is abnormal. This detected fault is stored as a code number in the ECU's memory, and it is this that is known as a 'fault code'. ECU mappers have built in a limited number of these fault codes into most modern systems. Each manufacturer has a different coding system and a different way of reading the codes, but they have some basic similarities.

Somewhere on the car's harness is a diagnostic socket which connects directly to the ECU. By connecting a second computer (the ACR Systems Tester, in this instance) to this socket we can send messages back and forth between the two computers and read any stored codes.

A simple code reader will show only the two or three digit code numbers, which you then need to look up in a book to find out what they mean. The Sykes-Pickavant ACR Systems Tester does much more, by READING, CLEARING, ACTUATOR TESTS and COMPONENT TESTS.

The ACR System Tester is able to read codes and display the meaning of each code. For example, if the coolant temperature sensor is faulty, some Bosch ECUs will store the fault code '15'. The SCR System Tester shows the following on its two line display:

15 COOLANT SENS
VOLTAGE LOW

CLEAR CODES

When this problem has been corrected - by first checking the wiring connectors for resistance-increasing corrosion, re-running the test and, if the fault persists, replacing the sensor - the stored fault code (15) must be removed from the ECU, or it will still think something is wrong. This is what the Clear Codes function is for. After clearing codes, the engine should be started and a final check for any stored fault codes made, just to be certain nothing else has occurred.

ACTUATOR TEST

This feature is a method of testing the ECU outputs when the engine is turned off. For example, testing the injectors involves sending a signal from the ACR System Tester to trigger a special program in the ECU which, in turn, causes the injectors to open and close every second. The injectors can be heard quite clearly 'ticking' each second, like a clock. This is clear proof that a signal is able to reach the injectors. We DO NOT YET KNOW if the injectors are blocked or gummed, so we do not know if any fuel can get into the engine - there are other ways of testing for that.

The final group of tests are components (this is another way of saying sensors) where we can show the voltage, time or angle of various sensors which provide input to the ECU.

IMPORTANT NOTES:
• The ECU can only measure what is happening to the car using the sensors it is connected to. If a part of the car has no sensor, its activity cannot be measured. This may seem obvious, but many people, due to the 'voodoo' mystique surrounding computers, mistakenly believe that such equipment is endowed with supernatural powers and is all-powerful.
• The ECU calculates the value that the ACR System Tester displays; the tester is NOT connected directly to the sensor. The display you see is what the ECU 'thinks' is the correct value. So if you see "Coolant Temperature 79 degrees Celsius" and the engine is stone cold, you know something is wrong but the ECU doesn't. This is a vital point to understand about code readers which connect to the diagnostic socket: they can only display what the ECU is programmed to show and what the ECU calculates from its input.

Section B: The ACR Tester - using.

STEP B1: The first job is to identify what you will need to use on the car you are testing. Look up the cable and plug-in pod numbers in the handbook entitled "ACR System Tester Applications", supplied with the kit. Out of the Sykes-Pickavant case you will need to take:

→ The main ACR unit.
→ The correct plug-in pod, containing the ROM of data relevant to the particular vehicle you are working on.

STEP B2: Find the location of the diagnostic plugs or sockets (as appropriate) on the model you are working

on. The plug-in pod's mini manual will show you where to look. On some models, the socket is under a lift-up panel on the centre console, beneath the parking brake handle (arrowed).

STEP B3: On other models, it is behind the fuse box cover, on the dashboard below and to the right of the steering column (arrowed).

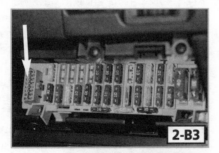

STEP B4: With the ignition OFF, push the plug into the socket. Some makes have twin or triple leads, but all Vauxhalls have a single lead.

STEP B5: Each ROM pod (and there are several - each one covering a range of makes and models) comes with its own small manual which 'walks' you through the setting up and running procedure.

→ When the unit comes to life, you select the type of system (e.g. Multec) on the menu offered on the readout. The ACR then communicates with the vehicle's ECU.
→ You then select from the menu which type of test you want to carry out, such as "Read Errors", "Clear Errors" or "Test Actuators".
→ If you selected "Test Actuators" for instance, the next menu allows you to choose which one - say, "Fuel Injector Pulse Width".
→ The ACR then interrogates the system for errors.
→ If an error is found, the error code number appears AND (unlike with many testers) is an abbreviated description of the error, such as "**INJ P/W 2.4 Ms**" appears on the screen.
→ Pressing the "**OK**" button on the **ACR** expands the abbreviation to "**INJECTOR PULSE WIDTH**" - useful until you are able to memorise the abbreviations.

STEP B6: A similar approach is followed - a walk-through by the manual, combined with step-by-step prompts and menus - allows the user to carry out most of the other functions

that would normally only be carried out by the main dealer, such as:
→ Reading and resetting Servicing warning lights.
→ 'Firing' individual components (Actuators in the ACR's jargon) to find out whether or not they work.
→ Resetting error warning lights ('Blink Codes')

IMPORTANT NOTES:
• Fault code readers that only supply code numbers can be misleading. This is because the code numbering system has changed and continues to evolve. It may not be possible to identify the (correct) fault from the code number alone.
• We have dealt in outline with the type of ACR pods and cables used for the models covered by this manual. Other vehicles and other manufacturer's model groups require their own specific setting-up procedures. In each case, the set-up is described in the Sykes-Pickavant manual supplied with the relevant ROM pod.

Section C: Multimeter Testing

The following items are typical of those that can be checked using a simple multimeter.

→ You can find the values for a specific model and year in CAPS, or compare the values with those taken from components known to be in good working order.

☐ **STEP C1: ENGINE SENSORS:**

→ Engine/Coolant ECU temperature gauge – resistance.

→ Engine speed sensor – resistance – air gap (non-ferrous gauge needed).

→ Camshaft position sensor – earthing, supply voltage.

→ Knock sensor – clean and re-torque – supply voltage – operational voltage (mV) resistance.

☐ **STEP C2: FUEL:**

→ Injectors – resistance.

☐ **STEP C3: INTAKE SYSTEM:**

→ Throttle position sensor – resistance and change as throttle opens – supply voltage.

→ Mass air flowmeter – Resistance – supply voltage.

→ Manifold absolute pressure (MAP) sensor – supply voltage – operational voltage variance with vacuum levels.

→ Intake air temperature sensor – resistance for given temperatures.

→ Idle control valve – resistance.

☐ **STEP C4: IGNITION:**

→ Coil supply voltage, coil primary and secondary resistance.

☐ **STEP C5: EMISSION CONTROL:**

→ Heated oxygen sensor – supply voltage – signal (mV).

→ Oxygen sensor heater – supply voltage.

→ Secondary air injection solenoid – supply voltage.

→ Secondary air injection pump – operation with terminals 2, 8 shorted.

→ Exhaust gas recirculation solenoid – resistance – supply voltage.

→ Evaporative emission canister purge valve – resistance – supply voltage.

☐ **STEP C6: CONTROL SYSTEM:**

→ Fuel pump relay – resistance and supply voltage.

→ ECU - supply voltage and earthing.

JOB 3:	**CONTACT BREAKER POINTS** - *check, replace, adjust.*

This applies only to certain cars with distributor-type ignition. Later cars have electronic ignition that doesn't use contact breaker points.

See *Safety First!* at the start of this chapter and see *Chapter 2, Safety First!* before carrying out any of the work described here.

FACT FILE
DISTRIBUTORS

• On the vehicles covered by this manual, several types of mechanical (non-electronic) distributor have been fitted, depending on the engine variant and year of manufacture. Despite detail differences, all contact breaker distributors are similar in layout and identical in operation.

• A lobe on the distributor driveshaft opens and closes the contact breaker points.

• When closed, the points complete a circuit that generates a magnetic field within the ignition coil primary winding. When the points open, the magnetic field collapses and so generates high voltage in the concentric secondary winding.

• On the end of the distributor drive shaft is a rotor, which directs the high voltage current from the coil to each of the sparking plug leads in turn, via a set of contacts in the distributor cap.

• The gap between the contact breaker points can be adjusted. This affects the timing of the spark and also its strength. The gap can be measured using feeler gauges or using a dwell angle meter, and setting the points gap is the first stage in setting the ignition timing.

• The second stage in setting the ignition timing involves turning the distributor body. See *Job 5*.

CONTACT BREAKER POINTS

The contact breaker points gap and the dwell angle (and amount of time that the points are closed) are interrelated – a change in either affects the other. It is possible to set the points gap by measuring it with feeler gauges or by using a meter capable of measuring the dwell angle, the latter being more accurate.

→ While a dwell meter can give an accurate measurement of the points gap, it gives little indication of the actual condition of the points themselves, so a physical examination is necessary.

→ The most common problem affecting CB points is the eroding of one contact and the build-up of metal (in the form of a 'pip') on the other. In its early stages this condition may well have little detrimental effect on the production of a spark and a dwell reading will take it into account, but as the pip increases in size the spark is liable to become unstable and/or unreliable. While removing the points and smoothing-away the pip with emery is feasible it is far more sensible, and cost-effective in the long-run, to simply renew the inexpensive points assembly as a whole.

→ CB points erosion and build-up is often a symptom of a faulty condenser.

TOP TIP!

• If the dwell angle reading fluctuates, this indicates wear in the distributor drive bearings.

☐ STEP 1:
Remove the two spring clips or screws securing the distributor cap and place it to one side – there is no

3-1

need to disconnect any of the High Tension leads at this point.

☐ STEP 2: Lift the cap away and the rotor arm will be revealed; pull the arm off – it is a push-fit – followed by the inner cover.

3-2

☐ STEP 3: The points will be found beneath a metal outer-bearing plate secured by two screws (arrowed); carefully remove the screws and lift off the plate.

3-3

☐ STEP 4: Lever the points (arrowed) apart with a small screwdriver and examine the faces of the contact areas, looking for signs of metal transfer between them (i.e. a 'pip' on one and a 'pit' on the other); if

3-4

present, the points need replacing. On some distributors, the centre of the cam rotor contains a small felt pad – apply a drop of oil to this, just enough to soak in without getting onto the points themselves.

☐ STEP 5: Apply a light smear of grease to the sides of the cam, but take care not to over-do it – prevent grease getting directly onto the contacts.

TOP TIP!

• A blue-ish colour on the contacts indicates overheating caused by a faulty condenser.
• Replacement is both easy and cheap.
• To test a condenser, open the points with the ignition switched on using a non-metallic implement such as a pencil. You should see a small electrical 'splash' across the points. If there is an audible crack and a bright spark, the condenser is faulty.

On reassembly, clean the rotor arm of any dust or grime, also the inside of the distributor cap, with a cloth. In winter, spray the outside of the cap with a very small amount of water repellent, such as Castrol DWF maintenance spray which helps prevent condensation forming and possibly 'tracking' the spark to earth/ground.

CHECK/ADJUST POINTS DWELL ANGLE

FACT FILE
POINTS GAP SETTING

• Check the points gap/dwell angle setting using a 'dwell angle' meter if possible.
• Dwell meters are used with the engine running and therefore automatically take into account small (but significant) deviations and wear patterns in the distributor components and points, and they also require no dismantling of the engine to make a measurement.

☐ STEP 6:
A dwell meter or multi-meter will allow the dwell to be measured simply by attaching three wires, to the

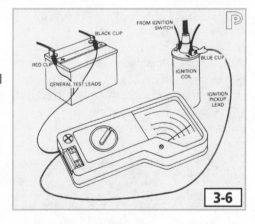
3-6

battery and coil, enabling an accurate reading to be taken without disturbing the distributor, and in a fraction of the time a physical measurement would take.
➔ Connect the dwell meter as per the maker's instructions, start the engine, and note the reading given.
➔ If necessary, stop the engine and adjust the points gap as follows.

• Checking and adjusting the points requires the engine to be turned slowly by hand - not a particularly easy proposition given the close proximity of the crankshaft pulley to the inner wing panel, which makes use of a spanner on the pulley bolt awkward.

• A far easier method is to apply the parking brake firmly, select second gear and jack-up the right-hand wheel until just clear of the ground. Secure the car on an axle stand.

• Now, when the road wheel is turned by hand, the engine will also turn, due to the car being in gear.

❑ **STEP 7:** It is important to measure the points gap when the heel of the contact arm is on the 'highest' part of the cam lobe - turning the engine (or road-wheel as above) will cause the cam to rotate to the required position. Insert a feeler gauge of the correct thickness in between the contact breaker points. When the gap is correct, the points will exert a very slight drag on the feeler gauge blade.

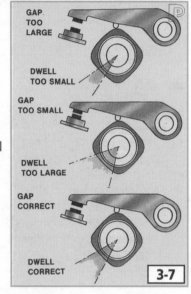

GAP TOO LARGE
DWELL TOO SMALL
GAP TOO SMALL
DWELL TOO LARGE
GAP CORRECT
DWELL CORRECT

3-7

➜ The points gap is adjusted in the same way for all distributors, though the precise position of the base plate screw varies.

➜ Note that reducing the gap increases the dwell, while increasing the gap reduces the dwell. (Illustration, courtesy Gunsons Ltd)

❑ **STEP 8:** Adjust the points gap by slackening the screw 'a' (arrowed) just sufficiently to allow the fixed point to move on the baseplate. Then by inserting the tip of a screwdriver

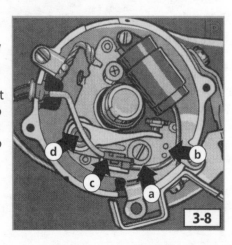

3-8

between the pips 'b' (arrowed) the fixed part of the points can be levered either towards or away from the sprung contact, thereby altering the gap between them.

RENEW CONTACT BREAKER POINTS

❑ **STEP 9:** After removing the distributor cap, replace the points by removing the screw 'a' – be careful not to drop it! Release the spring-steel arm of the

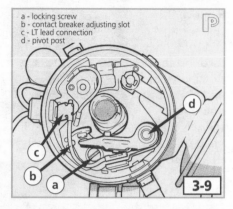

a - locking screw
b - contact breaker adjusting slot
c - LT lead connection
d - pivot post

3-9

moving point from its post 'd' and disconnect the wires at 'c'. Fit new points in the reverse order, but first clean the point faces of protective grease by wiping between them with spirit or solvent.

JOB 4: CONDENSER - *remove, replace.*

• This applies only to earlier cars fitted with contact breaker ignition.

• We recommend that you replace the condenser every time you renew the contact breaker points.

• The condenser ensures that there is no undue electrical arcing - which decreases the strength of the spark and erodes the points - across the points.

• The Delco-Remy distributor has its condenser inside the distributor body, on the same mounting baseplate as the contact points.

• The Bosch distributor has its condenser on the outside of the distributor body.

❑ **STEP 1:** Remove the distributor cap, rotor arm, condensation shield and, if applicable, bearing plate - see *Job 3*.

❑ **STEP 2:** Undo the condenser wiring connector (a).
➜ Remove the condenser mounting screw (b), and lift out the condenser (c).

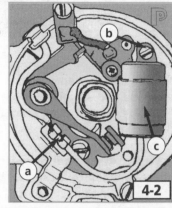

4-2

STEP 3: On some distributors, the condensor mounting bolt is found on the outside casing of the distributor.

→ Also, the low tension connection must be removed from the breaker points and, at the other end, from the coil, as the entire cable is supplied with the new condensor.

STEP 4: Reverse the sequence to refit the condenser.

JOB 5: IGNITION TIMING - *check, set.*

SAFETY FIRST!

• **THE IGNITION SYSTEM INVOLVES VERY HIGH VOLTAGES!**

• **All manufacturers recommend that only trained personnel should go near the high tension-circuit (coil, distributor and HT wiring) and it is ESSENTIAL that anyone wearing a medical pacemaker device does not touch any part of the ignition system.**

• **Also, stroboscopic timing requires the engine to be running - take great care that parts of the timing lights or parts of you don't get caught up in the moving parts! Don't wear loose clothing.**

IMPORTANT NOTE: Always check/adjust the points gap/dwell angle BEFORE checking or altering the timing.

FACT FILE

OLDER ENGINES NOT ORIGINALLY SPECIFIED FOR UNLEADED FUEL

• If the ignition timing has been changed to allow an engine to run on unleaded fuel, the timing marks referred to below, will not apply.
• It will then be necessary to adjust the timing using a stroboscopic timing light that is capable of being adjusted to give different readings from the timing marks.
• Vauxhall's recommendation is that cars converted to run on unleaded fuel should have their ignition timing retarded by 3 degrees.
• Not every model requires ignition retardation to run on unleaded fuel, however. Only models with contact breaker ignition, manufactured after 1985, require it.
• Pre-1985 contact breaker models, and models with breakerless distributor ignition need no adjustment.
• If you are unsure, or if a car is responding badly to unleaded fuel, consult a Vauxhall dealer for information.

TOP TIP!

• To check or adjust the ignition timing with the required degree of accuracy calls for the use of a stroboscopic timing light.
• Although a 'static' timing figure may be given in some workshop manuals, this is only a guide, mostly used to get an engine started after the distributor has been disturbed and the previous settings lost.

IMPORTANT NOTE: Some engines have extra requirements. Check *FACT FILE: ENGINES WITH ELECTRONIC IGNITION* at the end of this sequence.

STEP 1: Turn the engine by hand so that the timing marks (arrowed) can be highlighted with a dab of white paint or typist's correction fluid. This ensures the marks will be clearly visible when illuminated by the timing light.

STEP 2: These are the timing marks:
→ 'a' denotes the pulley mark, and 'b' the pointer on the engine.

STEP 3: Timing checks and adjustments are made with the vacuum pipe disconnected at the distributor end.
→ Plug with a suitable instrument, such as a small Phillips-type screwdriver or other suitable 'bung' such as the centre punch used here.

STEP 4: Run the engine (vehicle out of doors):
➜ Until the engine is at normal running temperature.
➜ If no temperature gauge is fitted, run until the cooling fan comes on and goes off twice.

STEP 5: Stop the engine. Connect a stroboscopic lamp (preferably with integral rev. counter) in accordance with the manufacturer's instruction.
➜ The timing light should always be connected to

No. 1 spark plug lead – the one nearest the timing belt end of the engine.
➜ Check that the stroboscopic timing light leads are clear of all engine bay moving parts.

STEP 6: Start the engine and run it at the engine's regular tick-over speed – see *Chapter 1, Facts and Figures*.

STEP 7: Point the flashing beam of the light at the timing marks.
➜ The strobe effect of the flashing beam 'freezes' the moving mark on the pulley and should appear stationary, adjacent to the static mark on the engine if the timing is set correctly.
➜ If the mark appears at the wrong place, the distributor will have to be slackened and turned to bring the marks into alignment, as follows:

SAFETY FIRST!

• Switch off the engine before slackening or turning the distributor.

• Very high voltages are present which can 'jump' through the insulation of HT leads and/or distributor cap, passing through your body to earth.

• This could be harmful or even fatal.

STEP 8: The distributor is retained by one bolt at the base of the body - shown with spanner fitted.
➜ Slacken the bolt but leave enough

'grip' to hold the distributor in position but enable it to move under firm pressure.

STEP 9: Turning the distributor clockwise increases the amount of 'advance', while anti-clockwise decreases it.

STEP 10: If the reading is 'out' stop the engine and slacken the clamp nut/s on the distributor so that the distributor can just be turned with firm hand pressure.
➜ Turn the distributor a very small amount, restart the engine and check again.
➜ Repeat the procedure until, with the engine running, the relevant timing marks line up.
➜ Retighten the nut/s securing the distributor to the engine.

STEP 11: Tighten the clamp bolts when the correct setting is achieved.
➜ In all cases, double-check the setting after the bolt has been tightened.

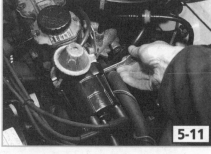

➜ If correct, disconnect the timing light, unplug and re-connect the vacuum tube to the distributor.

FACT FILE

ENGINES WITH ELECTRONIC IGNITION

• Connecting a tachometer (rev. counter) to some engines with electronic ignition can damage the ignition system. Check with the equipment instructions that the tachometer you will be using will be suitable for your engine.
• Never use a test bulb instead of a stroboscopic timing light. It might damage the ignition system.

JOB 6: IGNITION COIL - *remove, replace*.

There are several different types of coil, but the basic principles are similar for all of them.

❏ **STEP 1:** To remove the coil, disconnect the HT ignition lead (**a**) from the coil (**b**).

➜ Pull the wiring plug from the connector terminals (**c**).

➜ If the coil has spark plug leads attached direct to it, the plug positions will be shown on the coil body. Re-attach correctly.

➜ Undo the mounting nuts (**d**), and remove the coil, along with its bracket and cooling plate, from the side of the engine bay.

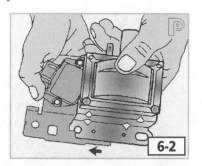
6-1

❏ **STEP 2:** Slide the coil, which comes complete with its mounting plate, off the bracket, in the direction shown.

6-2

❏ **STEP 3:** This type of coil simply unbolts from the inner wing panel after the HT lead and wiring plug have been removed.

6-3

JOB 7: DISTRIBUTOR - *remove, check, replace.*

❏ **STEP 1:** To make it easy to refit the distributor in the correct position, make a mark with typist's correction fluid across the joint between the distributor body and cylinder head.

7-1

IMPORTANT NOTE: On some models with fuel injection, the distributor has an alignment mark, which corresponds to a mark cast onto the camshaft housing.

• Because these variants require no ignition timing adjustment to run on unleaded fuel - see *Job 5 - Fact File*, these marks should always be aligned on refitting the distributor.

❏ **STEP 2:** Pull the HT leads from the distributor cap.

➜ Mark each lead, and its connection on the cap, with typist's correction fluid, to make sure that the leads are refitted in their correct locations.

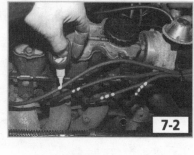
7-2

❏ **STEP 3:** Unscrew or unclip the distributor cap, as applicable. Remove it.

7-4

❏ **STEP 4:** To align the TDC mark see *Job 3, Step 9*.

➜ Turn the engine a little at a time until the rotor arm lines up with the TDC marker on the distributor body.

➜ This is in most cases a notch, as shown here. In some, however, it is an arrow, etched into the distributor body outer casing.

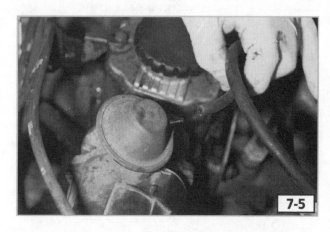
7-5

❏ **STEP 5:** If applicable, disconnect the vacuum advance pipe from the vacuum unit.

❏ **STEP 6:** Disconnect the LT (low tension) lead connector, or the lead to the Hall sensor, as applicable.

❏ **STEP 7:** Undo the mounting clamp nut or mounting bolts.

❏ **STEP 8A:** Remove the distributor from its mounting.

7-7

STEP 8B: If you are working on an OHC engine, remove the distributor from the camshaft housing.
➔ The distributor drive plate is arrowed.

7-8B

STEP 9: Remove, as applicable, the rotor arm, contact breaker points and the condenser - see *Jobs 3* and *4*.

STEP 10: If applicable, undo the screws securing the vacuum unit and unhook its operating arm from the distributor to remove it.

TOP TIP!

• Test the vacuum unit by sucking on the pipe nozzle. If it is functioning correctly, the operating rod should move towards you, then spring back when you release the suction.

STEP 11: Check the distributor cap - see *Job 3, Step 2*.
➔ Check for signs of heat damage or excess burning on the contacts.
➔ Look also for cracks in the cap.
➔ Rough-edged cracks with carbon inside indicate tracking. Scrape out the carbon as a temporary measure but renew the distributor cap at the earliest opportunity.
➔ If there is any doubt about its condition it should be renewed, along with the rotor arm.

STEP 12: Check the 'O' ring seal at the back of the distributor. If is damaged in any way, renew it.

STEP 13: Check the condition of the rotor arm bearings by moving the rotor spindle from side to side and pulling it up and down. If there is any play, the distributor should be renewed.
➔ Wear in the bearings causes the dwell angle - and hence the ignition timing - to fluctuate.

STEP 14: Refit the distributor.

TOP TIP!

• If you have replaced the distributor, you can approximately replicate the alignment mark you made before removing the old one.
• Put a length of masking tape around the body of the old distributor and mark the position of a definable point, such as a casting mark, or the TDC alignment notch/arrow.
• Then mark the position of the typist's correction fluid mark on the masking tape.
• Peel the tape off, and stick it in the same position on the new distributor.
• Use the equivalent of the correction fluid mark to (roughly) align the new distributor.

STEP 15: Turn the rotor arm to align with the TDC marker notch/arrow - see *Step 4*.

STEP 16: Carefully fit the distributor, then turn it until its alignment marks line up.
➔ Make sure, if applicable, that the contact breaker dwell angle is correctly set - see *Job 3*.

STEP 17: Fit the mounting clamp, if applicable, and the mounting bolts or nut, then tighten. Refit the HT and LT lead connections.

JOB 8: EMISSIONS CONTROL SYSTEM.

FACT FILE
LATER MODELS

• Later models feature complex emissions control systems.
• These have several sub-systems, including a fuel evaporative control circuit, and an exhaust gas recycling mechanism.
• Very little component maintenance can actually be done on this system. However, a fault can lead to running problems and faulty components can be replaced. See *Job 2*.

Depressurise the fuel system by either of the two methods described in *Job 12, Steps 1 and 2*.

Section A: Fuel evaporative loss control system.

STEP A1: This system centres on the carbon canister, which captures evaporating fuel vapour, mostly when the car is standing and stores it as fuel.
➔ When the engine is running fuel vapour is released from the canister, into the inlet manifold, to be used by the engine.

8-A1

STEP A2: To renew the canister:
➔ Raise the front of the car, and secure it on axle stands or raise it on a car hoist.
➔ Remove the right hand front wheel.
➔ Remove the wheel arch liner.
➔ Undo the clamp bolt, and lift the canister off its

bracket.
→ Disconnect the hoses from the top, noting their correct locations.
→ Reverse the process to refit the canister.

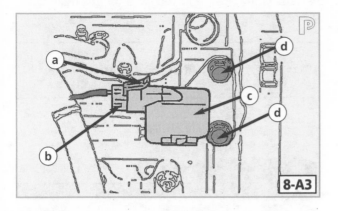

8-A3

❏ **STEP A3:** The canister purge valve is located:
→ **CAVALIER:** On top of the canister.
→ **CAVALIER X16SZ ENGINE:** On the camshaft housing.
→ **VECTRA 1.6:** On the left hand side of the inlet manifold
→ **VECTRA 1.8, 2.0:** On the left hand end of the cylinder head.
→ Depress the catch (**a**) and disconnect the electrical connector (**b**) from the valve body (**c**)...
→ ...undo the mounting bolts (**d**)...
→ ...and lift the valve off its mounting on the cylinder head, next to the camshaft housing.

❏ **STEP A4:** If the tank vent valve needs replacing:
→ Disconnect the wiring plug.
→ Disconnect the hoses from the valve.
→ Lift the valve upwards out of its retaining clip.
→ Refit as the reverse of removal.

Section B: Exhaust gas recirculation (EGR) system.

The EGR system returns a small amount of exhaust gases into the inlet manifold, which are then burnt. This lowers the nitrogen oxide levels in the ultimate emissions - which increases the exhaust gas temperature. As this in turn causes the oxidisation of more of the exhaust pollutants, 'cleaner' car exhaust emissions are the result.

❏ **STEP B1:** On most models, the EGR

8-B1

valve is situated on top of the inlet manifold.
→ On others the EGR valve is situated on the left hand end of the cylinder head.

❏ **STEP B2:** Disconnect the wiring plug from the valve.

❏ **STEP B3:** Undo the mounting bolts and remove the valve from the manifold or cylinder head, as applicable.
→ Remove any traces of old gasket from the mounting.
→ Refit as the reverse of removal, using a new gasket.

❏ **STEP B4:**
Some vehicles have an EGR valve control module, bolted to the left hand side of the engine bay.
→ Disconnect the wiring plug (**a**), vacuum hoses (**b**).

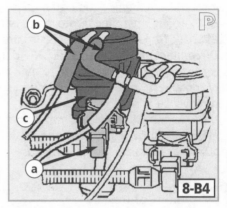

8-B4

→ Remove the valve (**c**) from its bracket.
→ Refit as the reverse of removal.

Section C: Secondary air injection system.

HOW IT WORKS...

❏ **STEP C1:** This system uses an electric pump to inject air into the exhaust manifold at certain times in the driving cycle. This raises the exhaust gas temperature, which, as explained in **Section B**, reduces pollutant emission. It also enables the catalytic converter' to reach working temperature more quickly.
• There are three principle components to the system - the secondary air injection pump, the secondary air injection switchover valve and the secondary air combination valve.
• To diagnose faults in the secondary air injection system, diagnostic equipment is necessary - see **Job 2**.

8-C2

☐ **STEP C2:** The secondary air injection pump (**a**) is located here, behind the battery.

→ To access the pump, remove the battery. See *Chapter 10, Electrical, Dash, Instruments, Fact File: Disconnecting the Battery* BEFORE doing so!

→ Depending on the model, to gain access to the pump and its mountings, you may also have to remove:

• The left front wheel and wing inner trim panel.
• The left upper cover from the air deflector.
• The coolant reservoir (**b**) and hose.

☐ **STEP C3:** Mark the position of the air hose (**a**) then disconnect it from the pump.

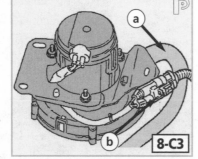

8-C3

→ Disconnect the wiring connector (**b**) from the pump.
→ Undo the pump bracket-to-body mounting nuts, and lift or lower the pump free, as appropriate.

☐ **STEP C4:** The secondary air injection switchover valve is situated on the left hand of the engine bay.

→ Remove anything preventing access to the valve.
→ Then disconnect the wiring plug from the valve.
→ Disconnect the vacuum hoses from the valve. Mark them first with typists' correction fluid, so you can be sure which hose goes where.
→ Undo the fixings from the valve. These can be either bolts or clips, depending on which vehicle you are working on.

☐ **STEP C5:** The secondary air injection valve is found adjacent to the exhaust manifold.

→ Mark the position of the valve.
→ Disconnect the vacuum and air hoses from the valve.

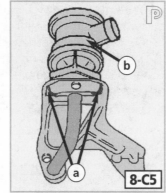

8-C5

→ Undo the mounting bolts (**a**) from the valve (**b**), and lift it off the air injection pipe.
→ Clean the mating surfaces of pipe and (if you are refitting it) valve.
→ Refit as the reverse of removal, using a new gasket.

JOB 9: CARBURETOR – *adjust*.

GENERAL

FACT FILE

CARBURETORS/LEGAL RESTRICTIONS

• It may be illegal in some territories for an unauthorised person to remove tamperproof plugs on carburetors or fuel injection systems.
• All modern Vauxhall cars use fuel injection, but many of the earlier variants covered by this manual have a carburetor.
• There are only two basic carburetor types – the Varajet II and Pierburg 2E3.
• It is more important to be able to identify the actual type of carburetor than the engine identification code or the model of car.
• Both the Varajet II and Pierburg 2E3, are of the twin-venturi type, in which two throttle valves open sequentially.
• Both units have an accelerator pump, to enrich the mixture if the driver 'floors' the throttle.

Setting the idle speed and mixture is not just a matter of making the car run smoothly and economically; it's also a question of allowing it to run within legal hydrocarbon (HC), Nitrous Oxide (NO) and carbon monoxide (CO) emission limits for the territory in which you live. However, a worn engine will fail even if the carburetor or injection system is set up correctly.

If meaningful results are to be obtained, gas testing equipment has to be used. If the car you are working on does not have one, a tachometer will also be needed.

The carburetor should be the last part of the engine tune-up procedure, as the general settings will be influenced by the condition of other engine components, such as the ignition system and basic service components, such as the air filter.

Familiarise yourself with the carburetor fitted to the car. Pay particular attention to the location of the 'idle speed' and 'mixture' adjustment screws.

FACT FILE
CARBURETOR IDENTIFICATION

STEP 1: This is the GM Varajet II, which was used in 1.3 and 1.6-litre engines.

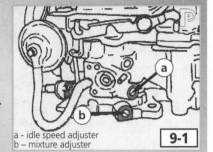

a - idle speed adjuster
b – mixture adjuster

9-1

STEP 2: This is the Pierburg 2E3, which was used in 1.4 and 1.6-litre engines.

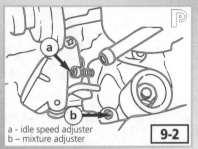

a - idle speed adjuster
b – mixture adjuster

9-2

SETTING THE CARBURETOR

STEP 3: Start the engine, and allow it to reach its normal operating temperature.

SAFETY FIRST!

• The exhaust gases contain toxic compounds. Run the engine only out of doors.

• Beware of hot exhaust components, which can burn badly - wear suitable industrial gloves.

• Watch out for the rotating cooling fan. Do not wear loose clothing or jewellery, and tie back long hair.

• Apply a strict No Smoking! rule whenever you are servicing your fuel system. Remember, it's not just the petrol that's flammable, it's the vapour as well.

STEP 4: Stop the engine, and remove the air filter assembly.

STEP 5: Connect the gas analyser and, if applicable, tachometer, according to the manufacturers' instructions.

9-5

STEP 6: Check the idle speed, using the tachometer. See *Chapter 1, Facts and Figures* to find out the correct speed for the car you are working on.

IMPORTANT NOTE: If the idle speed is incorrect, it is possible that the throttle cable is wrongly adjusted. Check it, and adjust it if necessary - see *Part C, Job 4*.

STEP 7: Using a flat-bladed screwdriver, turn the idle speed screw to set the idle at the desired level.
→ Turning

9-7

the screw clockwise increases the speed, while turning it anticlockwise reduces it.

STEP 8: Read the gas analyser to check the CO level. See *Chapter 1, Facts and Figures* to find out the correct level.

STEP 9: Adjust the mixture screw to set the CO at the correct level.

FACT FILE
TAMPER PROOFING

• Most of the carburetors covered by this manual originally had a tamper-proof seal, placed over the mixture adjustment screw.
• In some territories or countries it is illegal to remove a tamper-proof seal. Act accordingly.
• These seals are to prevent anyone unauthorised from altering the mixture and exhaust emissions.
• If the seal is a plastic cap placed over the adjuster screw, it can be broken off with pliers.
• If it is a plug within the screw recess, force it out with a sharp object, such as a small, flat-bladed screwdriver.

STEP 10: Turning the screw (arrowed) clockwise weakens the mixture, lowering the CO emission level.
Turning it

9-10

out, of course, enriches the mixture and raises the CO level.

STEP 11: Only small adjustments are normally necessary. Turn the screw a quarter turn at a time, waiting between turns for about 30 seconds before checking the new CO level, and making further turns as necessary.

TOP TIP!

• Generally, the newer the car, the more accurately the CO level can be set.
• Furthermore, it may be necessary to allow a small increase in CO emission in order to keep an older engine running smoothly.
• However, if the CO cannot both be set within the 'legal' range and around its recommended level, something is amiss mechanically.

STEP 12: When all the adjustments have been made, replace the air filter, disconnect the gas analyser and tachometer and take the car for a test drive, checking that it doesn't stall or overheat, and that it pulls smoothly throughout its rev range.
→ If such faults manifest themselves when the fuel system can be proved to be working correctly, other systems, such as the ignition circuit, should be tested for faults.

JOB 10: CARBURETOR
- remove, replace.

STEP 1: Before dismantling, disconnect the battery negative (-) earth/ground terminal. See *Chapter 10, Electrical, Dash, Instruments, Fact File: Disconnecting the Battery* BEFORE doing so!

STEP 2: Remove the air filter.

10-2

TOP TIP!

• Check for any vacuum hoses fitted in 'odd' places, such as the underside of the air filter housing.
• Disconnect and carefully lift the filter housing away.

STEP 3: Disconnect the accelerator and choke controls. See *Jobs 24* and *25*.

STEP 4: Disconnect the fuel lines from the carburetor body, and plug them.

10-5

TOP TIP!

• Ensure that, on models where both feed and return fuel lines are fitted, they are identified before you remove them, to ensure correct refitting.

STEP 5: Remove the carburetor mounting nuts and remove the unit from its studs on the inlet manifold (In this picture the carburetor and manifold have already been removed from the engine).

STEP 6: Clean the mating surfaces, fit a new gasket and refit/reconnect as the reverse of removal.

JOB 11: FUEL INJECTION BOSCH JETRONIC – test.

FACT FILE
LEGAL RESTRICTIONS

• It may be illegal in some territories for an unauthorised person to remove tamperproof plugs on carburetors or fuel injection systems.

GENERAL

→ Of the injection systems covered by this manual, only the Bosch Jetronic fitted to early 1.8 Cavaliers can be adjusted to affect both the idle speed and fuel/air mixture.
→ If the car is running badly, the obvious suspects such as filters and ignition HT components should first be checked.
→ Then diagnose individual component faults – see *Job 2*.
→ Multec and Simtec systems CAN be adjusted for idle speed and fuel/air mixture, but it is strongly

recommended that the vehicle is taken to a Vauxhall dealer with diagnostic equipment. An apparent fault can prove to be merely a symptom of a fault elsewhere, which a diagnostic tester will find.

❑ **STEP 1:** Test the Lambda probe with a Lambda tester - see *Job 21*. If the Lambda probe has failed, the ECU will switch the car to a 'map' which will keep it running, but not necessarily at optimum emissions levels and fuel efficiency.

❑ **STEP 2:** Test the CO levels using a CO meter. Start the engine, and allow it to reach its normal operating temperature.

SAFETY FIRST!

• **The exhaust gases contain toxic compounds. Warm the engine up and test it in the open air.**

• **Beware of hot exhaust components, which can burn badly - wear suitable industrial gloves.**

• **Watch out for the rotating cooling fan. Do not wear loose clothing or jewellery, and tie back long hair.**

• **Apply a strict No Smoking! rule whenever you are servicing your fuel system. Remember, it's not just the petrol that's inflammable, it's the vapour as well.**

❑ **STEP 3:** Stop the engine.

❑ **STEP 4:** Run the engine at idle. Check, using the car's tachometer, the engine's idle speed - see *Chapter 1, Facts and Figures* for the correct figure.

❑ **STEP 5:** If the idle speed is incorrect, it is possible that the throttle cable is wrongly adjusted. Check it, and adjust it if necessary - see *Part C, Job 4*.

❑ **STEP 6:** To adjust the tickover speed, slacken the locknut and adjust the screw

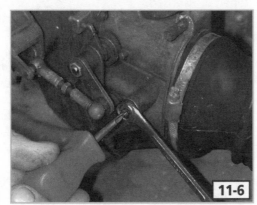

until the correct idle speed is achieved. Tighten the locknut.

❑ **STEP 7:** Connect the gas analyser according to the manufacturers' instructions. Read the gas analyser to check the CO level. See *Chapter 1, Facts and Figures* to find out the correct level.

❑ **STEP 8:** If the CO level is too high or low, stop the engine, remove the tamper-proof plug from the air flow sensor, adjust the Allen-head screw clockwise to enrich the mixture and vice versa.

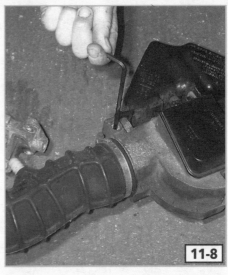

Lock the nut and re-test.

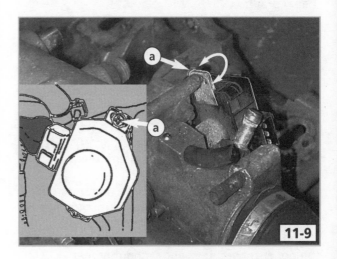

❑ **STEP 9:** Adjust the throttle valve switch. Slacken screw (a) and rotate the switch body anti-clockwise until resistance is felt. Tighten the screw.

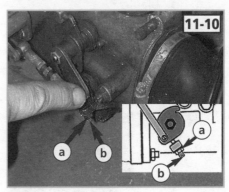

❑ **STEP 10:** With the throttle closed, slacken the locknut (**a**) and unscrew the stop (**b**) screw until it clears the cam, then screw it in until it touches the cam. Tighten the locknut. Release the operating rod locknuts and turn the rod until there is a 0.5mm gap between the cam and stop screw. Tighten the locknuts.

❑ **STEP 11:** If necessary, readjust the idle speed.

JOB 12: FUEL INJECTION UNIT OR RAIL - *dismantle, remove, replace.*

Section A: All types – essential preparation.

SAFETY FIRST!

• **Never work on the fuel system unless the engine is completely cold.**

• **Some fuel injection systems inject fuel at extremely high pressure.**

• **Pressure must be released in a controlled fashion to avoid any risk of a pressurised spray of fuel being produced.**

• **Residual pressure can remain in the fuel system for some considerable time, even if the engine has been switched off.**

IMPORTANT NOTES: There should be no major problem in working on fuel injection systems, provided that you follow the correct *Safety First!* procedures, and use a methodical, informed approach.
• No fault diagnosis of the fuel injection system is possible without the correct diagnostic equipment. You may wish to replace components, but we *strongly recommend* that you do not change components at random and without proper diagnosis of faults.
• There are two broad types of fuel injection unit: Single point, which injects fuel into the inlet manifold in place of the carburetor, and Multi-point, which injects fuel (more efficiently) directly into each cylinder.

❐ **STEP A1:**
Depressurise the fuel system by removing the fuel pump relay. Once the relay is removed, crank the engine on the starter for ten

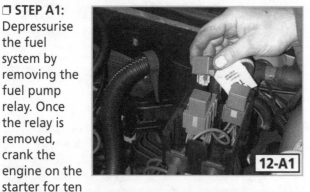
12-A1

seconds to remove fuel from the system.
➜ Do NOT touch ignition wires while engine is running or being turned on the starter.
➜ Switch the ignition OFF before disconnecting any components or wires.
➜ If the engine is to be turned on the starter

without starting, disconnect the centre HT lead from the distributor and fix it securely to a good earth/ground connection on the vehicle.

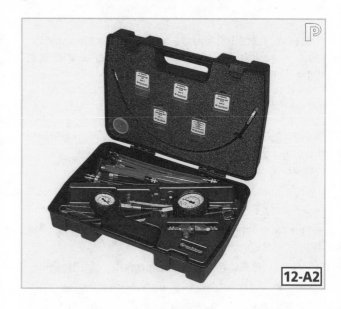
12-A2

❐ **STEP A2:** An alternative method is to use an injection pressure tester, such as the Sykes-Pickavant item shown here. Once attached, this has a relief valve, which can be used to safely depressurise the system.

❐ **STEP A3:** Disconnect the battery negative (-) terminal. If the car you are working on has a 'memory' radio security device or an on-board computer, arrange a bypass battery feed - see *Chapter 10, Electrical, Dash, Instruments, Fact File: Disconnecting the Battery* BEFORE doing so!

• IMPORTANT NOTE: Removing pressure from the fuel lines as described will not necessarily remove pressure from each of the components - they may still be pressurised.
• Before working on any part of the system it is necessary to relieve the pressure, as follows:
1. Disconnect the battery negative terminal. See *Chapter 10, Electrical, Dash, Instruments, Fact File: Disconnecting the Battery* BEFORE doing so!
2. Work out of doors and away from any sources of flame or ignition. Wear rubber or plastic gloves and goggles. Have a large rag ready.
3. Place a container beneath the filter to catch the fuel that is likely to be spilt.
4. Place your spanner on the first connection to be undone. Before undoing it, wrap the rag, folded to give several thicknesses, over the joint.
5. Undo the connection very slowly and carefully, allowing the pressure within the pipework to be let out without causing a dangerous jet of fuel.
6. Release the pressure from each of the pipes in the same way.
7. Mop up all traces of fuel and allow to dry thoroughly before starting any further work.

Section B: Single Point Injection.

The illustration numbers throughout this Section all relate to illustration *12-B1*.

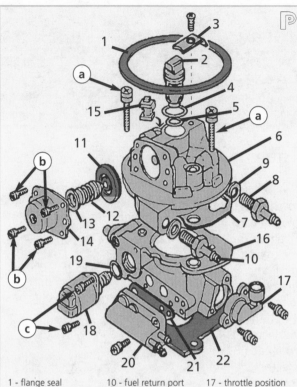

1 - flange seal
2 - injector
3 - injector retention clip
4 - injector upper sealing washer
5 - injector lower sealing washer
6 - injector unit upper housing
7 - gasket
8 - fuel inlet port
9 - fuel inlet port gasket
10 - fuel return port
11 - fuel pressure regulator diaphragm
12 - fuel pressure regulator spring
13 - fuel pressure regulator spring seat
14 - fuel pressure regulator cover
15 - wire connector rubber grommet
16 - injector unit lower, throttle body part
17 - throttle position sensor
18 - idle stabilisation valve
19 - O-ring
20 - vacuum connections flange
21 - vacuum connections flange gasket
22 - injection unit to inlet manifold gasket

12-B1

❐ **STEP B1:** These are the components of the Multec Single point injection unit.

❐ **STEP B2:**
Unbolt the air intake cover (**1**) from the injection unit. Move it to one side. You will probably have to unclip and

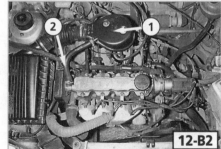

12-B2

disconnect the crank breather hose (**2**) to do this.

❐ **STEP B3:** Remove the wiring connector from the injector (**2**), then undo the retaining screw on the injector retention clip (**3**).

➜ The injector (**2**) can then be pulled from the injector unit.

❐ **STEP B4:** Note the sealing washers (**4, 5**), which must be replaced when the injector is refitted.

❐ **STEP B5:** Remove the rubber flange seal (**1**), and lever the injector wiring's rubber retaining grommet (**15**) from the edge of the flange.
➜ Disconnect the wiring connectors from the throttle position sensor (**17**) and idle control valve (**18**).
➜ Disconnect the vacuum hoses from the injector unit. If necessary, mark them so they can be replaced in the correct positions.

❐ **STEP B6:**
Undo the clip, and disconnect the MAP sensor hose from the back of the injector unit.
➜ Make sure nothing else is connected to the injector unit, remove the hexagonal head mounting

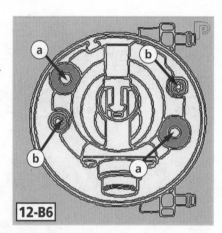

12-B6

screws (**a**) and lift the unit off.
➜ To separate the top half of the unit from the lower, throttle body part, remove the securing screws (**b**), then pull the sections apart.
➜ Remove the gasket, which must be renewed on reassembly.

❐ **STEP B7:** When reassembling, use new gaskets and use thread locking compound on the securing screws (**b**). It may be necessary to adjust the throttle free play - see *Job 11, Step 11*.

Section C: Multi-point injection.

❐ **STEP C1:** Those engines which feature multi-point injection have a more complex air induction system. It is straightforward to remove the fuel rail on early SOHC engine multi-point systems.

12-C2

❐ **STEP C2:**
Remove the air filter and duct.

❑ **STEP C3:** Disconnect the injector trigger wire.

12-C3

12-C4

❑ **STEP C4:** Disconnect the injector feed and return pipes.

❑ **STEP C5:** Remove the manifold heater pipe.

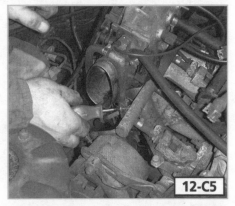

12-C5

❑ **STEP C6:** Remove the cold start auxiliary valve pipe.

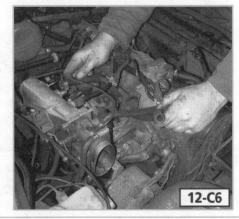

12-C6

12-C7

❑ **STEP C7:** Remove the injector rail.

❑ **STEP C8:** The Bosch LE-Jetronic injectors are fitted on the inlet manifold itself, and can be removed alone. To remove them:
→ Pull the wiring connector (**1**) from each injector.
→ Undo the mounting bolts (**2**) from the injector retention plate (**3**).
→ Pull the injector (**4**) from its housing (**5**).

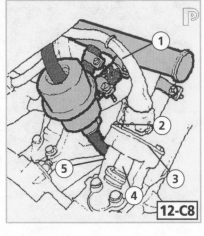

12-C8

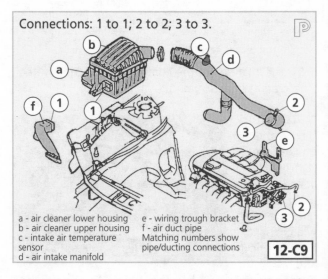

Connections: 1 to 1; 2 to 2; 3 to 3.

a - air cleaner lower housing
b - air cleaner upper housing
c - intake air temperature sensor
d - air intake manifold
e - wiring trough bracket
f - air duct pipe
Matching numbers show pipe/ducting connections

12-C9

❑ **STEP C9:** On later models, much of the multi-point injection is hidden from view. To access the fuel rail much of the system must be removed. These are its components.

STEP C10: Unbolt the engine cover plastic trim, then remove the oil filler cap. Lift the plastic trim away.

➡ Replace the oil filler cap, to prevent dirt getting into the head.

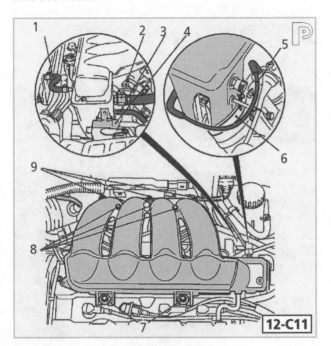

STEP C11: To remove the upper part of the intake manifold:
➡ Undo the retaining bolts (**7, 8**) and nuts (**9**) and lift the unit away.
➡ Disconnect the engine vent hose (**1**)...
➡ ...the brake servo vacuum hose (**2**)...
➡ ...and the manifold pressure sensor hose (**3**).
➡ Detach the fuel pressure regulator vacuum hose (**5**).
➡ Undo the clamp (**6**) on the throttle body connecting hose.
➡ Make sure nothing else remains attached.

STEP C12: Stuff the exposed lower inlet manifold tracts with clean rags, to prevent debris getting into them. Remember to renew the gasket when you refit the upper manifold.

STEP C13: Clear the way to the fuel rail by:
➡ Disconnecting the camshaft position sensor wiring connector (**1**).
➡ Unplugging the crankshaft position sensor wiring connector (**2**)...
➡ ...and, using a screwdriver, levering off the latter's retaining spring, then removing the plug from the bracket.

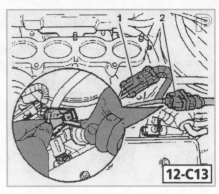

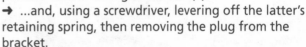

TOP TIP!

• If necessary, tag both parts of each connection with masking tape and write either a number, or the component name on them to identify their location.
• Check all connection plugs for bent or corroded connectors inside the plugs and sockets.
• ALWAYS pull on the connectors themselves, NEVER on the cable, when pulling them undone.
• Many electrical connectors have some kind of 'latch' to hold them in place - which must be lifted before the connector is pulled apart.
• Check vacuum pipes and connectors for deterioration, splits and air-tight connections.
• VERY MANY PROBLEMS ARE CAUSED BY POOR ELECTRICAL OR HOSE CONNECTIONS - don't assume immediately that expensive electrical components are at fault!

STEP C14: After closing them with hose clamps, disconnect the fuel feed line (**1**) from the fuel distributor pipe, and the fuel return line (**2**) from the fuel pressure regulator.

STEP C15: Unplug the injector rail wiring connector (**a**), then undo the nut (arrowed) and remove the wiring harness bracket from its mounting on the alternator bracket.

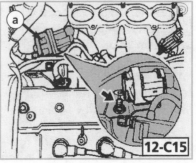

❒ **STEP C16:** Unplug the wiring connectors from the injectors, then undo the two mounting bolts holding the

12-C16

injector rail of the lower inlet manifold.
➜ Then lift the injector rail from the manifold.

❒ **STEP C17:** Disconnect each injector from the rail by removing the retaining clip (arrowed)...
➜ ...and pulling it free.

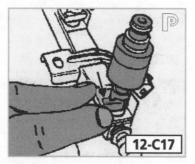

12-C17

❒ **STEP C18:**
Refitting the injectors and rail is the reverse of removal. However, remember the following:
➜ Replace the two sealing washers on the lower end of the injectors.
➜ Coat the injector with white silicone grease before refitting.
➜ To ensure the injector is in exactly the right position on the injector rail, align it with the metal tab on the rail. If you position it wrongly, an injector will deliver fuel incorrectly, making the engine run badly.

JOB 13: **FUEL PRESSURE REGULATOR -** *remove, replace.*

Section A: Single Point Injection.

The 'item' letters shown here refer to illustration *12-B1*.

❒ **STEP A1:** On Single point injection models, the fuel pressure regulator is situated on the injector unit (**item 14**).

IMPORTANT NOTE: Whenever the fuel pressure regulator cover is removed, the diaphragm (**item 11**) MUST be replaced!

❒ **STEP A2:** Depressurise the fuel system by either of the two methods described in *Job 12, Steps 1* and *2*.

❒ **STEP A3:** Remove the air cleaner and intake trunking.

❒ **STEP A4:** The serviceable parts of this fuel pressure regulator are the spring (**item 12**), the spring seat (**13**) and the diaphragm (**11**). To renew them:
➜ Remove the cover mounting screws (**b**).
➜ Take off the cover (**14**).
➜ Remove the spring and seating, and lever the diaphragm from its mounting.

❒ **STEP A5:** Refit the regulator components as the reverse of removal.
➜ Put a thin film of locking compound on the screws before fitting them.
➜ Tighten the bolts to the recommended torque setting 3 Nm (2lb ft).

Section B: Multi-Point Injection.

The fuel pressure regulator is fitted to the fuel rail in one of several different ways. The different fitting types include the following:

13-B1

❒ **STEP B1:** On earlier multi-point injection models, the fuel pressure regulator (arrowed) is between injectors three and four.

❒ **STEP B2:** On some later models, the regulator (arrowed) is fitted into the top of the fuel injection rail, secured by four Torx bolts.

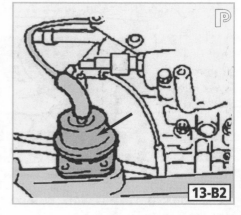

13-B2

13-B3

STEP B3: On others the regulator is fitted to the top end of the fuel rail and held by the clamp shown here.

STEP B4: Depressurise the fuel system by either of the two methods described in *Job 12, Step 1 and 2*.

STEP B5: ALL TYPES: The regulator is a non-serviceable unit, and must be replaced if it is not functioning correctly.
→ To remove it from SOHC engines, first remove the fuel rail as described in *Job 12*. If you are working on a DOHC engine, the fuel rail does not have to be removed.
→ **ALL TYPES:** Pull the vacuum tube from the regulator body.
→ Clamp the fuel feed and return hoses, as applicable, as close to the regulator as possible and undo their unions.

STEP B6: Undo the bolts or clamp holding the regulator to the fuel injection rail.

STEP B7: Refit, as the reverse of removal.
→ Put a thin film of locking compound on the mounting bolts (arrowed) before fitting them.

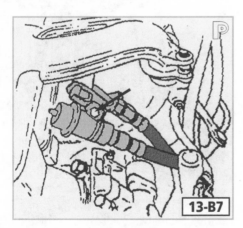
13-B7

→ Tighten the bolts (when used) to the recommended torque setting 8 Nm (6lb ft) for the older type of regulator; not specified for the later type).
→ If you are working on the later type of regulator, fit a new sealing O-ring to the fuel rail nozzle.

JOB 14: CAMSHAFT/CRANKSHAFT POSITION SENSORS - *remove, replace, check.*

FACT FILE
POSITION SENSORS
• The fuel injection and ignition systems on all later engines covered by this manual are governed by an Electronic Control Unit (ECU). These engines have a crankshaft position sensor, and in some instances a camshaft position sensor to 'tell' the ECU exact piston and valve positions.

Section A: Crankshaft position sensor.

STEP A1: The crankshaft sensor is fitted to a bracket on the oil pump housing. Undo the sensor mounting bolt (arrowed), which holds it in place on its bracket.

14-A1

STEP A2: Trace the wiring back from the sensor to the wiring harness plug (arrowed). If there are any clips holding the wiring in place on the way, release it from them.

STEP A3: Disconnect the crank position sensor wiring connection (arrowed). It is quite accessible, but the manifold in this diagram has been removed, to show it more clearly.
→ With a broad-bladed screwdriver, depress the two tags holding the connector to its mounting bracket, and slide it free.

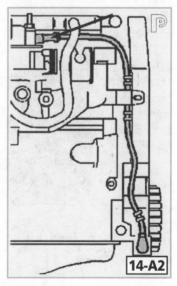
14-A2

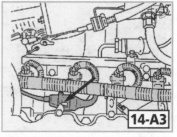
14-A3

STEP A4: Replace, as the reverse of removal.
→ Use a new O-ring seal.
→ Tighten the mounting bolt to its specified torque setting 8 Nm (6 lb/ft)

STEP A5: Using a feeler gauge, check the gap between the sensor and the ribbed crank, as shown by the arrows

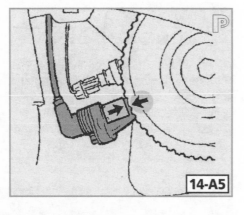

14-A5

in the diagram, is between 0.7 and 1.0mm
→ If not, Vauxhall recommends that the mounting bracket is replaced. It bolts to the oil pump housing.
→ In this case, its mounting bolts should be tightened to 10 Nm (7 lb/ft).

Section B: Camshaft position sensor.

The camshaft position sensor is situated either under the timing belt cover or under the ignition cable cover. In either case, the timing belt cover has to be removed.

STEP B1: Remove the upper outer timing belt cover - see *Chapter 6, Part C, Job 2*.

STEP B2: EITHER: Unbolt the sensor from the inner timing belt cover...

14-B2

STEP B3: ...OR: Unbolt the sensor fastening bolt, remove the ignition cable cover and remove the sensor, situated underneath.

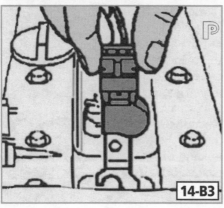

14-B3

STEP B4: Free the sensor wiring plug from its bracket.
→ Separate plug and socket.

STEP B5: Trace the wiring round to the timing cover, freeing it from any wiring clips.

STEP B6: Refit as the reverse of removal.
→ Tighten the mounting bolts to their specified torque settings 4 Nm (3 lb/ft).

JOB 15: IDLE CONTROL VALVE - *remove, replace.*

Refer to illustration *12-B1*.

The idle control valve is found on the throttle body or, in the case of Single point injection models, the injection unit. While removal and replacement is a similar process in all cases, the different architectures of these engines means that actually getting to the valve is a task of varying difficulty.

STEP 1: SINGLE POINT INJECTION MODELS: The idle control valve **(18)** is located on the face of the injection unit. To remove it, first remove the air intake ducting.
→ Remove the mounting bolts **(c)**.
→ Carefully pull the valve from the housing.
→ For replacing - see *Step 5*.

15-2

STEP 2: EARLIER MULTI-POINT INJECTION MODELS: The idle control valve is even more easily accessible, as it is on the rear side of the throttle body (arrowed).

STEP 3: To remove the valve from throttle body **(1)**:
→ Undo the wiring connection **(a)**.
→ Remove the two mounting bolts **(b)**.

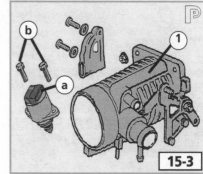

15-3

STEP 4: LATER MULTI-POINT INJECTION MODELS:

It is more difficult to get to the idle control valve (**a**).

➜ It is rather buried under the upper inlet manifold assembly (**b**)!

➜ For replacing - see **Step 5**.

TO ACCESS THE VALVE: Drain the coolant – see **Chapter 8, Cooling System**.

➜ Disconnect the fuel return line (**c**) from the fuel pressure regulation valve (**d**). Move the line aside.

➜ Disconnect the wiring connector from the idle control valve (**e**).

➜ Undo the two mounting bolts (**f**) that hold the valve to the throttle body (**g**).

➜ Pull the valve out of the throttle body.

15-4

ALL TYPES

STEP 5: Refit as the reverse of removal. If applicable:

➜ Use a new O-ring seal (**b**) on refitting.

➜ Smear a thin bead of locking compound on the bolt threads before refitting.

➜ Tighten the bolts to their recommended torque setting of 3 Nm (2 lb/ft).

➜ To prevent damage to the mounting, measure the distance (**a**) between, the valve piston and flange. It should be set to 28 mm for single and early multi-point injection models, and 33 mm for later multi-point examples.

➜ If the clearance is greater, carefully press the plunger (**c**) into the motor body.

15-5

JOB 16: THROTTLE POSITION SENSOR/THROTTLE VALVE SWITCH - *remove, replace*.

Section A: All models except Bosch LE-Jetronic variants.

STEP A1: SINGLE POINT INJECTION MODELS:
The throttle position sensor/potentiometer is mounted on the injector unit - see illustration **12-B1**.

STEP A2: SOHC MULTI-POINT INJECTION MODELS:
The sensor is on the throttle body next to the idle control valve – see illustration **15-3**.

STEP A3: DOHC MULTI-POINT INJECTION MODELS:
The sensor (arrowed) is screwed to the throttle body. However, this is more difficult to get to than on other models – some dismantling will be required to gain access to the sensor.

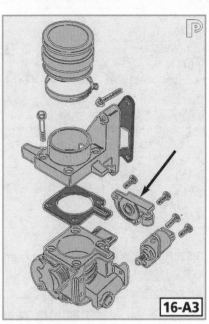

16-A3

ALL VERSIONS

STEP A4: To remove the sensor:
➜ Disconnect the wiring connection from the sensor.

➜ Remove the two mounting screws from the sensor, then remove it.

➜ Refit as the reverse of removal.

➜ Tighten the bolts to their recommended torque setting 2 Nm (1.5 lb/ft) on Single point injection models; 3 Nm (2 lb/ft) on all other versions.

Section B: Bosch LE-Jetronic injection system.

STEP B1:
Models with this injection system use a throttle valve switch instead of the later potentiometer-type position sensor. This is positioned on the throttle body - see **Job 11, Step 10**.

STEP B2:
To remove the switch, disconnect the wiring plug and remove the mounting screws.

➜ Pull the switch off the spindle of the throttle valve.

STEP B3:
Refit the switch as the reverse of removal, BUT, before you fully tighten the mounting screws you must adjust it correctly.

➜ Turn the switch unit anti-clockwise, until you feel a resistance.

➜ Tighten the screws.

➜ Check the adjustment is correct by opening the throttle a little. If you hear a 'click', it is correct. You should hear another as you close the throttle again.

➜ If the position is correct, tighten the mounting screws fully and reconnect the wiring plug.

JOB 17: MANIFOLD AIR PRESSURE (MAP) SENSOR - *test, remove, replace.*

☐ **STEP 1:** To remove the sensor:
→ Disconnect the wiring plug (**a**) from the sensor.
→ Pull the vacuum hose (**b**) from the sensor.
→ Undo the two mounting nuts.
→ Pull the sensor (**c**) off its mounting pegs.

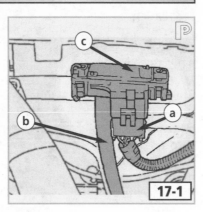

17-1

☐ **STEP 2:** Refit as the reverse of removal.

JOB 18: INTAKE AIR TEMPERATURE SENSOR - *remove, replace.*

18-1

☐ **STEP 1:** The intake air temperature sensor is plugged into the air intake trunking, upstream of the throttle body. To remove it, first disconnect the wiring plug by opening its locking tab and pulling it free.

18-2

☐ **STEP 2:** Pull the sensor out of the hose. It is a friction fit, so if it is difficult to remove, twist it first, to break its 'seal'.

JOB 19: AIR MASS SENSOR - *remove, replace.*

☐ **STEP 1:** The air mass meter (arrowed) is located in the inlet trunking.

19-1

☐ **STEP 2:** Remove the air filter box.
→ Disconnect the wiring plug.
→ Undo the retaining bolts.
→ Remove the mass meter from the throttle body end of the trunking.

☐ **STEP 3:** Remove the mounting screws (arrowed) and take the meter from the air filter box.

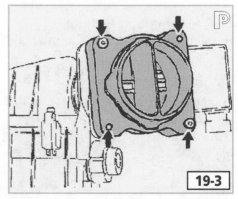

19-3

☐ **STEP 4:** The mass meter is not repairable. If a diagnostic test reveals it to be malfunctioning, it must be renewed. Replace it as the reverse of removal.

JOB 20: COOLANT TEMPERATURE SENSOR - *remove, replace.*

SAFETY FIRST!

• **DO NOT work on the temperature sensor if the engine is at all warm!**

☐ **STEP 1:** The coolant temperature sensor is located in various positions on the engine, depending on model.
→ On examples with Single point injection, it is screwed into the bottom of the inlet manifold (arrowed).
→ Bosch LE-Jetronic-

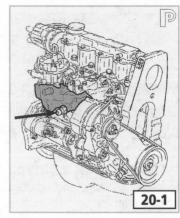

20-1

equipped cars have the sensor in the top of the inlet manifold.

STEP 2: On earlier multi-point injection models, the sensor (arrowed) is fitted to the thermostat housing...

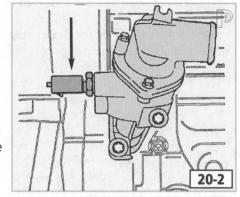

STEP 3: ...while on multi-point injection models, it is on the end of the crank, below the DIS module. Wherever it is, the

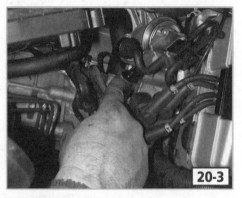

approach to removal is the same:

→ Remove the radiator cap, then replace it, to depressurise the cooling system.

→ If necessary to access the sensor, remove the air inlet trunking.

→ Disconnect the wiring plug from the sensor.

→ Unscrew the sensor from its mounting.

→ Remove it. Coolant will escape, until the coolant in the system reaches the level of the sensor. Catch it with a rag and a suitable container.

STEP 4: Replace the sensor, as the reverse of removal.

→ Use a new sealing ring.

→ Tighten to the recommended torque setting 14 Nm (10 lb/ft) in most cases, but 20 Nm (15 lb/ft) when bolted into the end of the head.

→ Refill the cooling system - see *Chapter 8, Cooling System*.

JOB 21: LAMBDA SENSOR - *remove, replace.*

SAFETY FIRST!

• Vauxhall recommend that the Lambda sensor should only be removed when the engine is at working temperature (but not running).

• Wear thick industrial gloves while removing the sensor and avoid burns from hot exhaust and engine parts!

Section A: Remove, refit

STEP A1: Run the engine until it is at working temperature - once the cooling fan has started and stopped twice.

STEP A2: Follow the wiring from the sensor and locate the connector. In this instance, it is clipped to a bracket next to the dipstick.

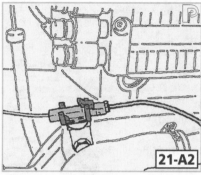

STEP A3: Unscrew the Lambda sensor (arrowed) from the exhaust manifold - or the catalytic converter, in the case of the X12XE engine.

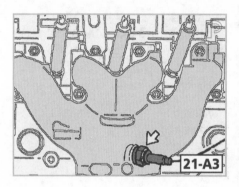

STEP A4: Refit as the reverse of removal.

→ Coat the threads of the sensor with an adhesive sealing compound.

Section B: Testing.

SAFETY FIRST!

• Beware of hot exhaust components, which can burn badly - wear suitable industrial gloves.

• Work only with the vehicle passing its exhaust gases directly to outdoors, never in an enclosed space.

STEP B1: You will need a Lambda sensor tester, such as this Sykes-Pickavant unit which is simply connected up to the sensor wiring as described in the tester manual.

STEP B2: Prepare for the test as follows:

➜ Run engine to normal full running temperature (fan turns on and off twice).

➜ If engine already hot, run for 30 seconds at 3,000 rpm to heat Lambda sensor and ensure exhaust gas at correct temperature.

➜ Allow engine to slow to normal tick-over. If testing is prolonged, run engine up to 3,000 rpm for 30 seconds at intervals.

STEP B3: A purpose-made tester will provide readings – green or red flashing lights, for instance – to enable

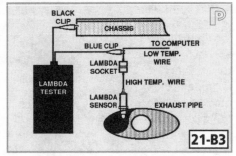

you instantly to judge the state of the Lambda sensor.

STEP B4: Alternatively, check the voltage at the sensor. The average should be about 0.45 to 0.5 volts but it's okay if it fluctuates between 0.2 and 0.8 volts. (Illustration, courtesy Sykes-Pickavant.)

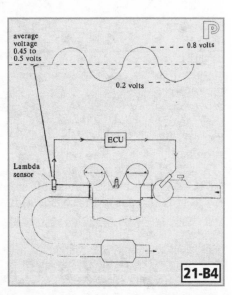

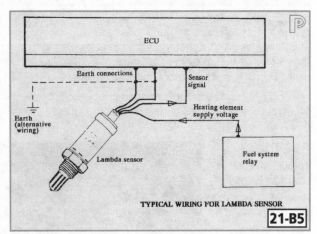

STEP B5: If readings are outside these limits:

➜ If the reading is consistently too high (mixture too strong), snap the throttle open several times and see if the level readjusts itself to within the acceptable range.

➜ If the reading is too low (mixture too weak), weaken the mixture - try running without air cleaner - and see if the level readjusts itself to within the acceptable range.

➜ If not, the Lambda sensor is probably faulty. (Illustration, courtesy Sykes-Pickavant.)

STEP B6: Some Lambda sensors also have a heating element. Typically, the supply is from the fuel system relay - check wiring diagrams - and the earth/ground will be via a separate cable, either to the ECU, or direct to an earth/ground point. To check the heating element, disconnect the plug and check the resistance across the heating element wires. A typical resistance is 5 to 10 ohms. A faulty element usually gives no reading (short circuit) or a very high reading (open circuit).

JOB 22: KNOCK SENSOR EQUIPMENT - *remove, replace.*

IMPORTANT NOTE: The knock sensor will only work properly if tightened to the correct torque. It is essential to do so.

STEP 1: The knock sensor is bolted into the back of the engine block, between the starter motor and the alternator. As a result, it is difficult to access but this is what it looks like.

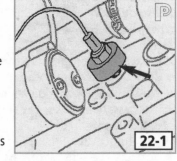

➜ Trace the wiring back from the sensor, until you come to the in-line connector plug.

➜ Separate the plug and socket.

STEP 2: Working from underneath the car, unscrew the sensor. Here, a 'split' socket is being used, as the cable cannot be removed from the sensor.

STEP 3: The knock module is mounted on a bracket on the LH side of the engine bay. Working from above, disconnect the wiring connector (**1**) from the knock sensor processing module

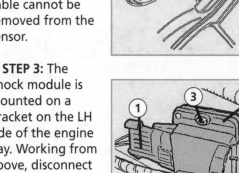

(2). Remove the arrowed bolts **(3)** and lift the module away from its mounting bracket **(4)**.

JOB 23: OCTANE NUMBER PLUG – *set as needed*.

☐ **STEP 1:** Some engines have a plug-in device, which tells the ECU the octane rating of the fuel the vehicle is using.
➜ It is located on the right-hand front suspension turret (***23-2***, arrowed) or the left side rear engine bay panel.
➜ Remove the plug assembly from its mounting bracket.

☐ **STEP 2:** Squeeze the sides of the plug **(a)**, and pull it out of the socket.
➜ Turn the plug until the desired rating (91 or 95 RON) is seen, printed on the uppermost side **(b)**.
➜ On later versions, the octane number appears in a window on the socket when the plug is pushed home.
➜ Reassemble and secure the unit in its mounting bracket.

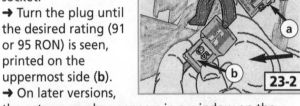

JOB 24: CHOKE CABLE - *remove, replace, adjust*.

☐ **STEP 1:** Some cars covered by this manual have a manual choke. There are several models of carburetor with choke - see ***Job 9***.

☐ **STEP 2:** The choke cable end fitting varies slightly between carburetor types.
➜ Remove the air filter.
➜ Undo the nut, Allen-head screw or screw clamping the inner cable.
➜ Undo the clamp screw to release the outer cable.

☐ **STEP 3:** Working inside the car and depending on the model:
➜ **EITHER:** Pull out the clip that holds the choke handle in place; **OR**, tap out the retaining pin and unscrew the choke control knob.
➜ Undo the choke knob retaining ring, **or** remove whatever under-dash trim necessary to access the back of the choke handle mechanism, and undo the nut clamping the outer cable to its mounting bracket.
➜ Tighten the nut or screw.
➜ Remove the drill.

➜ The cable is correctly adjusted when, if you push the control knob in fully then release it, it springs back about 1 mm.

☐ **STEP 4:** Replace the cable as the reverse of removal.

☐ **STEP 5:** To adjust the cable, ask an assistant to hold a 4 mm drill under the choke handle as shown...

☐ **STEP 6:** ...and working in the engine bay, loosen the inner cable clamp nut or screw **(1)**, and pull the inner cable **(2)** until there is just the slightest amount of free play.

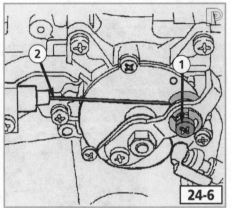

IMPORTANT NOTE: Where there is no inner cable adjustment, see ***Part C, Job 4*** for outer cable adjustment details.

JOB 25: MECHANICAL FUEL PUMP - *remove, replace*.

☐ **STEP 1:** Before dismantling, disconnect the battery negative (-) earth/ground terminal. See ***Chapter 10, Electrical, Dash, Instruments, Fact File: Disconnecting the Battery*** BEFORE doing so!

☐ **STEP 2:** Some of the models covered by this manual have a mechanical fuel pump. There are two types:
➜ One type is operated by a lever. Note the spacer and TWO gaskets between the lever-type pump and the engine block.
➜ The other (and more common type) is operated by

a plunger.

→ Some plunger-type pumps have a screw (arrowed) which enables the cover to be removed and internal components changed.

TOP TIP!

• If one of these fuel pumps is badly worn, it is usually best to fit a new pump, rather than attempt to overhaul the old one.

❏ **STEP 3:** Clamp shut, then disconnect the fuel lines from the pump. Be prepared with a clean rag, to absorb the fuel that will spill from the ends.

TOP TIP!

• Plug the ends of the pipes with rubber caps, if possible.
• Alternatively, leave the clamps in place.

❏ **STEP 4:** Undo the mounting bolts.

❏ **STEP 5:** Remove the pump, along with its spacer block.
→ Clean off any traces of gasket or sealant.

25-4

❏ **STEP 6:** Refit as the reverse of removal.

TOP TIP!

• Check the pump's oil seal. If engine oil seeps from the vent hole in the pump body, an internal seal has failed, and the pump must be replaced.
• If fuel is leaking from the vent, the diaphragm has failed, and the pump must be renewed.

JOB 26: ELECTRIC FUEL PUMP - *remove, replace.*

Depressurise the fuel system by either of the two methods described in *Job 11, Step 1 and 2.*

Section A: Inline pump.

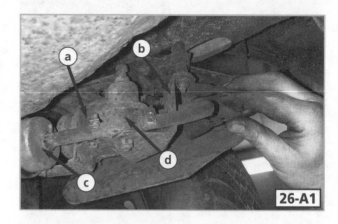
26-A1

❏ **STEP A1:** There are two types of electric fuel pump used in the models covered by this manual. The first is an inline unit (**a**), bolted to the underbody on a bracket (**b**), which also holds the fuel filter (**c**) and diaphragm damper (**d**). To remove this:

26-A2

❏ **STEP A2:** Clamp the pump's inlet and outlet hoses to prevent fuel escaping.
→ Catch any fuel that does escape in an absorbent rag.
→ Loosen the hose clamps, and pull the fuel lines off the input and output nozzles.

❏ **STEP A3:** Disconnect the wiring plug or connectors from the pump.
→ Undo the pump mounting nut.
→ Slide the pump out of the bracket.

❏ **STEP A4:** Refit as the reverse of removal.

Section B: Internal pump.

☐ **STEP B1:**
The other type is an electric pump, which sits inside the fuel tank. To access it, lift the rear seat squab, and lever out the

26-B1

plastic cover, shown being lifted away.

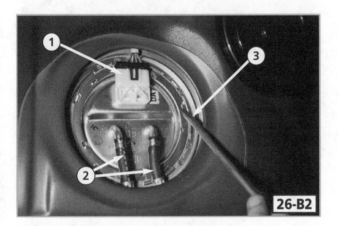

26-B2

☐ **STEP B2:** Disconnect the wiring plug (**1**) from the pump.
→ Disconnect the fuel lines (**2**).
→ Plug the lines (there is very little room to get pipe clamps onto them).
→ The pump is released by unclipping the retaining ring (**3**).

☐ **STEP B3:**
Preferably use a special tool similar to Vauxhall Part No. KM-797.

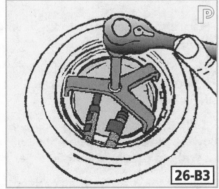

26-B3

☐ **STEP B4:** Alternatively, tap the ring round with a hammer and a non-metallic drift.

SAFETY FIRST!

• Do not use a metal drift to tap round the sealing ring. There is a risk of causing an explosion from sparks.

☐ **STEP B5:** You can now lift the pump from the tank.

☐ **STEP B6:** Refitting is the reverse of removal.
→ Always use a new sealing ring under the pump flange.

Section C: Internal pump.

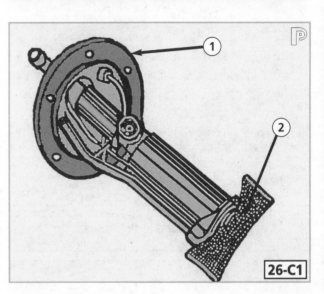

26-C1

☐ **STEP C1:** Some models had a slightly different internal pump, illustrated here. Removal is the same as in *Section B*, except that:
→ Instead of unclipping a retaining ring and locking tabs, the pump is unbolted from the top.
→ It is removed complete with a rubber sealing ring (**1**), which must be replaced on refitting.
→ It features a pre-filter (**2**), which slides onto the bottom of the pump. This must be renewed on refitting.

Part B: Diesel Engines

CONTENTS

FURTHER INFORMATION

1: DIESEL ENGINE FAULT FINDER MANUAL, produced in association with **Perkins Engines**, is the definitive guide to Diesel engine diagnostics, fault-finding, tune-up and overhaul. All Diesel engine types are covered.

If you have difficulty locating a copy of this book, please contact **Porter Manuals** at the address/telephone number shown near the start of this book.

JOB 1: SYSTEMS EXPLAINED.

SAFETY FIRST!

• Never work on the fuel system unless the engine is completely cool.

• All Diesel fuel injection systems inject fuel at extremely high pressure.

• Pressure must be released in a controlled fashion to avoid any risk of a pressurised spray of fuel being released.

• If spray from injectors or partly detached pipes penetrates the skin or hits the eyes, serious injury can result.

• Residual pressure can remain in the fuel system for some considerable time, even if the engine has been switched off.

• IMPORTANT NOTE: Removing pressure from the fuel lines, as described below, will not necessarily remove pressure from individual components - they may remain pressurised.

• Before working on any part of the system, it is necessary to relieve the pressure, as follows:

1. Before dismantling, disconnect the battery negative (-) earth/ground terminal. See *Chapter 10, Electrical, Dash, Instruments, Fact File: Disconnecting*

the Battery BEFORE doing so!

2. Work out of doors and away from any sources of flame or ignition. Wear rubber or plastic gloves and goggles. Have a large cloth ready to catch escaping fuel.

3. Place a container beneath the filter to catch the fuel that is likely to be spilt.

4. Place a spanner on the first connection to be undone. Before undoing it, wrap the cloth, folded to several layers, over the joint.

5. Undo the connection very slowly and carefully, allowing the pressure within the pipework to be let out without causing a dangerous jet of fuel.

6. Release the pressure from each of the pipes in the same way.

Mop up all traces of fuel and allow to dry thoroughly before reconnecting the battery, starting the car or taking it back indoors.

IMPORTANT NOTE: When working on the diesel fuel delivery system, cleanliness is vital. Clean all connections before stripping them.

JOB 2: DIAGNOSTIC - *carry out checks*.

The process of carrying out diagnostic electronic tests on the diesel-powered cars covered by this manual is exactly the same as for petrol/gasoline versions - see *Part A, Job 2: Diagnostic - carry out checks*.

JOB 3: HIGH PRESSURE PIPES – *remove, replace*.

❏ **STEP 1:** De-pressurise the system – see *Job 1*.

❏ **STEP 2:** Slacken the end fittings at the injectors and the pump.

3-3

❏ **STEP 3:** Lift the pipes together, taking care not to bend any.

❏ **STEP 4:** Refit as the opposite of removal.

JOB 4: DIESEL INJECTION PUMP - *remove, refit, set timing*.

SAFETY FIRST!

• Read and follow the *Safety First!* information at the start of this chapter, as well as *Chapter 2, Safety First!*.

Section A: Pump removal and refitting.

4-A1

❏ **STEP A1:** Remove the timing belt cover. See *Chapter 6, Part C, Job 2*. Turn the crankshaft until the timing mark on the injection pump sprocket (arrowed) aligns with the mark on the pump bracket (arrowed).

❏ **STEP A2:** Disconnect and remove the fuel return hose...

4-A2

4-A3

❏ **STEP A3:** .and the fuel feed hose.
➜ Hold the union to stop it from turning as the pipe is undone.
➜ Plug the ends of the hoses.

☐ **STEP A4:** Disconnect the accelerator cable and (when fitted - very early models only) the fast idle control cable.

4-A4

☐ **STEP A5:** Undo the fuel high pressure pipe unions at the pump and at the injectors.

☐ **STEP A6:** Unplug the wiring connector from the fuel feed solenoid valve.

☐ **STEP A7:** Loosen - but do not remove - the nut holding the injection pump sprocket in place.

4-A7

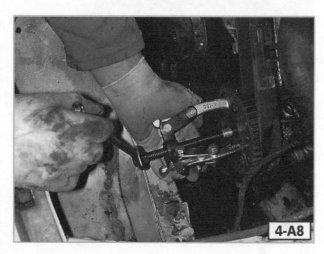

4-A8

☐ **STEP A8:** Using a puller, loosen the sprocket from the pump shaft. Remove the puller, unscrew the retaining bolt and take off the sprocket.
→ The sprocket can be very difficult to remove; if so, use an hydraulic puller.

☐ **STEP A9:** Remove the Woodruff key (arrowed) from the pump shaft.

4-A9

4-A10

☐ **STEP A10:** If the model you are working on has the oil filter directly behind the injection pump, drain the engine oil and remove the filter - see *Job 8*

☐ **STEP A11:** Remove the pump mounting bolts or nuts and, if the variant you are working on has a mounting bracket, take out the two bolts that hold it to the engine block.

4-A11

☐ **STEP A12:** Remove the pump.

☐ **STEP A13:** Replace the pump as the reverse of removal.

Section B: Engines with chain driven camshaft.

☐ **STEP B1:** Remove the intake manifold.
→ Remove the fuel injection lines and then the upper manifold.
→ DO NOT separate the exhaust gas recirculation valve from the manifold – if either is defective, replace as a set.
→ Disconnect hoses and wiring and then the lower manifold.

❐ **STEP B2:** Lock the engine at No. 1 cylinder TDC.

❐ **STEP B3:** Remove the upper timing chain – see *Chapter 6, Part C, Job 1*.

❐ **STEP B4:** Remove the fuel injection lines.

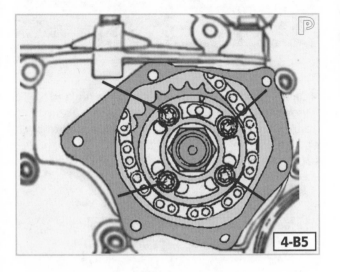

4-B5

❐ **STEP B5:** Undo the injection pump sprocket bolts (arrowed).

❐ **STEP B6:** Remove the injection pump bracket.

❐ **STEP B7:** Remove the injection pump.

❐ **STEP B8:** Replace as the reverse of removal.

Section C: Check/set injection timing (1.7L only).

IMPORTANT NOTE:
• On 2.0 litre models the injection timing is controlled by the ECU and adjustment is not possible without special diagnostic equipment.
• If a fault occurs, the ECU switches to a special mode that allows the engine to run and illuminates the instrument panel warning light.
• If the warning light illuminates, the vehicle should be taken to a Vauxhall specialist for diagnostic tests.

❐ **STEP C1:** Set the engine to TDC. See *Chapter 6, Part C, Job 3*.

❐ **STEP C2:** Now the crankshaft pulley TDC mark will be in line with the pointer on the engine block (No.1 cylinder is at TDC).
➜ Turn the engine anti-clockwise - in the opposite direction to that of normal rotation - by 60 degrees.
➜ Either use an angle gauge...
➜ ...or use a protractor to measure the angle, and make a suitable mark on the cambelt cover to which the mark on the pulley can be aligned.

4-C3

❐ **STEP C3:** A dial gauge now has to be fitted to the pump body.
➜ Remove the central plug from the back end of the pump. This will expose the threaded hole for fitting the dial gauge.

IMPORTANT NOTE: On some Lucas/CAV pumps, the plug is situated on the top of the unit.

4-C4

❐ **STEP C4:** Fit the dial gauge to the pump.
➜ The plug (see *Step C3*) is shown in the right hand.
➜ Turn the movable dial on the gauge so the pointer is set to zero.
➜ At this point the fuel injection pump piston (NOT the engine!) is at TDC.

FACT FILE
ADJUSTMENT TYPES

• You will find one of two types of timing adjustment used on these engines.
• On some, Vauxhall have kindly fitted a pump sprocket held in place with three bolts passing through slotted holes in the sprocket. This enables the timing to be adjusted without moving the pump.
• On others, there is no such facility. The pump has to be adjusted in the traditional way, by slackening its mountings and rotating it.

❒ **STEP C5: EARLY MODELS WITH COLD START KNOB:** Slide a screwdriver or metal dowel, through the holes in the cold start lever, to immobilise it.

❒ **STEP C6:** Turn the crankshaft, in the normal direction of engine rotation (i.e. clockwise at the crankshaft pulley), until the TDC mark on the crankshaft pulley is in line with the pointer on the engine block.

→ In other words, the engine will now be at TDC again.

→ The reading now shown on the dial gauge is a measure of the fuel injection pump piston's travel at engine TDC.

→ Check that the reading indicated on the dial gauge is in the range shown below:

ENGINE CODE DEFLECTION (READING) 15D 0.85-0.95mm 15DT / X15DT 0.63-0.73mm 16D 0.95-1.05mm 16DA 0.85-0.95mm 17D 0.60-0.70mm X17D 0.60-0.70mm

If adjustment is necessary, do one of the following:

❒ **STEP C7A: PUMP SPROCKET WITH THREE SLOTTED HOLES:**
→ Slacken the bolts (a).
→ Turn the mounting flange (b) at the back of the sprocket to turn the shaft on the pump until the reading on the dial gauge is correct.
→ Retighten the bolts (a).

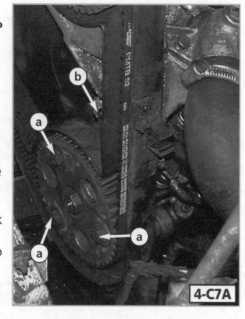
4-C7A

❒ **STEP C7B: PUMP SPROCKET WITHOUT SLOTTED HOLES:**
→ Slacken the pump's mounting bolts - see *Section A, Step 11*.
→ Rotate the pump as shown until the reading on the dial gauge is correct.
→ Retighten the mounting bolts.

4-C7B

❒ **STEP C8:** Remove the dial gauge.
→ Replace the central plug, using a new sealing washer.
→ Reattach the fuel line/s and electrical cables.
→ Free the cold start lever, if fitted.

Section D: Speed adjustments (1.7L only).

IMPORTANT NOTE:
• On 2.0 litre models the engine maximum and tickover speeds are controlled by the ECU and adjustment is not possible without special diagnostic equipment.
• If the engine speed needs adjustment, the vehicle should be taken to a Vauxhall specialist for diagnostic tests.

To accurately check and adjust the diesel's idle speed you need a diesel-specific rev counter. If you're happy with the idle speed, leave it well alone. But if you do wish to adjust the speed:
→ Run the engine to normal operating temperature.
→ Check that there is some free play in the accelerator cable before proceeding.
→ Turn OFF all electrical components.
→ Ensure the cold-start knob (if fitted) is not pulled.
→ **MAXIMUM SPEED SETTINGS:** Do not attempt to set the maximum speed unless you are sure that the general engine condition, and in particular the condition of the timing belt, are in sufficiently good condition. Otherwise, running the engine at maximum speed may cause a catastrophic breakdown!

FACT FILE
SPEED SETTINGS

IDLE SPEED:
→ 17 DTR and 17 DTL when ambient temp. is below 20 degrees C: 1,200 rpm.
→ 17 DTR and 17 DTL when ambient temp. is above 20 degrees C: 850 rpm.
→ **ALL OTHER ENGINES:** The correct idle speed for all other diesel engines is 830-930rpm.

MAXIMUM SPEED: The correct maximum speeds for Vauxhall diesel engines are as follows: ENGINE CODE MAXIMUM SPEED 15D 5800rpm 15DT / X15DT 5600rpm 16D 5600rpm 16DA 5600rpm 17D, DR and DTL 5500-5600rpm

BOSCH INJECTION PUMP

STEP D1: IDLE SPEED: Slacken the lock-nut of the adjuster screw (**1**), then turn the screw inwards to increase the speed - or outwards to decrease.

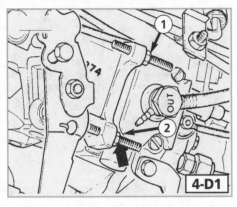

`4-D1`

Re-tighten the lock-nut when the speed is correct.

➔ **MAXIMUM SPEED:** Slacken the lock-nut of the adjuster screw (**2**), then turn the screw inwards to increase the speed - or outwards to decrease. Re-tighten the lock-nut when the speed is correct.

LUCAS/CAV INJECTION PUMP

STEP D2: IDLE SPEED: Slacken the lock-nut of the adjuster screw (**1**), then turn the screw inwards to increase the speed - or outwards to decrease.

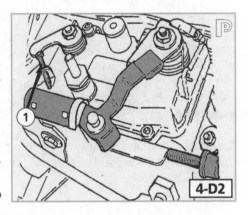

`4-D2`

Re-tighten the lock-nut when the speed is correct.

➔ **MAXIMUM SPEED:** Slacken the lock-nut of the adjuster screw (**1**), then turn the screw inwards to increase the speed - or outwards to decrease. Re-tighten the lock-nut when the speed is correct.

JOB 5: FUEL INJECTORS - *check, replace*.

☞ SAFETY FIRST!

• Read and follow the *Safety First!* information at the start of this chapter, as well as *Chapter 2, Safety First!*.

Section A: Injector faults.

The following are some of the symptoms of defective injectors:

➔ Misfiring.
➔ Knocking on one or more cylinders (difficult to discern!).
➔ Engine overheating, loss of power, higher fuel consumption.
➔ Excessive blue smoke when started from cold.
➔ Excessive black smoke when running.

Section B: Indirect injection injectors - remove, overhaul, replace.

IMPORTANT NOTE: This applies to all eight valve engines.

STEP B1: After depressurising the system, clean thoroughly around each injector and pipe end to prevent dirt entering as you remove pipes and injectors.

STEP B2: Remove the high pressure pipes - see *Job 3*.

STEP B3: Unscrew the injector using a deep socket. A purpose-made injector removal socket is by far the best tool for the job.

`5-B3`

STEP B4: Remove the injector.

`5-B4`

STEP B5: The injector (**a**) is fitted with pipe seal (**b**); sealing washer (**c**); corrugated washer (**d**) and heat sleeve (**e**).
➔ All seals must be renewed on reassembly.

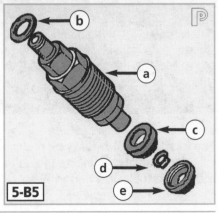

`5-B5`

STEP B6:
Once the injector is out, take a look at the nozzle end and see if it is caked with carbon – even a tiny amount of the stuff can upset that critical fuel spray pattern.

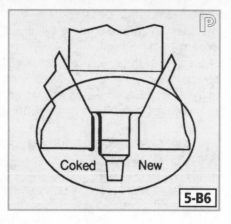

(Illustration, courtesy V L Churchill/LDV Limited)

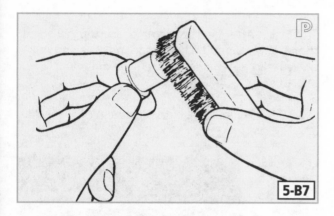

STEP B7: There's certainly no harm in brushing around the pintle of an indirect-injection type injector (the pintle is the central protruding tip) or the holes of an indirect-injection engine with a brass brush (a suede shoe brush is ideal for this).
→ You can even poke a strand of brass wire into the holes to ensure they're clear.
→ But DON'T ever be tempted to use a steel brush or strands on any part of the injector.

TOP TIP!

• Professional injector cleaning kits are available.
• These kits include useful probes and scrapers for internally de-gumming nozzles, but we really don't recommend that any DIYer become involved with injector dismantling.
• Remember, you don't have to renew an injector if it's misbehaving, you can take it to a Bosch or Lucas agent for re-nozzling and calibration, and this needn't be very expensive.

The injector should ideally be removed from the cylinder head and connected to an injector tester for a check on fuel spray pattern.

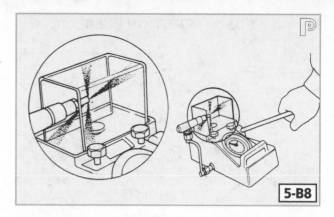

STEP B8: An injector tester is essentially a bench-mounted, hand-operated high-pressure pump containing diesel fuel, to which the injector is connected.
→ It is fitted with a pressure gauge so that verification can be made of the pressure at which the nozzle opens (the break pressure).
→ If the break pressure is incorrect, engine operation will be impaired. (Illustration, courtesy V L Churchill/LDV Limited)

STEP B9: The injector bore should always be cleaned carefully before an injector is refitted as any dirt can cause leakage of cylinder compression, as can the use of an old, flattened seal, and leakage can lead to expensive erosion of the injector by hot, escaping gases.
→ When refitting screw-type injectors, observe the specified tightening torque, or tighten firmly without using excessive force or too long a drive lever on the socket.

STEP B10: Refit the injector, tightening it to its specified torque setting.

TOP TIP!

• Run the engine at a fast idle initially to clear air from the high pressure side of the fuel system.

Section C: Direct injection injectors - remove, overhaul, replace.

STEP C1: Remove the injectors as described in *Chapter 6, Part C, Job 5.*

STEP C2: The copper seal ring and rubber seal must be renewed.

STEP C3: Check and clean the injector - see *Steps B6 to B9.*

STEP C4: Refit, referring to *Chapter 6, Part C, Job 5.*

JOB 6: TURBOCHARGER - *remove, replace.*

SAFETY FIRST!

• DO NOT, on any account work on the turbocharger unless the engine is completely cold.

• Turbochargers, and the pipework associated with them, reach extremely high temperatures when working.

Unless specified otherwise, the items referred to throughout this Job relate to illustration **6-1**.

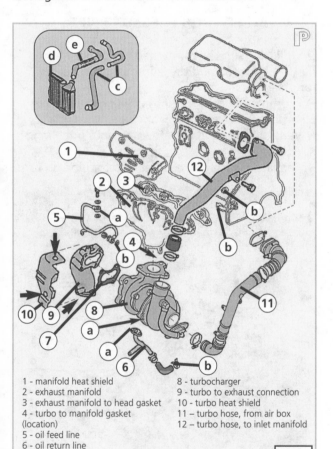

1 - manifold heat shield
2 - exhaust manifold
3 - exhaust manifold to head gasket
4 - turbo to manifold gasket (location)
5 - oil feed line
6 - oil return line
7 - turbo to exhaust connection gasket
8 - turbocharger
9 - turbo to exhaust connection
10 - turbo heat shield
11 – turbo hose, from air box
12 – turbo hose, to inlet manifold

6-1

❏ **STEP 1:** This is the layout of the turbocharging system.

❏ **STEP 2:** Before dismantling, disconnect the battery negative (-) earth/ground terminal. See *Chapter 10, Electrical, Dash, Instruments, Fact File: Disconnecting the Battery* BEFORE doing so!

❏ **STEP 3:** Drain the engine oil and the coolant.

❏ **STEP 4:** Remove the radiator fan, along with its shroud.

❏ **STEP 5:** Remove the exhaust manifold heat shield bolts, and remove the shield (**1**).
→ If fitted, remove the glow plug bus straps (2.0 litre only).

6-6

❏ **STEP 6:** Undo the clamps to the charge air and intake pipes (**11** and **12**), and remove the pipes (arrowed).

❏ **STEP 7:** Remove the turbo heat shield mounting bolts (**item 10**, arrowed), and lift the shield away.

❏ **STEP 8:** Unbolt the exhaust front pipe (**item 9**, position arrowed) to the turbocharger connecting piece (**9**).

❏ **STEP 9:** Disconnect the following pipes from the turbocharger (**8**):
→ Oil supply (**5**) and return (**6**) hoses. Remove both hoses.
→ Be ready to catch oil as it drains out!
→ Coolant hoses.

❏ **STEP 10:** Unbolt the coolant pipe bracket from the cylinder head.

❏ **STEP 11:** Unbolt and remove the turbo support from the engine block.

❏ **STEP 12:** Undo the exhaust manifold mounting bolts:
→ Remove the manifold along with the turbocharger.
→ Unbolt the turbocharger and remove from the manifold.

❏ **STEP 13:** Clean all traces of gasket from the joints.
→ Refit as the reverse of removal.
→ Use new gaskets throughout.
→ Tighten the mounting bolts to their recommended torque settings.

❏ **STEP 14: TURBOCHARGED ENGINES WITH INTERCOOLER:** Removal of the turbocharger with intercooler is similar to the above, except that the charged air pipes (**c**) runs from the turbocharger to the intercooler (**d**) and then to the inlet manifold via the pipe (**e**).

JOB 7:   COLD START CABLE – *remove, replace.*

❒ **STEP 1:** To remove the cable:
➜ Begin at the injection pump end of the cable, by releasing the cable clip from the cold start lever at the injection pump.
➜ Pull the cable, along with its rubber grommet from its 'U'-shaped bracket mounting.
➜ Thereafter, removal is exactly the same as for the choke cable in a petrol-engined car - see *Part A, Job 26*.

JOB 8:   FUEL SYSTEM AND FILTERS - *bleed, change filters.*

SAFETY FIRST!

• Read and follow the information given in *Safety First!* at the start of this chapter, as well as *Chapter 2, Safety First!*

• Do not allow Diesel fuel onto your skin. Wear impermeable gloves.

IMPORTANT NOTES:
• NEVER slacken a high pressure connection before you have wrapped several layers of cloth around the connection. This should prevent jets of fuel under high pressure being produced. Pressurised fuel is capable of penetrating the skin with, potentially, extremely dangerous consequences.
• If the engine stalls because of lack of fuel, if low pressure fuel lines have been disconnected, or the fuel filter has been changed, bleed the pump as shown below.
• Many later systems are self-bleeding and will purge themselves of air as the starter motor is turned over.
• If you need to 'spill' Diesel fuel, make sure that you have plenty of rags and containers, to catch fuel and keep it from polluting the ground or drains.
• Do not let Diesel fuel get onto rubber components, such as hoses or mountings. If you do, wipe it off immediately.

Section A: Bleeding the fuel system.

TOP TIP!

• This job is made much easier with the help of an assistant to turn the engine over on the starter when needed.

BLEED FILTER HOUSING

❒ **STEP A1:** Loosen the vent screw (**1**) on the top of the fuel filter by two turns - filter still attached; NOT as shown here! Exact position of vent screw may vary between models.

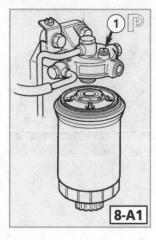

8-A1

❒ **STEP A2:** Crank the engine on the starter. Fuel will bubble from the vent bolt. When it flows free of air bubbles, stop the engine and retighten the vent bolt.

BLEED HIGH PRESSURE PIPES

❒ **STEP A3:** With the car parked out of doors, the parking brake on and the transmission in neutral, run the engine.
➜ Loosen one of the unions fastening the delivery pipe to an injector.
➜ Vent the line until fuel flows without bubbling. Retighten.
➜ Repeat on each of the other injector pipes in turn.

Section B: Draining water from the Diesel fuel filter.

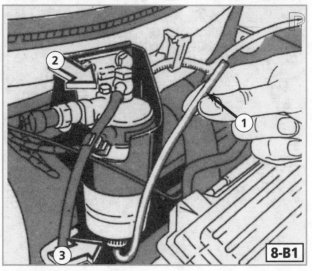

8-B1

❒ **STEP B1:** Over a period of time, water builds up in the fuel filter. To drain it out:
➜ Fit a piece of plastic pipe to the drain stub on the bottom of the filter (**1**).
➜ Undo the vent screw (**2**) by two turns.
➜ Loosen the knurled drain plug (**3**).
➜ Drain out the contaminated fuel - usually about a cupful - into a suitable container.
➜ Remember to retighten both drain plug and vent screw.

Section C: Changing the Diesel fuel filter.

❏ **STEP C1:** If the filter you are working on is the type with disposable body:

8-C1

→ Detach the temperature switch and heating wiring connectors, when fitted.

→ Clamp the fuel hoses in case of residual pressure.

→ Drain the filter - see *Section B*.

→ Use a strap wrench to remove the filter.

→ Fill the new filter with diesel fuel before fitting.

→ Coat the new filter's sealing ring with diesel fuel and tighten it by hand - not with the strap wrench! - that would distort the seal.

→ Refit all connections.

JOB 9:	**EXHAUST GAS RECIRCULATION (EGR) SYSTEM** - *repair*.

If the Diesel-powered car you are working on has an EGR system, refer to *Part A, Job 8* if the diagnostic process suggests that the EGR valve needs replacing.

JOB 10:	**ACCELERATOR CABLE –** *remove, replace*.

❏ **STEP 1:** Lever the ball end fitting from the accelerator lever.

10-1

→ Access is improved if you first remove the accelerator damper, also a ball and socket fastening.

❏ **STEP 2:** Lever the cable outer grommet from the pump bracket.

❏ **STEP 3:** Compress the spring on the end of the cable and ease it from the slot at the top of the pedal assembly.

❏ **STEP 4:** Release the cable from any clips along its length, noting the cable route as you do so.

❏ **STEP 5:** Refit as the reverse of removal.

❏ **STEP 6:** Adjust the cable by moving the clip along the cable sheath.

→ There should be a very small amount of play left in the cable inner when the cable is correctly adjusted.

Part C: Both Engine Types
CONTENTS

SAFETY FIRST!

• Read *Chapter 2, Safety First!* before carrying out this work!

• Disconnect the battery earth lead before starting work on the fuel system.

• When siphoning fuel into a metal container, attach a wire (a battery jump lead is ideal) between the tank and the container to eliminate the danger of static electric sparks jumping between the two.

• We strongly recommend that you carry out all of this work outdoors.

• If this is not possible, make sure your workshop is very well ventilated!

• Always work away from all sources of ignition or heat.

• DO NOT smoke.

• Make sure you have a container that is big enough to take all the fuel remaining in the tank.

• Store the fuel in an approved, safe container (or containers).

• Stuff rags into the tank apertures to prevent petrol vapour being given off, and store the tank in a well-ventilated place well away from any possible sources of ignition.

• Wear gloves and goggles - petrol can be harmful.

JOB 1: FUEL TANK - *remove, refit.*

❏ **STEP 1:** If the vehicle has fuel injection, de-pressurise the system – see *Part A Job 12*.
➜ Jack the vehicle up high enough to give sufficient clearance, and support it using axle stands or a vehicle hoist.
➜ Siphon off as much fuel as possible. The less fuel there is in the tank, the easier it is to manoeuvre.
➜ Disconnect the wiring plugs from the fuel sender unit and, where fitted, the integral fuel pump. See *Chapter 9, Part A, Job 26*.

❏ **STEP 2:** With the rear of the vehicle off the ground:
➜ Remove the exhaust rear section – see *Job 3*.
➜ Remove the parking brake cables – see *Chapter 12, Job 18*.

❏ **STEP 3:** If the vehicle you are working on has an electric fuel pump, undo the electric connector.
➜ Clamp the fuel hose(s), and remove them/it from the pump cover.

❏ **STEP 4:**
If not, clamp the flexible fuel hoses as close as possible to the tank.
➜

1-4

Disconnect the fuel hoses at the tank connections.

❏ **STEP 5:** Support the fuel tank on a trolley jack, with a piece of wood to spread the load.

❏ **STEP 6:** There are two mounting straps holding the tank to the body. These have hinges at the front end, and a single mounting bolt each at the rear.

❏ **STEP 7:** Remove the fixing nuts.

❏ **STEP 8:** Remove the fuel filler cap, and undo the screw retaining the filler tubing.

1-7

❏ **STEP 9:**
Lower the tank to the floor, complete with the filler tubing, as seen here.
➜ You may have to manoeuvre the tank a little on the way down, in order to free the filler tubing.

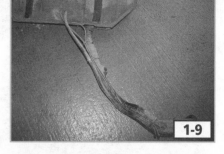

1-9

➜ The help of an assistant is invaluable here, as the tank assembly is very unwieldy.

❏ **STEP 10:** Pour any remaining fuel from the tank into an approved container, use a funnel to avoid spillage.
➜ See SAFETY FIRST for details of tank storage.

❏ **STEP 11:** Refit as the reverse of removal.

JOB 2: FUEL GAUGE SENDER UNIT - *remove, refit.*

SAFETY FIRST!

• See *Chapter 2, Safety First!*
• **Disconnect the battery earth/ground lead before starting work on the fuel system.**
• **Work only out of doors.**
• **Ensure there are no potential sources of ignition in the vicinity before starting.**
• **When siphoning fuel into a metal container, attach a wire (a battery jump lead is ideal) between the tank and the container to eliminate the danger of static electricity creating sparks between the two.**

FUEL GAUGE AND SENDER UNIT COMBINED

See *Part A, Job 26 - Electric Fuel Pump - replace*. This shows how to remove and replace the type of unit where the pump and sender unit are combined.

SEPARATE SENDER UNIT

❏ **STEP 1:** Drain the fuel tank to below the level of the sender unit.

❏ **STEP 2A:** Disconnect the

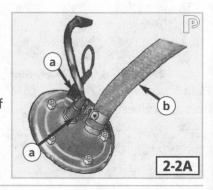

2-2A

wiring connectors (**a**).

➜ On models with fuel injection, clamp the fuel hose (**b**) and disconnect it.

➜ At this point, a small amount of fuel may be lost. Be ready with a suitable container and rags!

❑ **STEP 2B:** On other types, the sender unit has a bayonet fitting. Turn it anti-clockwise to free it from the tank. You will

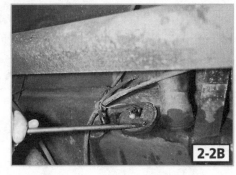

have to use a non-metal drift to do this – don't risk causing a spark!

➜ On some, the sender unit is held in place with a ring of nuts – see illustration **2-2A**.

❑ **STEP 3:** Refit as the reverse of removal.

➜ Always use a new sealing gasket.

JOB 3:	EXHAUST SYSTEM
	- remove, refit.

SAFETY FIRST!

• **NEVER work on an exhaust system when the engine is anything but cold!**

TOP TIP!

• It pays to clean exposed threads with a wire brush and spray on releasing fluid before tackling the exhaust.

❑ **STEP 1A:** Exhaust systems are similar in principle – though they differ greatly in appearance, depending on whether they incorporate a catalytic converter and where in the system it is situated. On some models, the catalytic converter is integrated into the exhaust manifold...

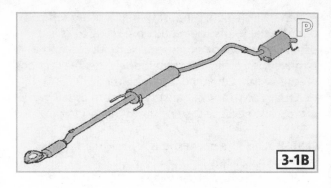

❑ **STEP 1B:** ...with separate centre and rear silencer sections, each incorporating silencer boxes.

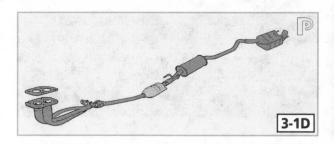

❑ **STEP 1C:** On other models, the catalytic converter is in the front (downpipe) section, in between the manifold and centre section of a three piece system.

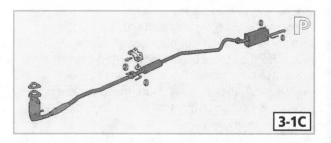

❑ **STEP 1D:** Some models have a four piece system in which the catalytic converter is in a run of pipe between the downtube and the centre silencer section...

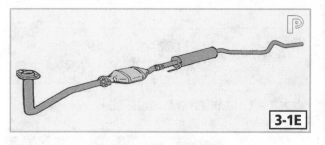

❑ **STEP 1E:** ...or the catalytic converter can be mounted directly via flanges to the downpipe and rear silencer box section in a three-piece system.

❑ **STEP 2:** Clean the exposed manifold stud threads with a wire brush then apply WD40 or similar and leave it to soak in. It's easy to shear off studs otherwise, which means the manifold has to come off.

→ Removing an old exhaust system is often difficult because the joints seize solid. In such cases, it is often best to cut or chisel scrap sections off, and to remove their remnants from the salvageable sections on the bench, where heat may be used.

→ Repairs to rusted exhaust systems are usually very temporary, because they rot from the inside out.

❏ **STEP 3:** To fit a new exhaust system, carry out work in this order:

→ Use all new gaskets and new rubber hangers when replacing an exhaust system.

→ Start at the manifold, but don't tighten the manifold nuts (or any other fastenings) until the whole system is in place.

→ Apply exhaust jointing paste, assemble the components and fit the pipe clamps loosely.

→ Ensure that the exhaust is mounted evenly, with adequate clearances.

→ Fit new rubber straps. Ensure no straps are under more tension than others by realigning the pipe joints as necessary.

→ Shake pipe to check for clearances. Only now should the pipe clamps be tightened.

→ Start and run the engine to make sure there are no rattles.

TOP TIP!

• Make doubly sure that the exhaust will not foul the heat shields nor any part of the bodywork. Ensure that no wires are hanging near the exhaust – a hot exhaust can melt insulation and start an electrical fire!

IMPORTANT NOTE: Catalytic converters are fairly fragile. Avoid dropping them on the ground, as this can cause internal damage that will cause a great rise in exhaust emissions and render the vehicle unfit for use on the road because it will fail any emissions test. Catalytic converters are very expensive to replace.

JOB 4:   ACCELERATOR CABLE - remove, replace, adjust.

❏ **STEP 1: CARBURETOR ENGINE:** Start by removing, as necessary:

→ The air filter assembly.

→ **SINGLE-POINT FUEL INJECTION:** The air intake trunking.

4-1

→ **MULTI-POINT INJECTED VEHICLES:** You may need to remove the upper inlet manifold - see *Part A, Job 12, Steps B1 to B4.*

❏ **STEP 2:** Once you have gained access to the carburetor, injector unit or throttle body, disconnect the cable balljoint from the throttle linkage.

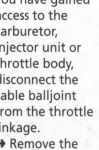

4-2

→ Remove the retaining clip (inset) off the ball using long-nosed pliers.

→ Using a flat-bladed screwdriver, lever off the ball, as shown.

❏ **STEP 3:** The outer cable is retained in a push-fit bracket.

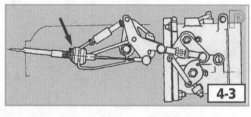

4-3

This is where it is found on earlier Multi-point systems.

→ Pull the rubber grommet (arrowed) from the clip to release the cable.

❏ **STEP 4:** Release the cable from any clips along its length, noting the cable route as you do so.

→ On some models the cable is held in a bracket on the exhaust manifold.

❏ **STEP 5:** This is an exploded view of a typical accelerator assembly.

→ Working in the driver's footwell, unhook the cable end from the pedal.

→ On some models the cable grommet is secured by a circlip – remove this.

→ Push the cable up and back, through the bulkhead.

→ Lift the cable out of the engine bay.

4-5

❏ **STEP 6:** Refit as the reverse of removal.

❏ **STEP 7:** Adjust the cable by moving the clip along the cable sheath.

→ There should be a very small amount of play left in the cable inner when the cable is correctly adjusted.

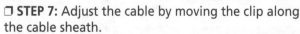

CHAPTER 10: ELECTRICAL, DASH, INSTRUMENTS

*Please read **Chapter 2 Safety First** before carrying out any work on your car.*

CONTENTS

SAFETY FIRST!

• Never smoke, use a naked flame or allow a spark near the battery.

• Never disconnect the battery with battery caps removed – or with engine running, which will damage electronic components.

• BATTERY TERMINALS: Always disconnect earth/ground FIRST and reconnect LAST.

• If battery acid comes into contact with skin or eyes, flood with cold water and seek medical advice.

• Don't top up the battery within half an hour of charging it – electrolyte may flood out.

FACT FILE

DISCONNECTING THE BATTERY

• If you disconnect the battery, you might find the alarm goes off, the ECU loses its 'memory', or the radio needs its security code.

• You can ensure a constant electrical supply with a separate battery, protected with a 1 amp fuse.

• In some cases, you might need to disconnect the battery completely. For instance, if you need to disable the air bags.

• When the battery DOES need to be disconnected, you MUST make sure that you've got the radio security code before disconnecting it.

JOB 1: ALTERNATOR - *remove, refit.*

FACT FILE

ALTERNATOR TYPES

Although the alternator is found in different positions on the engine, there are only two types.

TYPE ONE is driven from the crankshaft pulley, by its own V-belt.

• The belt is tensioned by swinging the alternator body away from the engine along a support bracket.

TYPE TWO appears on later DOHC engines.

• It is driven by a ribbed serpentine belt, along with other auxiliary equipment such as the power steering pump.

• This belt has its own tensioner, so there is no adjustment via the alternator mountings.

Section A: Both types.

❏ **STEP A1:** Before dismantling, disconnect the battery negative (-) earth/ground terminal. See *Fact File: Disconnecting the Battery* BEFORE doing so!

❏ **STEP A2:** Remove ancillary components mounted over the alternator.
➜ Depending on the model, these may include the air cleaner (arrowed) and intake hose.

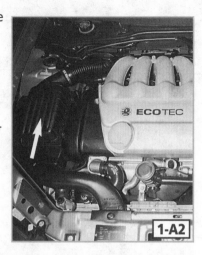

1-A2

❏ **STEP A3:** Undo the wiring connectors from the alternator.
➜ Follow the rest of the instructions relevant to the alternator type.

❏ **STEP A4A:** This is typical of the 1.3 and 1.4 engines without power steering...

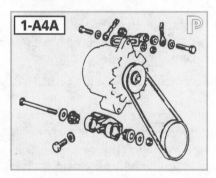

1-A4A

❏ **STEP A4B:** ...and this is a typical with-power-steering set-up.

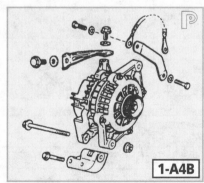

1-A4B

❏ **STEP A4C:** This is typical of 1.6D engines...

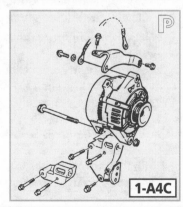

1-A4C

❏ **STEP A4D:** ...and this of other diesel engine types.

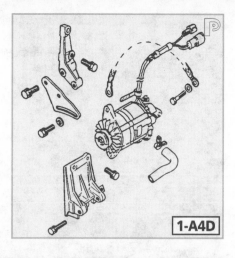

1-A4D

Section B: V-belt driven alternator.

❏ **STEP B1: DIESEL ENGINES:** If the brake servo vacuum is attached to the alternator, disconnect the vacuum hose (3) and oil feed (1) and return (2) hoses from the pump.

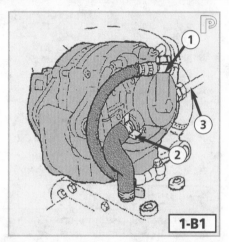

1-B1

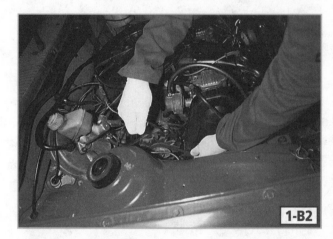

1-B2

❏ **STEP B2:** Undo the upper mount bolt a little, then slacken the lower mount bolt.
➜ Swing the alternator body towards the engine, and slip the V-belt off the alternator pulley.
➜ Remove the upper mount bolt.
➜ If necessary, remove the upper mounting bracket completely.

☐ STEP B3:
Undo the lower mount bolt, which is, in most cases, a long through bolt (**1**), which fixes the alternator to its mounting bracket (**2**).

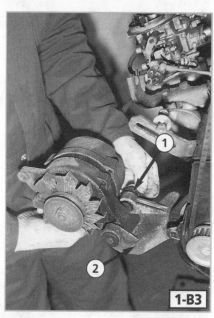

1-B3

1-B4

☐ STEP B4: Take the alternator from the engine and recover the mounting bracket bushes from the bracket.

☐ STEP B5: Refit as the reverse of removal, using new mounting bushes if necessary.

Section C: Ribbed belt-driven alternator.

☐ STEP C1:
With a spanner or, if available, Vauxhall's special tool (KM-959 or KM-6017), release the tension on the ribbed belt by turning the tensioner

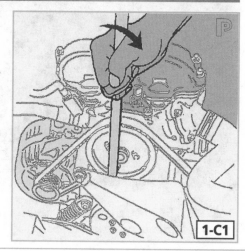

1-C1

clockwise.
➔ Slip the belt off the alternator pulley.

☐ STEP C2: Undo the upper mounting bolt.
➔ Remove the alternator support from the inlet manifold.
➔ Undo the lower mounting bolt.

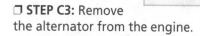

1-C2

☐ STEP C3: Remove the alternator from the engine.

☐ STEP C4: Refit as the reverse of removal.

JOB 2: STARTER MOTOR - *remove, refit.*

☐ STEP 1: Raise the front of the vehicle, and support it off the ground.

☐ STEP 2: Before dismantling, disconnect the battery negative (-) earth/ground terminal. See *Fact File: Disconnecting the Battery* BEFORE doing so!

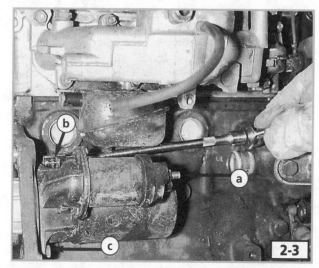

2-3

☐ STEP 3: Disconnect the electrical connections from the starter motor (**a**).
➔ Undo the earth/ground cable from the upper starter bolt (**b**).
➔ Remove the upper and lower (**c**) mounting bolts.

☐ STEP 4: Remove the starter motor

☐ STEP 5: Refit as the reverse of removal.

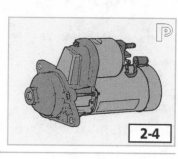

2-4

JOB 3: DASHBOARD AND INSTRUMENT PANEL - *remove, refit.*

FACT FILE

DASHBOARD TYPES

• There are many common features to the dashboards fitted to the models covered by this manual.
• In all cases, the steering wheel and column shroud have to be removed.
• Along with the control stalks - unplug the wiring connectors and undo the retaining bolts.
• You often have to remove trim or finishers (especially near the outer sides) to get at fixings, and a number of screws are concealed behind trim plugs, which should be levered out carefully.
• If some fixings can't be found, try removing as many as possible, then carefully twisting and moving the component. Look for points of non-movement as likely locations for hidden fixings.
• The drawings shown here are mainly of left-hand drive vehicles. Right-hand drive variants are the same, but transposed.

SAFETY FIRST!

MODELS WITH AIR BAGS:
• Read and make sure you understand the whole of *Chapter 14, Job 9* before starting to carry out any of the work described in this Job.
• Carry out all of the precautions and preparations described in *Chapter 14, Job 9*.

Section A: General.

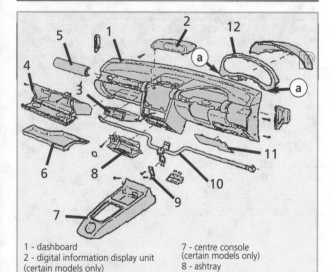

1 - dashboard
2 - digital information display unit (certain models only)
3 - stereo head unit aperture
4 - glove compartment
5 - crash padding (certain models only)
6 - storage shelf (certain models only)
7 - centre console (certain models only)
8 - ashtray
9 - support bracket
10 - reinforcement
11 - fuse box cover
12 - instrument binnacle surround.

3-A1

☐ **STEP A1:** This is a typical dashboard assembly.

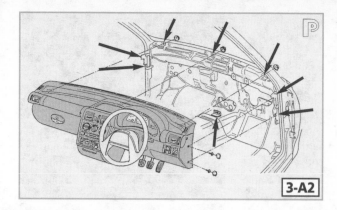

3-A2

☐ **STEP A2:** These are the main dashboard mounting points (arrowed).

☐ **STEP A3: ALL TYPES:** Before dismantling, disconnect the battery negative (-) earth/ground terminal. See *Fact File: Disconnecting the Battery* BEFORE doing so!

SAFETY FIRST!

INTERIOR TRIM AREAS:

• Before carrying out any work on any part of the interior, read *Chapter 14, Jobs 9, 10 and 11.*

• There are very important safety hazards attached to working on, OR IN THE VICINITY OF these components.

Section B: Instrument panel, removal.

☐ **STEP B1:** On some models, you will have to lever out switches before you can remove the instrument panel surround. Disconnect their wiring plugs and keep them aside.

3-B2

☐ **STEP B2:** Remove the screws securing the instrument panel surround.

STEP B3: You can now remove the surround.

3-B4

STEP B4: Take out the instrument panel screws (see *3-A1, items a*).

STEP B5: Pull the top of the instrument panel towards you.
→ Reach around the back of it, and release the speedometer

3-B5

cable by pressing the retaining spring and pulling the cable backwards.

STEP B6: Remove the panel.

Section C: Dashboard, removal.

SAFETY FIRST!

• Read and make sure you understand the whole of *Chapter 14, Job 9* before starting to carry out any of the work described in this Job.

• Carry out all of the precautions and preparations described in *Chapter 14, Job 9*.

STEP C1: Remove the steering column shroud, by removing its mounting screws and lifting it away in two halves.
→ Then remove the control stalks.

3-C1

STEP C2: Remove the steering wheel. See *Chapter 11, Job 2*.

STEP C3: Remove control knobs.
→ This is the headlight switch.
→ Slide a small screwdriver into the hole in the

3-C3

knob's underside, and push up the locking tab. The knob will then slide free.

3-C4

STEP C4: In order to access the main dashboard fixing bolts, it is necessary to remove surface panels and mouldings. These will vary according to the model.
→ On some models it will be necessary to remove the fuse carrier and side air vents.
→ This is the fog light panel. Remove these panels as their fixings become accessible...

STEP C5: ...unclipping wiring plugs as you go...

3-C5

STEP C6: ...and revealing dashboard or, as in this case, upper fascia mounting screws.

3-C6

STEP C7: Remove the centre console - see *Chapter 14, Job 7*.

STEP C8: The centre section is usually a separate panel.
➔ Remove any control knobs then hunt for fixing screws.

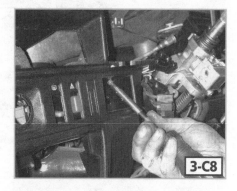

3-C8

STEP C9: To access this fixing, the demister knob must be lowered.

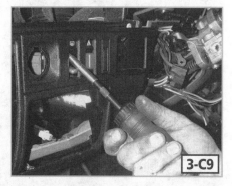

3-C9

3-C10

STEP C10: Lift the centre (or lower dash) panel out and undo all the wiring plugs..

STEP C11: This reveals yet more dashboard fixings.

3-C11

3-C12

STEP C12: Undo the glove compartment lower fixing screws.

STEP C13: The glove compartment upper fixings will be hidden away inside the compartment.

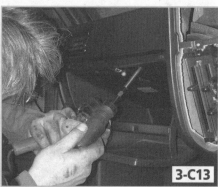

3-C13

3-C14

STEP C14: Lift the glove compartment out.
➔ Disconnect the interior light wire.

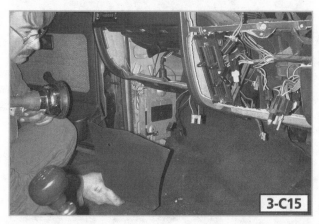

3-C15

STEP C15: Remove the footwell side trim panels.

□ **STEP C16:** It should now be possible to gain access to the main dashboard fixing bolts...

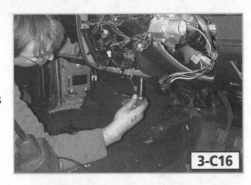

3-C16

□ **STEP C17:** ...including some behind the glove compartment...

3-C17

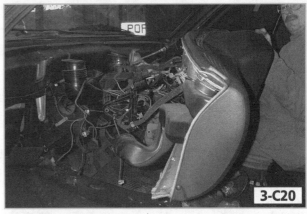

3-C20

□ **STEP C20:** When all the fixings have been removed, lift the dashboard up, slowly and carefully and out of the vehicle.
➔ Look out for any electrical connections you have missed, and, if necessary, unplug them.

Section D: Refitting.

3-C18

□ **STEP C18:** ...and others behind the instrument and other external panels.

□ **STEP C19:** Leave the central bolts until last.

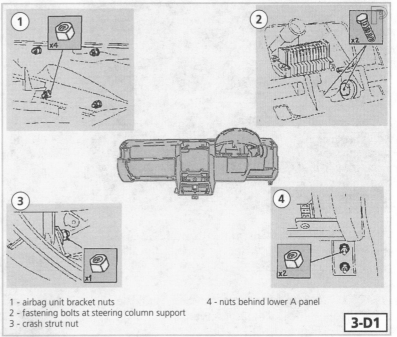

1 - airbag unit bracket nuts
2 - fastening bolts at steering column support
3 - crash strut nut
4 - nuts behind lower A panel

3-D1

3-C19

□ **STEP D1:** Generally, refitting is the reverse of removal.
➔ Fit the dashboard assembly with the airbag, then fit the airbag unit bracket and the wiring connector.
➔ The airbag unit bracket nut torque is 16 lbf.ft (22 Nm).
➔ The instrument panel to bulkhead nut torque is 16 lbf.ft (22 Nm).
➔ The lower bracket to floor torque is 4.5 lbf.ft (6 Nm).

☐ **STEP D2:** Referring to illustration *3-D1*, tighten the bolts at the steering column support (2), the crash strut nuts (3) and the nuts behind the lower panel (4).

SAFETY FIRST!

• Take extreme care not to short out any electrical connections, or to pinch any cables when reconnecting.

• Have a fire extinguisher ready when you reconnect the battery and stand by to disconnect the battery instantly if necessary.

• An electrical fire can easily be caused by careless reassembly!

JOB 4: SPEEDOMETER - *remove, refit.*

IMPORTANT NOTE: This Job does not apply to vehicles with an LCD instrument panel.

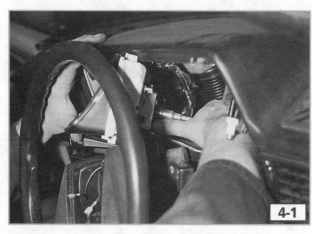

☐ **STEP 1:** Remove the instrument panel and unclip the speedometer cable from the back of the speedometer - see *Job 3, Section A*.

☐ **STEP 2:** Working from the engine bay, pull the cable through the bulkhead/firewall.

☐ **STEP 3:** Unscrew the cable from its connection on the transmission casing.

☐ **STEP 4:** Replace as the reverse of removal.
➜ Make sure the rubber grommet is seated correctly in the bulkhead.

JOB 5: WINDSCREEN WIPER MOTOR - *remove, refit.*

☐ **STEP 1:** Before dismantling, disconnect the battery negative (-) earth/ground terminal. See *Fact File: Disconnecting the Battery* BEFORE doing so!

☐ **STEP 2:** Mark the position of the wiper blades on the windscreen.

TOP TIP!

• Stick masking tape to the screen, indicating where the wiper blades lie when parked.

☐ **STEP 3:** Lift the end caps, remove the retaining nuts and lift the wiper arms off their spindles.

☐ **STEP 4:** Undo the plastic scuttle trim nut.

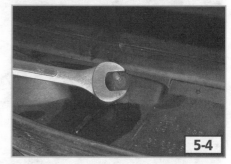

☐ **STEP 5:** If the model you are working on has a plastic scuttle trim, remove it then remove the plastic tray underneath. Here, a clip is being levered out with a specialist trim clip removal tool.

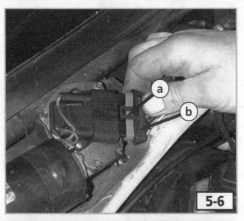

STEP 6: Lift the lock tab (**a**) and disconnect the wiring plug (**b**) from the motor.

5-6

5-7

STEP 7: Undo the three mounting bolts and remove the motor.
➔ The motor can now be disengaged from the linkage (arrowed) by removing the three screws.

EARLY CAVALIER

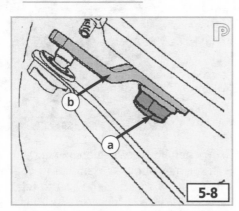

5-8

STEP 8: Undo the bolt (**a**) connecting the crank mechanism (**b**) to the motor.

STEP 9: Lift the lock tab (**a**) and disconnect the wiring plug (**b**).
➔ Unclip the wiring harness.
➔ Remove the three fixing bolts (arrowed) and remove the motor.

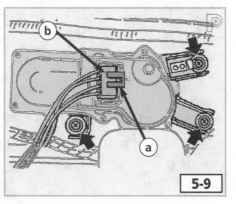

5-9

REPLACEMENT

STEP 10: Refit as the reverse of removal.
➔ Replace the wiper arms in the correct positions, according to the masking tape marks you put on the screen.
➔ Operate the wipers (with screen wet), to make sure the sweep of the wiper blades is correct.

JOB 6:	TAILGATE WIPER MOTOR - *remove, refit.*

STEP 1: Before dismantling, disconnect the battery negative (-) earth/ground terminal. See *Fact File: Disconnecting the Battery* BEFORE doing so!

STEP 2: Mark the position of the wiper blade on the rear screen.

TOP TIP!

• Stick masking tape to the screen, indicating where the wiper blade lies when parked.

STEP 3: Lift the end cap, remove the retaining nut and lift the wiper arm off its spindle.

6-3

STEP 4: Remove the rubber or plastic grommet from the wiper spindle.

6-4

STEP 5: Remove the spindle securing nut.

6-5

STEP 6: Lift the tailgate, and remove the inner trim panel – see *Chapter 14, Interior, Trim*.

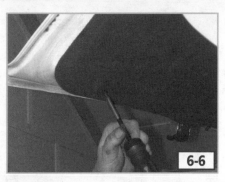
6-6

TOP TIP!

- The best tool to use here is a trim clip removal tool.
- A screwdriver often snaps or distorts clips.

STEP 7: If the model you are working on has a contact switch on its tailgate, undo the wiring connections for the wiper motor from it.

6-7

→ If not, undo the wiring connector on the motor.
→ Unclip the motor wiring wherever it is secured to the tailgate.

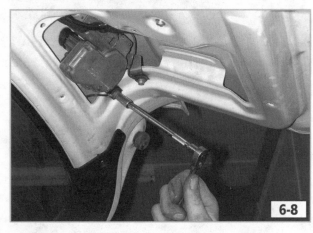
6-8

STEP 8: Undo the motor's mounting bolts.
→ Lift the motor from the tailgate.

STEP 9: Replace as the reverse of removal.

JOB 7: RADIO AERIAL/ANTENNA - *remove, refit.*

STEP 1: Before dismantling, disconnect the battery negative (-) earth/ground terminal. See *Fact File: Disconnecting the Battery* BEFORE doing so!

STEP 2: FRONT WING-MOUNTED AERIAL/ANTENNA: Pull the radio from the dash, and disconnect the radio aerial input plug from the back.

7-2

STEP 3: Remove the relevant wheel and remove the inner wheelarch liner.
→ If the aerial is being renewed, cut through the cable near to the aerial.
→ Securely fix the radio-end of the new cable to the aerial-end of the old one.
→ As the old cable is withdrawn, the new one can be carefully fed into place.

STEP 4: Inside the engine bay pull the aerial cable through its rubber grommet in the bulkhead. Unclip it along its run through the engine bay.

IMPORTANT NOTE: Depending on which model you are working on, you may have to remove the plastic scuttle trim.

STEP 5: Remove the aerial mounting nut(s).
→ If the aerial is an electric-powered model, there will be more fixings behind the wheelarch liner.

7-5

→ There may also be a clamp bracket around the aerial stem. If so, undo the bolt to release the aerial.

STEP 6: Draw the aerial down from the wing cavity, and pull the aerial lead through the inner wing.

STEP 7: ROOF MOUNTED AERIAL/ANTENNA: Remove the trim from the rear quarter pillars.

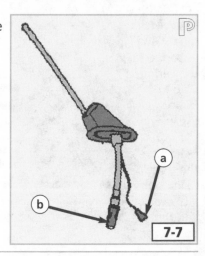
7-7

→ Pull down the rear edge of the headlining.
→ Disconnect the antenna amplifier power supply (a) and the aerial cable connection (b).
→ Undo the mounting bolt, and remove the aerial.

7-8

7-9

☐ **STEP 8: REAR WING MOUNTED ARIEL/ANTENNA:** Remove the rear side trim panel - see *Chapter 14, Interior, Trim*.

☐ **STEP 9:** Undo the aerial lower end bolt.
→ Disconnect the earth and co-axial leads.

☐ **STEP 10:** The aerial may now be lifted out.

☐ **STEP 11:** Replace as the reverse of removal.

IMPORTANT NOTE: Some vehicles have a power-retractable aerial, mounted on the rear wing. This should be

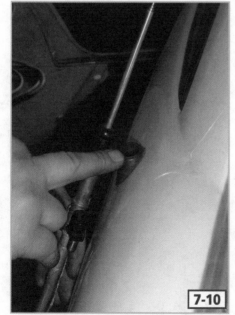

7-10

treated as a roof-mounted unit, as it unplugs from its power and aerial connections. It is accessible after removing the left-hand rear light cluster - see *Job 8*.

JOB 8: LIGHTS AND FUSES - *replace*.

CONTENTS

Section A: Headlight and sidelight bulbs - replacement.
Section B: Front indicator bulbs – replacement.
Section C: Indicator side repeater - replacement.
Section D: Headlight and front indicator units - replacement.
Section E: Front foglight bulbs - replacement.
Section F: Rear light bulbs and units - replacement
Section G: Number plate light.
Section H: Interior and luggage compartment lights.
Section I: Dashboard bulbs.
Section J: Fuses and relays.

TOP TIP!

• Whenever a light fails to work, check its fuse before replacing the bulb.
• A blown bulb often causes a fuse to blow at the same time.
• See *Section J: Fuses and relays*.

There are detail differences between the light units fitted to the various models and variants covered by this manual, but the principles of removing and replacing them are the same in all instances.

IMPORTANT NOTE: After removing and replacing any headlight components, have the headlights aligned with a beam adjuster before using the vehicle at night.

SAFETY FIRST!

• Before dismantling, disconnect the battery negative (-) earth/ground terminal. See *Fact File: Disconnecting the Battery* BEFORE doing so!

• Beware! A bulb that has recently been ON may be extremely hot and cause a burn.

Section A: Headlight and sidelight bulbs - replacement.

TOP TIP!

• If you touch a halogen headlight bulb with bare fingers you will shorten its life, so handle it with a piece of tissue paper.
• If the bulb is accidentally touched, wipe it carefully with methylated (mineralised) spirit.

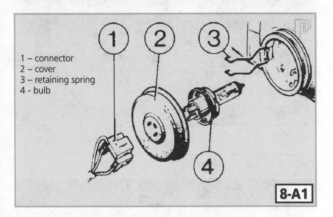

1 – connector
2 – cover
3 – retaining spring
4 - bulb

8-A1

❑ **STEP A1:** These are the components of a typical headlight bulb assembly.
➜ Pull off the headlight electrical multi-plug connector.

❑ **STEP A2:** Pull the rubber cover (**a**) from the back of the headlight unit.
➜ On early vehicles, turn the retaining ring anti-clockwise to release it, then remove the bulb.

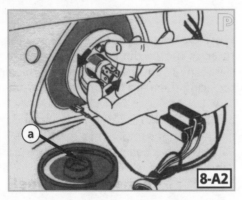

8-A2

❑ **STEP A3:** On later vehicles, squeeze the bulb retaining spring (arrowed), and swing it out of the headlight - one side is hinged.

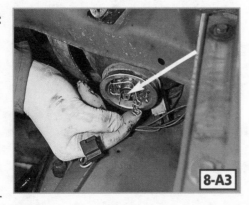

8-A3

➜ Take the bulb out of the reflector.
➜ When fitting the new bulb, make sure the locating lug on the bulb fits correctly in its recess in the reflector.

IMPORTANT NOTE: Most of the vehicles covered by this manual have halogen headlight bulbs. In fact, even older vehicles not specified with them originally are likely to have been converted since. DO NOT touch the bulb glass of halogen bulbs!

8-A4

❑ **STEP A4:** To remove the side light bulb, twist the holder to the left to release it from the reflector.
➜ Push and twist the bulb to release it.

Section B: Front indicator bulbs – replacement.

MODELS WITH BULB ACCESSED FROM INSIDE ENGINE BAY

8-B1

❑ **STEP B1:** On models from 1991 and up to 1984 the front indicator bulb holders, located next to each headlight, are simply twisted anti-clockwise to release them.
➜ Remove the bulb by pressing down slightly and twisting.

STEP B2:
On 1984 to '91 vehicles, disengage the bulb holder from the light by squeezing the long

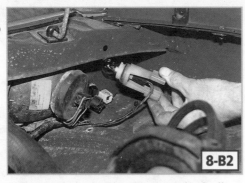

8-B2

'ears' and turning anti-clockwise. Remove the bulb from the holder by pushing slightly and twisting.

MODELS WITH SEPARATE FRONT INDICATOR ASSEMBLY

STEP B3: The front indicator assembly is released from inside the engine bay by pulling on the looped retaining clip...

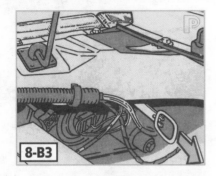

8-B3

STEP B4: ...which allows the assembly to be pulled forward; the bulb holder is turned anti-clockwise to release it, while the bulb is of the 'push-and-turn' type.

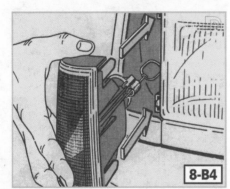

8-B4

Section C: Indicator side repeater – replacement.

STEP C1: There are two variants of side repeater used on models covered by this manual.
→ The early type has a cover which is removed by being rotated a quarter-turn to the left anti-clockwise and pulled free.

8-C1

TOP TIP!

STEP C2: The push-fit bulb is quite deeply recessed in the holder, leaving little on which to grip – if the bulb is defective and has to be replaced, apply a

8-C2

slip of masking tape to the bulb-glass, squeeze the excess into a small 'handle', as shown, and use this to withdraw the bulb.

STEP C3: The later type clips in. Slide a flat, thin blade under the cover, and push back the locking tab (arrowed).
→ Pull the repeater free of the wing.

8-C3

STEP C4: Pull the lens off the bulb holder to access the bulb.
→ To remove the bulb, turn anti-clockwise and pull free.

8-C4

Section D: Headlight and front indicator units - replacement.

STEP D1: Remove the radiator grille - see **Chapter 13, Bodywork**. On some of the models covered by this manual, the grille is integral with the front bumper - which should be removed.

STEP D2: Disconnect the wiring both to the headlight bulb and the sidelight - see **Section A**.

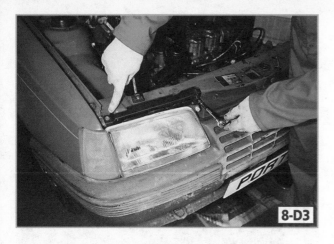

8-D3

STEP D3: Remove the mounting bolts.

STEP D4:
Some models have mounting bolts on the lower edge of the unit.
➜ This typical later vehicle requires the

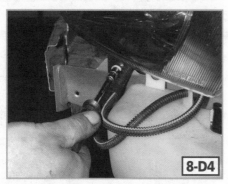

8-D4

removal of the bumper to remove the light unit.

STEP D5: Others have a clip location, which may require persuasion with a lever.

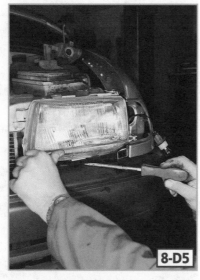

8-D5

STEP D6: Lift the headlight unit away from the front panel. Some come as a dedicated light unit...

8-D6

STEP D7:
...while others come with the indicator unit either attached or integral.
➜ In this case, remove the indicator bulb as you lift the headlight unit free.

8-D7

STEP D8: If you are replacing an indicator unit which is separate from the headlight, remove the bulb holder - see *Section C*.
➜ Then remove the mounting screw through a hole in the wing channel.

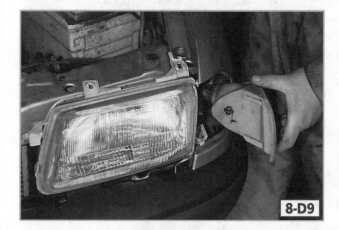

8-D8

STEP D9: ...then pull the unit from the front panel.

STEP D10: If the indicator unit is integral to the headlight, then the entire unit will have to be replaced to renew it.

8-D9

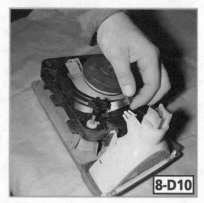

8-D10

→ But if it merely was removed attached to the headlight, lay the assembly flat, and remove the securing clip...

→ ...before sliding the two units apart.

❏ **STEP D11:** Refitting is the reverse of removal.

Section E: Front foglight bulbs - replacement.

Several types of foglight assembly are fitted to the models covered in this manual.

→ In most cases, the bulb is accessed from the rear of the unit.

❏ **STEP E1:** Where fitted, disconnect the wiring plug from the back of the foglight.

❏ **STEP E2:** Turn the rear cover anti-clockwise to release it.

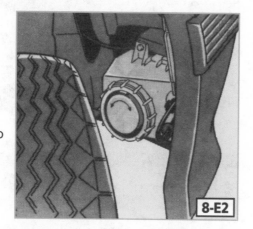

8-E2

❏ **STEP E3:** Squeeze the clip and open it to release the bulb. Disconnect the bulb's cable connection from the cover and remove the bulb.

❏ **STEP E4:** Reverse this process to refit the bulb.

Section F: Rear light bulbs and units - replacement.

❏ **STEP F1:** Open the access panel in the rear luggage compartment trim, to reach the light unit.

→ Release by turning a small button-catch, in most cases.

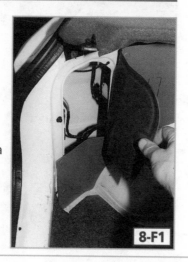

8-F1

❏ **STEP F2:** Unclip the bulb carrier from the lens body...

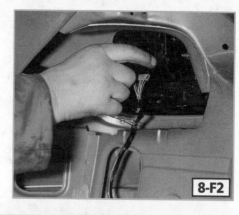

8-F2

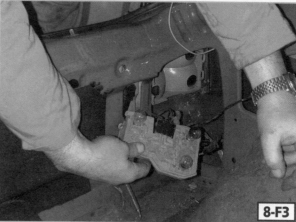

8-F3

❏ **STEP F3:** ...and pull it into the luggage bay.

❏ **STEP F4:** Locking tabs can be in a number of positions, depending on models.

→ This one is in the centre of the bulb holder cluster.

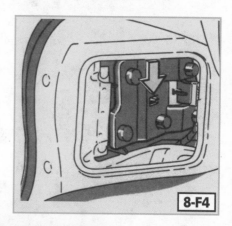

8-F4

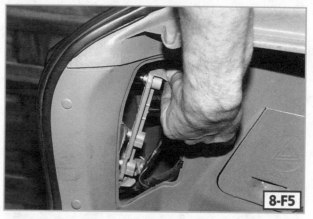

8-F5

❏ **STEP F5:** Where there is a clip top-and-bottom, it's easiest to release the top clip then angle the holder before lifting it out.

STEP F6: If any bulbs require replacement, twist and push, then pull them free of the carrier.

8-F6

LIGHT UNIT REPLACEMENT

STEP F7: Undo the mounting nuts from inside the aperture.

8-F7

STEP F8: Pull the lens body from the aperture.

STEP F9: Refitting is the reverse of assembly.

8-F8

Section G: Number plate light.

STEP G1: Carefully lever the number plate light holder out of the bumper.
→ If you are working on a model with steel bumpers, reach

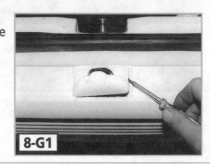

8-G1

underneath the bumper, and squeeze the retaining tabs together. Then push the light unit upwards out of the bumper blade.

STEP G2: Disconnect the wiring connection from the light unit.

8-G2

STEP G3: Lift the light body from the lens casing, by pressing the tab.
→ Change the bulb.

8-G3

Section H: Interior and luggage compartment lights.

STEP H1: There are, depending which model you are working on, lights in any of the following places: cabin, luggage bay, under the bonnet, glovebox. All the lights are of the same design: although there are variations in detail, they are all approached in the same way.

8-H1

STEP H2: Lever the light lens from the panel it sits in or diffuser.

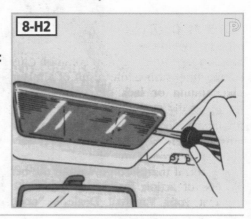

8-H2

STEP H3: Pull the bulb from its holder, and replace it if necessary.

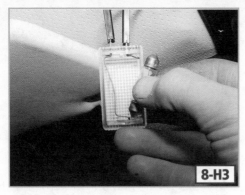
8-H3

STEP H4: If you are renewing the lens, undo the wiring connections.

STEP H5: Replace as the reverse of removal.

Section I: Dashboard bulbs.

IMPORTANT NOTE: This applies to both the analogue and LCD instrument displays.

STEP I1: Remove the instrument binnacle - see *Job 3*.

STEP I2: Each bulb is fitted into a small, black bulb holder. Twist the holder anti-clockwise and remove it from the binnacle's rear face.
➔ Change the bulb and replace the plug.

8-I2

Section J: Fuses and relays.

TOP TIP!

• If a complete 'set' of bulbs fails to operate (and especially if other electrical components fail at the same time) check the fuses before suspecting any other fault.
• If a replacement fuse blows, you probably have a circuit fault.
• If the fuse protecting the circuit has not blown, it may be that a relay had failed.

SAFETY FIRST!

• Make sure that all electrical circuits are switched off before removing or replacing a fuse.

• Never try to 'cure' a fault by fitting a fuse with a higher amperage rating than the one specified.

STEP J1: The fuses are colour coded, brown for 7.5 amp, red for 10 amp, blue for 15 amp, yellow for 20 amp, neutral or clear for 25 amp and green for 30 amp.
➔ A 'blown' or

8-J1

defective fuse is indicated by a break in the wire (arrowed) which may appear melted.

STEP J2: The fuse box is located to the right of the steering column, beneath a panel on the lower dashboard; the cover is retained by

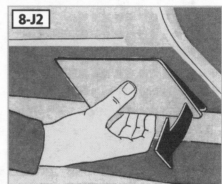
8-J2

plastic lugs and simply pulls off.

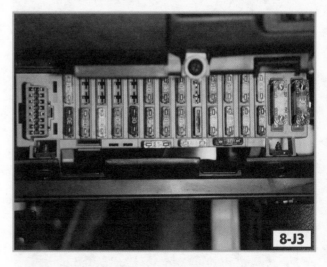
8-J3

STEP J3: The underside of the cover gives a pictorial description of the circuits protected by each fuse, which are numbered.

STEP J4: Relays are either placed adjacent to the fuses, or on the bulkhead, in the engine bay – see *Chapter 15, Wiring Diagrams*.

JOB 9: CENTRAL LOCKING - *replace components.*

SAFETY FIRST!

• Before working on any part of the electrical system, disconnect the battery's negative (-) terminal.

Section A: Typical layout.

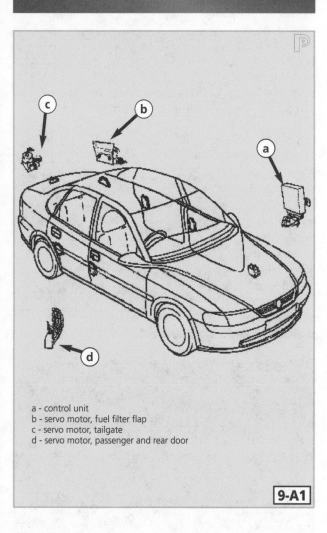

a - control unit
b - servo motor, fuel filter flap
c - servo motor, tailgate
d - servo motor, passenger and rear door

9-A1

☐ **STEP A1:** This is the arrangement of central locking components in a Vauxhall vehicle.
→ Each lock has its own servo motor, connected to a central control unit.
→ The driver's door lock motor is activated by a microswitch, clipped to the back of the door handle/lock mechanism. This door has 'command status' in the system.
→ The other motors are directly connected to the lock they serve.

Section B: Control unit – replacement

IMPORTANT NOTE: All of the illustration letters in this Section refer to illustration *9-A1*.

☐ **STEP B1:** Remove the right-hand kick panel trim - see *Chapter 14, Job 2*.

☐ **STEP B2:** Unclip and pull back the insulation, which will, depending on which model you are working on, either reveal the central locking control unit (**a**)...

☐ **STEP B3:** ...or the ECU. If this is the case, unclip the ECU from its support bracket and lay it carefully aside.
→ Then unclip the bracket from the panel, and remove it, along with the insulation behind it, to reveal the central locking control unit.

☐ **STEP B4:** Remove the control unit mounting bolts.

☐ **STEP B5:** Lift the unit out, and disconnect its wiring connector.

☐ **STEP B6:** Replace as the reverse of removal.

Section C: Servo motors – replacement.

FACT FILE
SERVO MOTORS.

• On Cavaliers to 1988 and all Calibra, separate front and rear door servo motors were used.
• From 1988, the servo motors were attached to the door locks.
• On the Vectra, the servo motors are integral with the door locks.

CAVALIERS TO 1988/CALIBRA

☐ **STEP C1:** Remove the trim panel from the door, tailgate or luggage bay, as appropriate - see *Chapter 14, Job 2*.

9-C1

STEP C2: Unscrew the motor (**c, d**) from its mounting (tailgate illustrated).
→ Remove the electrical plug (**a**).
→ Disconnect (**b**) from lock and remove.

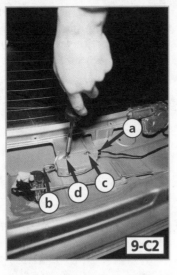

9-C2

STEP C3: Refit as the reverse of removal.
→ Door lock motors should be adjusted on refitting, so that the driver's door lock lever is flush against its stop.
→ The passenger and rear door lock levers, on the other hand, must be 2 mm off their stops.

STEP C4: Once the locks are correctly adjusted, tighten the motor mounting bolts and replace the trim panel.

CAVALIERS FROM 1988

STEP C5: Remove the door lock – see *Chapter 13, Job 8*.

STEP C6: The servo motor is held by two bolts.

STEP C7: Refit as the reverse of removal.

JOB 10: **DIESEL GLOW PLUGS -** *check, replace*.

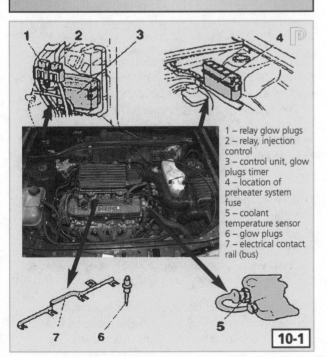

1 – relay glow plugs
2 – relay, injection control
3 – control unit, glow plugs timer
4 – location of preheater system fuse
5 – coolant temperature sensor
6 – glow plugs
7 – electrical contact rail (bus)

10-1

STEP 1: The glow plugs are the active components in the preheater system, the locations of components are set out in the diagram.
→ On certain models, the relay is situated inside the passenger compartment behind the right-hand kick panel.
→ On other models, it is adjacent to the left-hand suspension turret.
→ The third of the most common positions is in the rear, right-hand corner of the engine bay.

IMPORTANT NOTE: Unless you are experiencing trouble starting the engine, the glow plugs should demand no more attention than a clean-up when the injectors are checked.

STEP 2: Disconnect the battery negative (-) terminal. Undo the retaining nuts at the top of each glow plug.

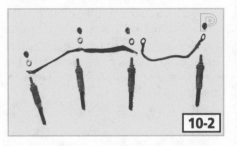

10-2

→ Remove the connecting rail/bus – sometimes with a flexible section, as here.

STEP 3: Clean away any dirt from around the plugs, then fully unscrew and remove them.
→ Examine the condition of each plug by wiping soot away and examining for erosion of the element sheath.
→ Check the internal resistance of each glow plug with an ohmeter, as shown. The reading should, if the plug is healthy, be 0.4 ohms.

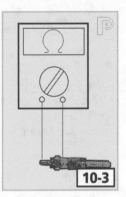

10-3

STEP 4: 11V GLOW PLUGS ONLY: To check the plug is functioning, connect it as shown to a 12 Volt battery. It should quickly begin to glow at the end.

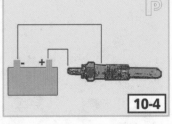

10-4

→ DO NOT prolong this test, or you will damage the plug.
→ Fit a complete set of new plugs if any one is in poor condition.
→ Refit as the reverse of removal, but note that overtightening a glow plug can damage it!
→ NOTE: More recent engines are fitted with 5V glow plugs. These cannot be tested without specialised equipment.

JOB 11: AIR CONDITIONING EQUIPMENT.

SAFETY FIRST!

• DO NOT open the pressurised pipework on the air conditioning system.
• The air conditioning system contains pressurised gas and specialist equipment is needed to extract it. It is illegal to discharge the gas to the air.
• The gas used in air conditioning systems is extremely harmful to the environment and may cause injury to the person carrying out the work if released in an uncontrolled manner.
• If air conditioning units are in the way of carrying out other work, whenever possible move units to one side with pipes still attached.
• If this is not possible, or there is a risk of damage to pipes or connections, have an air conditioning specialist de-pressurise the system so that you can dismantle it.

☐ STEP 1: • If the system needs to be drained for any reason, or if you suspect a leak, take the vehicle to a Vauxhall dealer or an

11-1

air-conditioning specialist such as Autoclimate - see *Specialists and Suppliers* at back of manual.
• After it has been re-assembled, the system will have to be tested (usually with inert nitrogen) and recharged by a specialist.

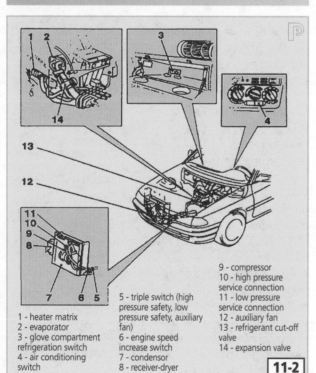

1 - heater matrix
2 - evaporator
3 - glove compartment refrigeration switch
4 - air conditioning switch
5 - triple switch (high pressure safety, low pressure safety, auxiliary fan)
6 - engine speed increase switch
7 - condensor
8 - receiver-dryer
9 - compressor
10 - high pressure service connection
11 - low pressure service connection
12 - auxiliary fan
13 - refrigerant cut-off valve
14 - expansion valve

11-2

IMPORTANT NOTE: After gas has been removed from the air conditioning system by a specialist, seal all open connections to stop any contamination from entering the system.

☐ STEP 2: You need to be able to identify the parts of the air conditioning system so that you can differentiate them from cooling or heater system components.
→ Bear in mind that certain items may be on the opposite side of a right-hand drive vehicle.
→ Neither the condensor, the compressor, the receiver nor the evaporator should be detached from their pipes.
→ If the condensor (7) should have to be moved to change the coolant radiator, support it carefully out of the way while the work is being carried out, taking care not to damage the tubes or fins.
→ If the compressor has to be taken from its bracket, do not disconnect any of the pipes. There are several different types of compressor and bracket.
→ Make a careful note of the way in which the compressor is mounted and take equally careful note of the system of bushes, shims and bolts which are used in the mounting and adjusting system.

JOB 12: ANTI-THEFT EQUIPMENT.

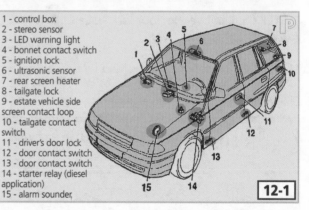

1 - control box
2 - stereo sensor
3 - LED warning light
4 - bonnet contact switch
5 - ignition lock
6 - ultrasonic sensor
7 - rear screen heater
8 - tailgate lock
9 - estate vehicle side screen contact loop
10 - tailgate contact switch
11 - driver's door lock
12 - door contact switch
13 - door contact switch
14 - starter relay (diesel application)
15 - alarm sounder,

12-1

☐ STEP 1: Some of the models covered by this manual are fitted with an alarm/immobiliser security system. These are the principal components.
→ The system is activated when the driver's door is locked, and deactivated when a door is unlocked.
→ It is triggered if a door, the luggage bay or the bonnet are opened, if the stereo head unit is tampered with, or if breaking glass triggers an ultrasonic sensor.

☐ STEP 2: The alarm system contains a self-diagnostic function.
→ As a result, all servicing on this system should be referred to a Vauxhall dealer, which will have the necessary equipment to diagnose faults, and reset the system when they are rectified.

IMPORTANT NOTE: Non-Vauxhall diagnostic equipment, the use of which is discussed in *Chapter 9*, is not equipped to deal with this system.

CHAPTER 11: STEERING, SUSPENSION

*Please read **Chapter 2 Safety First** before carrying out any work on your car.*

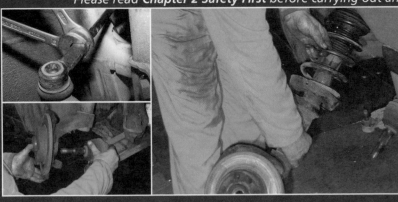

CONTENTS

JOB 1: SYSTEMS EXPLAINED.

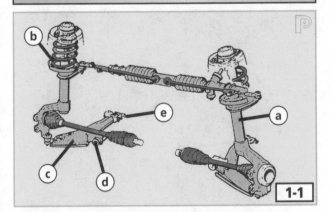

1-1

❒ **STEP 1:** The models covered by this manual all have independent front suspension of the McPherson strut type.

➜ There are, however, two fundamental differences between variants.

➜ Shown here is the arrangement fitted to Cavalier models from 1981 to 1988.

➜ Note that the strut (**a**) is a cast unit incorporating steering arms, into which the damper (**b**) is inserted.

➜ Note also the lower arm (**c**), which mounts to the chassis longitudinal members with a rearward-facing bolt (**d**) and a U-shaped clamp (**e**).

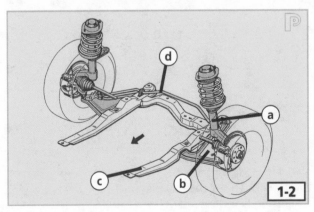

1-2

❒ **STEP 2:** On later Cavalier, all Vectra and Calibra models, however, the strut (**a**) and lower arm (**b**), although essentially the same as its predecessors, are mounted on a subframe (**c**).

→ On these later models, the subframe also carries mounting points for the anti-roll bar (**d**).

→ Although the illustration, *Step 1*, does not show an anti-roll bar, earlier Cavalier models equipped with one have it mounted on the bulkhead instead – see *Job 12*.

→ Note that the arrow in this diagram indicates the direction of the vehicle's front-end.

❏ **STEP 3**: The Vectra configuration differs in two ways:

→ The strut is composed of a platform spring/damper assembly, bolted to a hub carrier.

→ The steering arm is part of the hub casting.

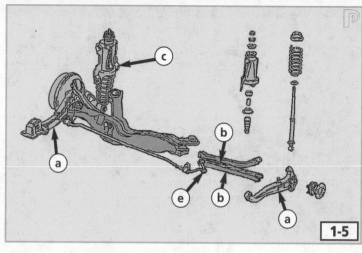

1-5

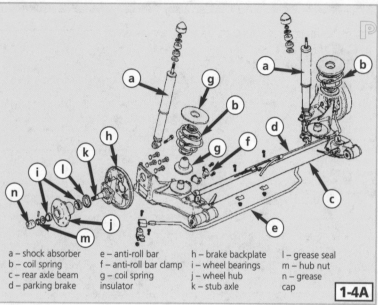

a – shock absorber
b – coil spring
c – rear axle beam
d – parking brake
e – anti-roll bar
f – anti-roll bar clamp
g – coil spring insulator
h – brake backplate
i – wheel bearings
j – wheel hub
k – stub axle
l – grease seal
m – hub nut
n – grease cap

1-4A

❏ **STEP 5**: This is the Vectra rear suspension.

→ The rear trailing arm (**a**) is attached to upper and lower arms (**b**).

→ The rear dampers are held in brackets (**c**).

❏ **STEP 4A**: This is the Cavalier and Calibra semi-independent suspension layout.

→ This type is fitted to vehicles with SOHC engines.

1-4B

❏ **STEP 4B**: This is the Cavalier and Calibra fully independent, semi-trailing arm suspension.

→ This type is fitted to vehicles with DOHC engines.

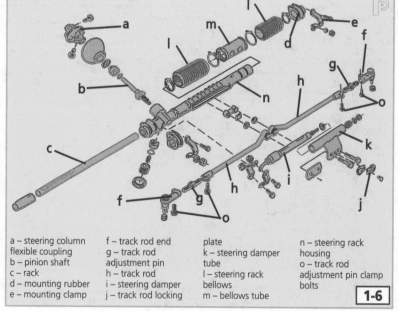

a – steering column flexible coupling
b – pinion shaft
c – rack
d – mounting rubber
e – mounting clamp
f – track rod end
g – track rod adjustment pin
h – track rod
i – steering damper
j – track rod locking plate
k – steering damper tube
l – steering rack bellows
m – bellows tube
n – steering rack housing
o – track rod adjustment pin clamp bolts

1-6

❏ **STEP 6**: These are the components of the 'sliding sleeve' steering system of the Cavalier and Calibra.

STEP 7: This is the conventional rack and pinion steering system of the Vectra.

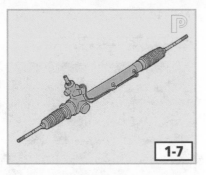

1-7

JOB 2: STEERING WHEEL – *remove, refit*.

STEERING WHEELS WITH AIR BAGS: See *Chapter 14, Interior and Trim*.

SAFETY FIRST!

• Read and make sure you understand the whole of *Chapter 14, Job 9* before starting to carry out any of the work described in this Job.

• Carry out all of the precautions and preparations described in *Chapter 14, Job 9*.

STEP 1: Before dismantling, disconnect the battery negative (-) earth/ground terminal. See *Chapter 10, Electrical, Dash, Instruments, Fact File: Disconnecting the Battery* BEFORE doing so!
→ **VEHICLES WITH DRIVER'S AIRBAG:** Follow the instructions in *Chapter 14, Job 11*, then move to *Step 4*.

STEP 2: Turn the steering wheel until the road wheels face forwards.

STEP 3: With a broad-bladed screwdriver, lever out the steering wheel centre.
→ On models with the horn button in the steering wheel centre, disconnect the wires.

2-3

STEP 4: Holding the steering wheel still, undo the large retaining nut.
→ Some models have a lock washer with tabs, which have to be bent back before tackling the nut.

2-4

STEP 5: Lift out the lock washer. If possible, this will be renewed when you put the steering wheel back on. Also remove the spring, if fitted.

STEP 6: Mark the steering wheel's position on the steering column with typists' correction fluid.

2-7

STEP 7: Pull the steering wheel from the column.

TOP TIP!

• The steering wheel is a tight fit on the column splines. It can usually be released after a few upwards blows (towards you) on alternate sides of the wheel.
• If, however, it proves stubborn, a twin-legged puller will work - there may be suitable holes in the steering wheel centre.

STEP 8: On the bottom of the steering wheel is either a horn contact ring or an integral horn ring/indicator cancellation unit (depending on model).
→ Remove the ring and check that it is not worn.
→ Replace it if necessary.

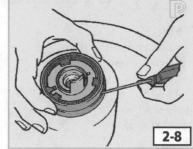

2-8

STEP 9: Refitting is the reverse of removal. Refit the air bag assembly, if necessary – see *Job 2* and *Chapter 14, Interior and Trim*.

JOB 3: TRACK ROD END - *replace*.

SAFETY FIRST!

• Whenever you disconnect the track rod end balljoint, its retaining locknut should be renewed. If you are renewing the joint, the replacement kit should include this item.

STEP 1: Raise the vehicle, support it on axle stands and remove the relevant road wheel.

VECTRA

STEP 2: The internally-threaded joint is fixed in position by a locknut – use one spanner to turn the nut and another to

3-2

hold the square-ended joint, working them in opposing directions.

TOP TIP!

• So that the 'tracking' of the front wheels can be retained, it is only necessary to slacken the locknut just enough to free the joint – half a turn of the 'nut' is usually sufficient.

• However, if the locknut has to be removed to allow a steering gaiter to be fitted, mark its position by applying a dab of paint, or wrap a piece of tape around the steering arm so that the nut, when replaced, will return to its original position.

• Alternatively, count the number of turns when removing the track rod end.

CAVALIER/CALIBRA

STEP 3: Cavalier models have no locknut securing the track rod end.

→ Instead, the joint **(a)** is retained on the adjustment pin

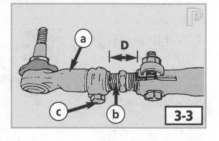

3-3

(b) by a clamp bolt **(c)**, acting in a captive nut on the clamp. This bolt must be undone, and the clamp worked loose.

→ Measure distance 'D' with a Vernier calliper to retain tracking settings.

ALL MODELS

STEP 4: Remove the nut securing the tapered pin to the forged steering arm of the hub assembly.

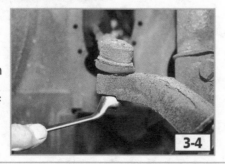

3-4

→ Clean up the exposed thread with a wire brush.

TOP TIP!

If the nut seizes on the thread:
• Try releasing fluid.
• It may be necessary to cut the nut from the pin using a hacksaw or nut-splitter.
• Alternatively, it is sometimes possible to use a ball joint removal tool in reverse, so it compresses rather than opens the joint. This can create enough friction in the joint to enable you to remove it.

STEP 5: A ball-joint removal tool is required to separate the tapered pin from its seat in the steering arm.

3-5

→ DO NOT try to free the pin by hammering it – damage to the pin and seat may result.

STEP 6: With the balljoint freed from its seat, unscrew the joint from the steering rack arm.

→ If you are working on an Vectra model,

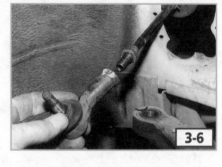

3-6

count the number of exposed threads between the central flats of the adjustment pin and the track rod end. This number will help you replace the joint in approximately the right position.

→ Unless the steering rack gaiter is being removed, do not disturb the locknut on Cavalier models.

CAVALIER/CALIBRA

STEP 7: If the locknut is to be moved:

→ Wrap a length of tape around the steering arm immediately behind and abutting the

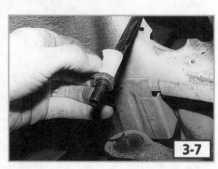

3-7

nut before unscrewing it.

→ This will provide a 'mark' up to which the nut can be replaced.

→ The 'tracking' of the front wheels will then be

okay to allow the vehicle to be driven to a specialist, who can set the tracking accurately.

→ Before screwing-on a new joint, lightly grease the threads on the arm to make the joint easy to remove in the future.

SAFETY FIRST!

• It is important to have the 'tracking' or wheel alignment checked after disturbing the steering joints.

• This is to ensure the handling of the vehicle remains safe and also to avoid uneven tyre wear.

JOB 4: STEERING RACK GAITER - *replace*.

The gaiters on the steering rack are vulnerable to wear. If they fail, road dirt and water will quickly enter and cause expensive, wear to the rack.

The two different forms of steering rack covered by this manual - see *Job 1, Steps 5 and 6* - require different approaches to replacing the rubber gaiters.

Section A: Vectra.

❏ **STEP A1:** The gaiters (there are two: left and right-hand) are located on the bulkhead behind the lower half of the engine;
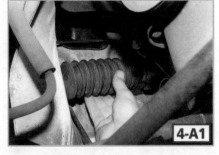
that shown is on the drivers' side of the vehicle, viewed from beneath the wheel arch.

→ With the steering turned on full lock check the gaiter, looking for splits, chafing, perishing etc..

→ Now turn steering fully in the opposite direction and check the other gaiter in the same way.

❏ **STEP A2:** Make sure the gaiter securing clips (arrowed) are secure and doing their job.

→ Clips can be metal bands either sprung or twisted into position, metal clips, or plastic

bands commonly known as 'cable-ties'.

If one or both of the gaiters are defective, replace them as follows:

→ First, slacken the bolts of the wheel nearest the defective gaiter, before raising and supporting the front of the vehicle.

→ Remove the wheel completely.

→ Remove the track-rod-end as described in *Part B* of this Job.

❏ **STEP A3:** The clips securing each end of the gaiter can now be removed, the small outer clip of the

type shown can be released with pliers and passed over the gaiter.

❏ **STEP A4:** The larger tensioned metal-band type can be undone by levering the free-end of the clip

upwards with a screwdriver, while the 'cable tie' type can simply be cut off with wire cutters.

❏ **STEP A5:** The gaiter can now be slid-off along the steering arm, and a new one slipped into place.

→ Note that no lubrication of the inner mechanism of the 'rack' is necessary on this particular model but take care not to allow dirt or grit to settle on the exposed parts.

STEP A6: Fit new clips, which should have been supplied with the new gaiter:
→ Replace the track-rod-end.

4-A6

→ Fit the wheel and lower the vehicle to the ground.
→ Finally, tighten the wheel bolts.

IMPORTANT NOTE: The steering 'track' will need to be checked by a specialist, otherwise poor steering response and rapid tyre wear can result.

Section B: Cavalier and Calibra.

Each track rod bolts to the steering rack, through a sliding collar with rubber gaiters on the outer end of it. The bolts which hold these on are retained by a locking plate, which must be replaced as you re-attach the track rods.

STEP B1: Raise up the passenger side front wheel and remove the wheel.

STEP B2: Disconnect the track rod arm:
→ Remove the locking plate(s).
→ Undo the bolts securing the track rod to the steering rack.
→ Remove the spacer plate and washers beneath.

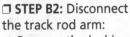

4-B2

STEP B3: If the vehicle has power steering, remove the fluid pipes from the passenger side of the steering rack assembly.
→ Seal the ends with plastic and tie them out of the way to the bulkhead, facing upwards.

STEP B4:
Remove the clamp securing the steering rack to the bulkhead, on the passenger side.
→ Slide the rubber bush that sits underneath it off the end of the rack assembly.

4-B4

STEP B5: Release the outer clips securing both rubber gaiters.

TOP TIP!

STEP B6: • Slide the sliding collar (**a**) and both rubber gaiters (**b**) from the passenger side end of the steering rack assembly.
• It may help to lubricate the part that was covered by the rubber bush with silicone lubricant.

a   b

4-B6

STEP B7: Remove the remaining securing clip(s), and replace the damaged rubber gaiter(s).

STEP B8: Refitting is the reverse of removal.
→ Renew bulkhead bolts, tie rod to steering gear lock plate and the power steering pipe 'O' rings (if fitted).
→ REMEMBER to top up the power steering fluid reservoir and bleed the system if fluid was lost during removing and reattach the power steering pipes (if present). Refer to *Job 7*.

JOB 5:	STEERING RACK – *remove, replace*.

On the models covered in this manual, two different forms of steering rack are used.

Section A: Vectra.

STEP A1: Raise up the passenger side front wheel and remove the wheel.

STEP A2: Remove items that prevent easy access to the steering rack.
→ Depending on model, these can include: air filter assembly, air box, intake piping and air mass meter housing, radiator expansion tank.
→ Secure pipework and wiring out of the way using string or cable ties.

STEP A3: Separate the track rod end balljoints from the hub carriers, using a balljoint breaker.

STEP A4:
Undo the pinchbolt on the clamp securing the steering column to the steering rack.
→ Depending on model, you may have to remove under-dash trim to do this.

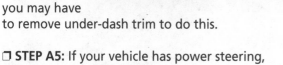
5-A4

STEP A5: If your vehicle has power steering, remove the fluid pipes from the passenger side of the steering rack assembly.
→ Cover the ends of these with plastic and tie them out of the way to the bulkhead with string or cable ties, facing upwards.

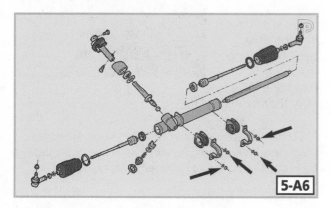
5-A6

STEP A6: Remove the four nuts and/or bolts (arrowed) securing the steering rack clamps to the bulkhead.
→ Remove the clamps.

STEP A7: Pull the steering rack forwards, then remove it from the engine bay sideways, through the large aperture in the right hand side inner wing.
→ This may have a large, flexible gaiter on it.

STEP A8: Replacement is the reverse of removal.
→ HOWEVER, remember to set the steering wheel and road wheels straight before tightening the pinch clamp on the steering column.
→ Use new nuts/bolts on the steering rack-bulkhead clamps, tightening them to the recommended torque.
→ You must have the front wheel track professionally set after replacing the steering column.
→ Renew bulkhead bolts, tie rod to steering gear lock plate and the power steering pipe 'O' rings (if fitted).
→ REMEMBER to top up the power steering fluid reservoir and bleed the system if fluid was lost during removing and reattach the power steering pipes (if present). Refer to *Job 7*.

Section B: Cavalier and Calibra.

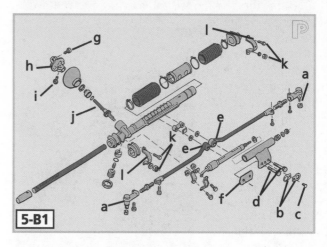
5-B1

STEP B1: These are the steering rack and track rod components. The letters in this Section relate to this drawing.

STEP B2: Remove items that prevent easy access to the steering rack.
→ Depending on model, these can include: air filter assembly, air box, intake piping and air mass meter housing, radiator expansion tank.
→ Tie pipework and wiring out of the way.

STEP B3: Separate the track rod end balljoints from the steering arms – see *Job 3*.

STEP B4: Remove the locking plate(s) by removing the screw (**c**).
→ Undo the bolts (arrowed and **d**) securing the track rods to the steering rack.
→ Remove the spacer plate (**f**) and washers from beneath.

5-B4

STEP B5: If the vehicle has power steering, remove the fluid pipes from the steering rack assembly.
→ Cover the pipe ends and tie them to the bulkhead, facing upwards.

STEP B6: Working inside the footwell:
→ Undo the pinchbolt

5-B6

(arrowed and **g**) on the clamp securing the steering column flexible coupling (**h**) to the steering column.
→ Also, remove the nut securing the flexible coupling fixing bolt (**i**) to the steering rack spindle (**j**). Depending on model, you may have to remove under-dash trim to do this.
→ Push the flexible coupling off the steering rack spindle and up the column.

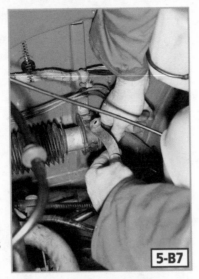

☐ **STEP B7:** Remove the four nuts and/or bolts (**k**) securing the steering rack clamps to the vehicle body. Remove the clamps (**l**).

5-B7

☐ **STEP B8:** Pull the steering rack forwards, then remove it from the engine bay sideways, through the large aperture in the right-hand side panel. This may have a large, flexible 'mud flap' on it.

☐ **STEP B9:** Refitting is the reverse of removal.
→ Set the steering wheel and road wheels straight before tightening the pinch clamp on the steering column.
→ Use new nuts/bolts on the steering rack-bulkhead clamps, tightening them to the recommended torque.
→ Renew the track rod bolt locking plates.
→ Have the front wheel track set on appropriate wheel tracking equipment.
→ Renew bulkhead bolts, tie rod to steering gear lock plate and the power steering pipe 'O'-rings (if fitted).
→ REMEMBER to top up the power steering fluid reservoir and bleed the system if fluid was lost during removing and reattach the power steering pipes (if present). Refer to *Job 7*.

JOB 6:	POWER STEERING PUMP – *remove, refit.*

The power steering pump on Vauxhall vehicles cannot be serviced. If it goes wrong it must be replaced, or exchanged for a reconditioned item from your Vauxhall dealer or specialist.

☐ **STEP 1:** Raise and secure the front of the vehicle off the ground. Remove the driver's side wheel.

☐ **STEP 2:** If necessary to access the pump, remove

the air box and air intake piping.

☐ **STEP 3:** It's inevitable that some fluid will escape from the pipework, no matter how careful you are.
→ Have a suitable receptacle at hand to catch spillage.

☐ **STEP 4:** Slacken then remove the auxiliary drive belt which drives the power steering pump.

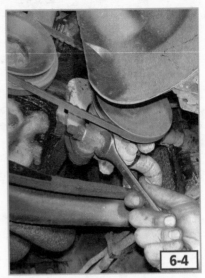

☐ **STEP 5:** Remove the drive pulley from the power steering pump.
→ Hold it still with a chain wrench, or tightly gripped drive belt, as you undo the three mounting bolts.
→ Withdraw the pulley.

6-4

☐ **STEP 6:** If necessary to access the power steering pump mounting bolts, remove the lower outer timing belt cover, referring to *Chapter 6, Part C, Job 1.*

☐ **STEP 7:** Clamp the fluid supply and feed hoses with brake pipe clamps or small G-clamps, to minimise the amount of fluid lost from the system.

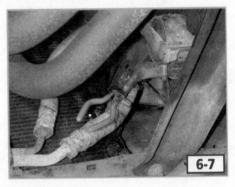

6-7

☐ **STEP 8:** Disconnect the fluid supply and feed hoses from the pump.
→ Seal the ends and tie out of the way with string or cable ties, with the ends pointing upwards if possible.

6-8

STEP 9:
Undo the pump mounting bolts.

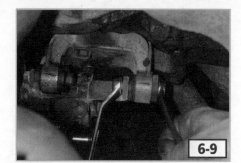

6-9

STEP 10: Remove the pump.

STEP 11: Refitting is the reverse of removal.
→ Replace the 'O' ring seal on the feed pipe union before attaching it to the pump.
→ Bleed and top up the power steering fluid – see *Job 7*.

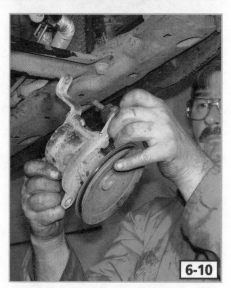

6-10

JOB 7: POWER STEERING SYSTEM – *filling, bleeding*.

STEP 1: Check the level of fluid in the power steering fluid reservoir. With the engine cold, it should be at the 'MIN' marking.
→ At full working temperature, it should be at the 'MAX' marking.

7-1

TOP TIP!

STEP 2: • Get the person helping you to start the engine. The fluid level will drop quickly, so you will have to quickly pour in enough to keep the 'MIN' mark covered.
• If the level drops below this, the pump will draw air into the system, taking it much longer to bleed!

STEP 3: Once the system has stopped drawing in fluid, turn the wheel slowly several times from lock to lock and return to the centre.

→ Do so several times, until air no longer bubbles into the fluid reservoir.

STEP 4: When there is no air left in the system, stop the engine. Make sure the fluid level is correct.

JOB 8: FRONT STRUT - *remove, refit*.

On the models covered in this manual, two different forms of front strut are employed.
→ The Vectra has a hub carrier bolted to the bottom of the damper/spring platform, making up a composite strut.
→ Cavalier and Calibra, on the other hand, have a single-piece strut, where the damper unit is an insert in the hub carrier casting.

SAFETY FIRST!

• Do NOT remove the centre nut at the top of the strut, inside the engine bay.

• Follow the instructions carefully and only remove the nuts holding strut to bodywork.

Section A: Vectra.

Refer to *Job 1, Steps 1-3* for an overview of the components covered here.

STEP A1: These are the components of the front strut assembly.
→ Some models may have slight differences, especially in components fitted above the coil spring. Take note of their positions when dismantling.

STEP A2: If the model you are working on has ABS, unclip the electrical cable for the wheel sensor from the strut.

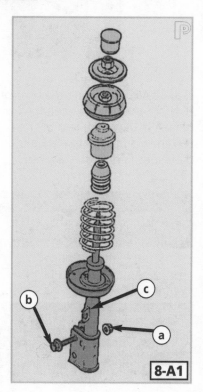

8-A1

STEP A3: Refer also to illustrations *8-A1*.
→ Undo the two nuts (**a**) and bolts (**b**) holding the strut assembly (**c**) to the hub carrier.

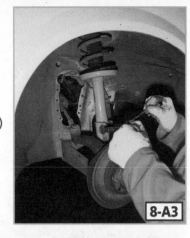
8-A3

STEP A4: Inside the engine bay, undo the two nuts holding the strut to the suspension turret.
→ Hold the strut with one hand as you do so, to prevent it from falling.

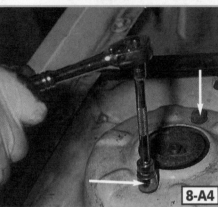
8-A4

STEP A5: Remove the strut assembly.

8-A5

STEP A6: Refitting is the reverse of removal.
→ NOTE that the bolt heads (illustration *8-A1, items* **b**) face towards the rear of the vehicle.
→ Renew all the nuts and bolts, and tighten them to their specified torque settings.

Section B: Cavalier/Calibra.

TOP TIP!
• The hub nut has to be undone – and it is extremely tight!
• Always loosen it while the vehicle is still on the ground.

STEP B1: Apply the parking brake, and loosen the nut with an extension bar, having removed the split pin from it first.

8-B1

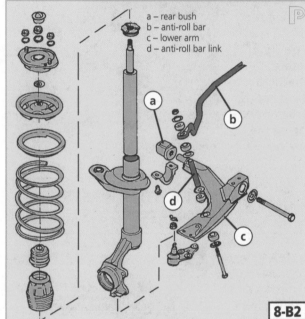
a – rear bush
b – anti-roll bar
c – lower arm
d – anti-roll bar link
8-B2

STEP B2: These are the components of the front strut assembly.
→ Some models may have slight differences, especially in components fitted above the coil spring. Take note of their positions when dismantling.

STEP B3: If the vehicle you are working on has ABS (anti-lock braking), disconnect the sensor, and unclip its wiring loom from the strut.
→ Tie it out of the way with string or a cable tie.

STEP B4: Remove the brake caliper from the hub carrier - see *Chapter 12, Brakes*. Tie it out of the way with wire or a strong cable tie.

STEP B5: Undo the nut on the track rod end balljoint connection to the steering arm.

8-B5

STEP B6: Using a balljoint breaker, remove the track rod end from the steering arm.

8-B6

STEP B7: Remove the hub nut and washer. See *Job 11*.

STEP B8: Lift the circular part of the locking clip (if applicable) and pull it out.
→ Undo the nut holding the bottom balljoint to the hub carrier. See *Job 14*.
→ Remove the bolt and lever the lower arm down, to clear the eye in the hub.

8-B8

STEP B9: Remove the outboard end of the driveshaft from the hub.
→ Use a soft-faced hammer, and don't hit it too hard, or the driveshaft CV joint will be damaged. See *Chapter 7, Transmission, Clutch*.

8-B9

STEP B10: Tie the driveshaft up, using wire or a strong cable tie.
→ If you let the driveshaft hang free the CV joint will be damaged.

STEP B11: Inside the engine bay, undo the two nuts holding the strut to the suspension turret.

8-B11

STEP B12: Hold the strut as you remove the last nut to prevent the strut from falling.

8-B12

STEP B13: To refit:
→ Have a helper offer the strut up into the suspension turret.
→ Locate it properly, then fit renewed nuts to secure it.

8-B13

❐ **STEP B14:** Grease the splines of the driveshaft, and refit the end into the hub.
→ Fit a new washer and nut, and tighten the nut.

❐ **STEP B15:**
Refit the lower arm balljoint to the hub carrier and tighten the nut to its specified torque

8-B15

settings. Secure it as follows:
→ If a self-locking nut, renew every time the nut is removed.
→ If a castellated nut, use a new split pin.
→ If a Vauxhall spring clip (arrowed) fit a new replacement.

❐ **STEP B16:** Refit the track rod end balljoint to the hub carrier, using a new self-locking or castellated nut, as applicable. The castellated nut may have a retention clip, which must be put in place. See *Step 12*.

❐ **STEP B17:** Refit the brake caliper - see *Chapter 12, Brakes*.

❐ **STEP B18:** If the vehicle has ABS brakes, refit the sensor to the hub carrier, and clip its wiring loom back onto the strut. See *Chapter 12, Brakes*.

❐ **STEP B19:** Fit the wheel, and lower the vehicle.

❐ **STEP B20:** Tighten the hub nut to its specified torque setting, and fit a new split pin.

SAFETY FIRST!

• **It is essential to have the tracking checked and adjusted by your Vauxhall dealer or wheel and tyre specialist, to avoid dangerous handling and uneven tyre wear.**

JOB 9: FRONT STRUT - *overhaul*.

SAFETY FIRST!

• **Leaking dampers guarantee MoT failure. Moreover, tired suspension components can have a marked effect on a vehicle's handling and compromise its safety.**

• **Dampers and springs MUST be replaced in pairs, front and rear. If only one side is renewed, it will have a significant effect on the vehicle's handling, and make it extremely dangerous to drive!**

The front dampers are integral parts of the front struts of the Vectra. If a damper is faulty then the whole strut assembly should be renewed. The front dampers fitted to the Cavalier and Calibra can be removed from the struts and renewed.

FACT FILE

INSPECT AND PREPARE DAMPERS

• You can tell if a damper requires replacement by pushing the rod up and down a couple of times. If you feel any jerkiness to its movement, or hear a bubbling sound, or it leaks fluid, it must be renewed. On the Vectra this entails renewing the strut assembly.

• New damper units need to be 'primed' before fitting. This is a simple process, involving depressing the damper rod and pulling it back out about five times. This draws oil into the reservoir and seals, ready to do its job.

• Pay attention to the top mounts. If there is play in the bearing, replace the unit. These take a hard pounding, and it is very annoying to go to the effort of replacing dampers and springs, only to find 'clunking' noises still coming from the front end once the job is done!

Section A: Vectra.

❐ **STEP A1:** Remove the front spring/damper assembly - see *Job 8*.

❐ **STEP A2:** Holding the spring/damper assembly firmly in a vice, compress the spring using coil spring compressors, such as the Sykes-Pickavant tool shown here.

9-A2

SAFETY FIRST!

• **If used incorrectly, spring compressors can cause serious injury.**

• **Place the compressor on opposite sides of the spring and covering as many coils as possible.**

• **Tighten and release them evenly on both sides, a couple of turns at a time.**

• **Stop when the spring comes clear of the top mount.**

STEP A3:
Remove the damper rod nut using a ring spanner, holding the rod itself still, using a spanner or Allen key, as applicable.

9-A3

STEP A4: Remove the upper stop (**a**), support bearing and nut (**b**), upper spring seat (**c**) and buffer (**d**).
→ Remove the spring and compressors.

STEP A5: Release the compressors evenly, a couple of turns on one and the same amount on the other.
→ Carefully release the compressors from the springs.
→ Remove the spring when the tension is fully released.

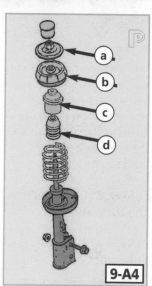

9-A4

STEP A6: Renew the struts, springs, bump stops or top mount assemblies as necessary.

STEP A7: Replace the buffer. Fully extend the damper rod.

STEP A8: Compress the spring if necessary, and carefully put it back in place on the damper unit.

STEP A9: Replace the top mount assembly.

STEP A10: Always fit a new damper rod locknut.

STEP A11: Release the spring compressors:
→ Make sure that the spring is seated correctly against the raised stops on the spring seats.

STEP A12: Refit the spring/damper assembly – see *Job 8*.

Section B: Cavalier/Calibra.

STEP B1: Remove the front strut – see *Job 8*.

STEP B2: Holding the strut firmly in a vice, compress the spring using coil spring compressors.

9-B2

STEP B3: Inside the engine bay, lever off the plastic cover (if fitted) from the top of the front suspension turrets.

STEP B4:
Remove the damper rod nut using a ring spanner.
→ Hold the rod from turning with a spanner or an Allen key, as applicable.

9-B4

STEP B5: Remove the top mount assembly and the spring top plate, along with its supporting ring.
→ Alternatively, remove the assembly complete, along with the spring, as shown here.

9-B5

TOP TIP!

• If the spring is not to be renewed, it can stay in the compressors until it is replaced but ONLY if it is to be immediately refitted.
• However, store the compressed spring extremely carefully and out of the way, as it is effectively a 'loaded weapon'.

STEP B6: Carefully release the compressors from the springs. Remove the spring.

STEP B7: Pull the rubber dust shield and bump stop off the damper rod.

9-B7

STEP B8: If the damper insert is to be renewed, the cover nut that holds it inside the strut has to be undone.
→ There are several ways to achieve this.
→ Either turn anti-clockwise, using a Stilsons wrench, or turn the strut upside down, clamp the nut and lever the strut round in the correct direction, using a tyre lever or similar between the strut shaft and a screwdriver in the hole where the track-rod-end balljoint is mounted.

9-B8

→ If you invert the damper, be aware that a leaking damper will probably lose oil. Be prepared to catch this or mop it up.

9-B9

9-B10

STEP B9: Remove the cover nut.

STEP B10: Remove the damper insert. If it is very badly worn, there may be oil in the tube.
→ If so, drain it out by turning the strut upside down over a suitable container.

STEP B11: Put the new insert into place and refit the cover nut.

STEP B12: Replace the bump stop and dust shield. Fully extend the damper rod.

9-B13

9-B14

STEP B13: Compress the spring if necessary, and replace it on the damper unit.

STEP B14: Replace the top rubbers and top mount assembly.
→ The spring top plate has a small hole in its upper face.
→ This should line up with the outboard mounting points, where the hub carrier is attached.

STEP B15: Replace the damper rod nut, fit a new one.

9-B15

STEP B16: Release the spring compressors, making sure that the spring is seated correctly in the contours of the spring bed and spring top.

STEP B17: Refit the spring/damper assembly – see *Job 8*.

9-B16

JOB 10: FRONT HUB CARRIER - *replace.*

Because the Cavalier's hub carrier is an integral part of its front strut, this Job applies only to the Vectra.

STEP 1: The hub nut is extremely tight, so loosen it while the vehicle is still on the ground.
➜ Apply the parking brake, and loosen the nut with an extension bar, having removed the split pin from it first.

10-1

STEP 2: Raise the vehicle, remove the wheel and remove the hub nut and its washer.

STEP 3: If the vehicle you are working on has ABS, disconnect the sensor, and unclip its wiring loom from the strut. Tie it out of the way.

10-3

STEP 4: Remove the brake caliper from the hub carrier - see *Chapter 12, Job 2*. Tie it out of the way with wire or a strong cable tie.

STEP 5: Remove the brake disc - see *Chapter 12, Job 3*.

10-5

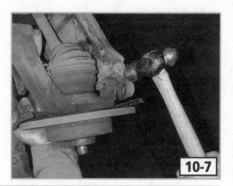

10-6

STEP 6: Undo the pinch bolt holding the bottom balljoint to the hub carrier.

STEP 7: If necessary, use a ball joint splitter or strike the hub carrier balljoint surround as shown to release the balljoint taper.

10-7

➜ Striking the surround distorts the hole and frees the taper.

❒ **STEP 8:** Remove the bolt and lever the lower arm down, to remove the balljoint spindle.

10-8

❒ **STEP 9:** Undo the nut on the track rod end balljoint's connection to the hub carrier.
➜ Using a balljoint breaker, remove the track rod end from the hub carrier.

❒ **STEP 10:** Undo the two nuts and bolts holding the spring/damper assembly to the hub carrier.
➜ NOTE that the bolts face towards the rear of the vehicle.

TOP TIP!

❒ **STEP 11:** Pull the hub outwards, off the driveshaft.
• If it is unwilling to leave the driveshaft

10-11

splines – it is often tight - tap the end of it gently with a soft-faced hammer but take great care not to damage threads or splines.

❒ **STEP 12:** Tie the driveshaft up, using wire or a strong cable tie. If you let it dangle, its CV joint will be damaged.

❒ **STEP 13:** Before reassembly, grease the splines of the driveshaft, and refit the end into the hub.
➜ Fit a new washer and nut, and tighten the nut.

❒ **STEP 14:** Refit the lower arm balljoint to the hub carrier and tighten the new castellated nut to its specified torque settings.
➜ Secure it with a new split pin.

❒ **STEP 15:** Refit the two nuts and bolts holding the spring/damper assembly to the hub carrier and tighten them to their specified torque settings.

❒ **STEP 16:** Refit the track rod end balljoint to the hub carrier, using a new nut. Tighten it to its specified torque setting.

❒ **STEP 17:** Refit the brake caliper, referring to *Chapter 12, Brakes*.

❒ **STEP 18:** If the vehicle has ABS, refit the sensor to the hub carrier, and clip its wiring loom back onto the strut.

❒ **STEP 19:** Fit the wheel, and lower the vehicle.

JOB 11: **FRONT WHEEL BEARINGS -** *replace.*

In this Job, 'hub carrier' refers, if you are working on an Cavalier or Calibra model, to the integral front strut assembly.

❒ **STEP 1:** Remove the hub – see *Job 10*.

❒ **STEP 2:** Remove the brake disc from the hub - see *Chapter 12, Job 3.*

❒ **STEP 3:** Place the hub carrier in a vice or press, with wooden blocks (arrowed) supporting the hub.
➜ Position a drift, such as a suitably sized socket against the inboard rim of the hub.

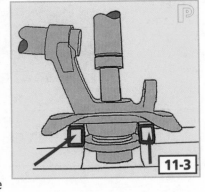
11-3

➜ Close the vice, to press the hub out of the hub carrier.
➜ The inner bearing race outer half will come away with the hub.

❒ **STEP 4:** Remove the inner bearing race outer half with a bearing puller or carefully hammer it out with a suitable drift.

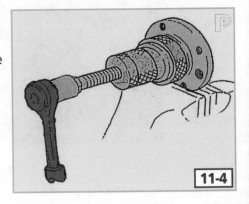
11-4

STEP 5: Undo the mounting screws and remove the brake disc shield from the hub carrier.

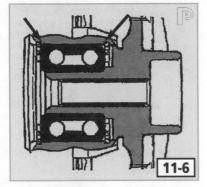

STEP 6: Using circlip pliers, remove the two circlips holding the bearing in the hub carrier (arrowed).

11-6

STEP 7: Either remove the bearing from the hub carrier with a bearing puller, or drive it out in a vice, using a suitably sized tube to drift it out, in the direction of the arrow.

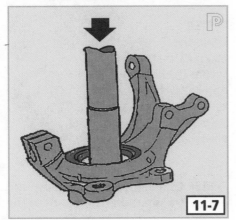

11-7

STEP 8: Clean both the hub and hub carrier. Check them for wear.
→ If either are worn or damaged, replace.

STEP 9: To reassemble:
→ Put a new circlip in place in the hub carrier.
→ Position it so that its 'open' part aligns with the groundward side of the hub carrier.
→ Oil the outside of the bearing, to make it easier to fit.

STEP 10: Using a suitably sized tube (1, 2) and a press or large vice, press the bearing into place, in the direction of the arrow. Be very careful that the bearing goes in straight!

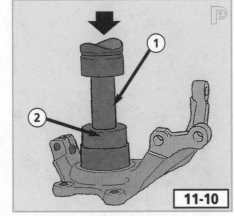

11-10

STEP 11: With the bearing butted up against the circlip, put a new inner circlip in place, again aligning its 'open' section with the groundward side of the hub carrier.

STEP 12: Refit the brake disc shield.

STEP 13: Press the hub into the hub carrier. Place a suitably sized socket or tube over the inner track of the bearing before placing the assembly in a vice - this supports it, preventing the pressure from driving the inner track out, as the hub is a tight fit.

If you are working on a Cavalier or Calibra model, it is a great help if someone can hold the strut as you do this, leaving you to concentrate on making sure everything aligns correctly.

STEP 14: Refit the brake disc, as described in **Chapter 12, Brakes**.

STEP 15: Refit the hub carrier or front strut assembly, as described in **Jobs 8** and **10**.

> **JOB 12:** **FRONT ANTI-ROLL BAR AND BUSHES** – *replace*.

On the models covered in this manual, most have anti-roll bars.
→ On the Vectra, The anti-roll bar is clamped to the subframe and is attached via drop links to the strut support tube.
→ The Cavalier has an anti-roll bar which runs between the lower suspension arms, behind the engine.
→ The two types of anti-roll bar need different approaches to replacing them, so we will deal with them separately.

Section A: Vectra.

STEP A1: Remove the front subframe – see **Job 16**.

STEP A2: Mark the drop links (a) so that they can be replaced on the same side, then unbolt them.
→ Use a spanner on the flats to prevent the drop link turning.

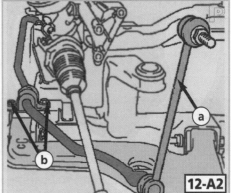

12-A2

□ **STEP A3:** Undo the clamp bolts (*12-A2, items b*).
→ The anti-roll bar can now be lifted away.

□ **STEP A4:** Regardless of their condition, it is worth renewing the bushes while you have the opportunity!
→ The cuts in the bushes must be facing forward when fitted to the anti-roll bar.

□ **STEP A5:** Replace as the reverse of removal

Section B: Cavalier pre-1988.

□ **STEP B1:** Raise the front of the vehicle.

□ **STEP B2:** Disconnect the anti-roll bar drop links from the suspension control arms.

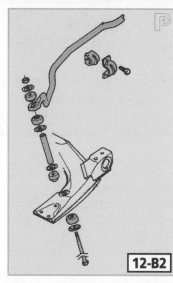

□ **STEP B3:** Undo but do not remove the control arm support bolts.
→ This allows clearance for removing the anti-roll bar.

□ **STEP B4:** The anti-roll bar can now be withdrawn.

□ **STEP B5:** Reassembly is the opposite of removal.
→ Distance 'A' should be 38 mm.

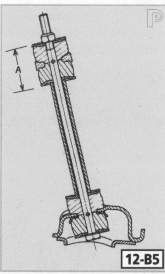

Section C: Cavalier 1988-on and Calibra.

In order to remove the anti-roll bar it is necessary to lower the rear end of the subframe. This necessitates supporting the engine and transmission from above, using an engine crane or support as described in *Chapter 7, Job 1.*

□ **STEP C1:** Support the engine and raise the front of the vehicle.

□ **STEP C2:** Undo the locknuts that secure the anti-roll bar to the lower suspension arms.

□ **STEP C3:** Remove the nuts securing the engine/transmission rear mount to the subframe.

□ **STEP C4:** Take the weight of the subframe using a jack and length of timber.

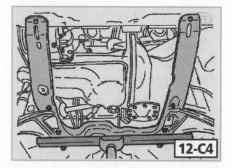

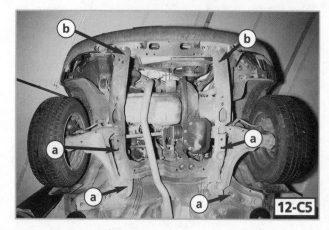

□ **STEP C5:** Remove the centre and rear subframe mounting bolts (**a**) and slacken the front bolts (**b**).

STEP C6: Lower the rear end of the subframe enough to access the anti-roll bar clamp bolts (arrowed) and remove them.

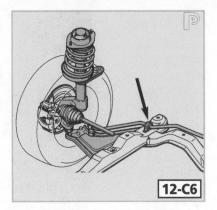

12-C6

STEP C7: The anti-roll bar can now be removed through the right-hand wheelarch.

12-C8

STEP C8: Refitting is the reverse of removal.
→ Fit new bushes and locknuts.

STEP C9: Tighten the anti-roll bar ends fitting until distance (A) is 39mm.

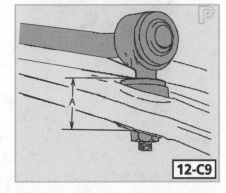

12-C9

JOB 13: FRONT LOWER ARM, BUSHES AND BALLJOINT - *replace*.

All the models covered by this manual have similar front suspension layouts, in principal.
→ McPherson struts join a lower arm at a ball joint.
→ The lower arm's inner pivot point is its mounting either on the bodyshell or, in later models, on a subframe which doubles as an engine/transmission cradle.
→ The lower arm balljoint is secured by rivets. These have to be drilled out.
→ Replacement balljoints are supplied with

alternative bolts and locknuts.
→ If only the hub carrier balljoint is to be renewed, it is possible to do this with the lower suspension arm in situ. See *Section C, Step C4*.

> ### TOP TIP!
> • The hub nut has to be undone – and it is extremely tight!
> • Always loosen it while the vehicle is still on the ground

FACT FILE

LOWER SUSPENSION ARM TYPES:

STEP FF1: This is the lower suspension arm fitted to Cavaliers 1988-on and all Calibras.

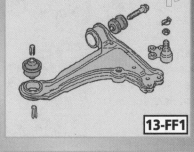

13-FF1

STEP FF2: This is the lower suspension arm fitted to the Vectra. It is similar to that fitted to pre-1988 Cavaliers.

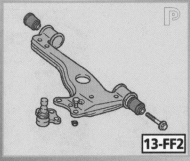

13-FF2

Section A: Lower suspension arm removal and refit.

STEP A1: Disconnect the hub carrier from the lower suspension arm – follow *Job 10, Steps 1-8*.

STEP A2: On pre-1988 Cavaliers, the lower suspension arm bushes locate in a pressing bolted to the chassis.
→ The lower suspension arm and its carrier pressing can be removed together.

STEP A3: Remove the front lower suspension arm bolt...

13-A3

☐ **STEP A4:** ...and withdraw it.

13-A4

☐ **STEP A5:** Remove the rear fixing bolt.
➔ This is the vertical bush/bolt fitted to late Cavaliers and all Calibras.
➔ The rear bush/bolt arrangement of pre-1988 Cavaliers and all Vectras is horizontal.

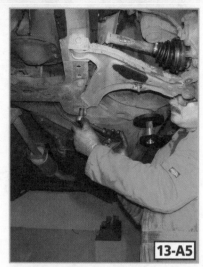

13-A5

☐ **STEP A6:** Pull or lever the lower suspension arm off...

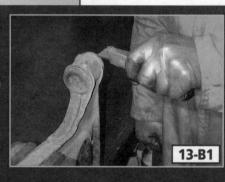

13-A6

☐ **STEP A7:** ...and retrieve the bush.

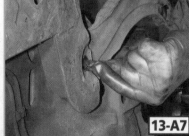

13-A7

TOP TIP!

• If they appear worn or damaged, take the opportunity to renew the suspension bushes. See *Job 9*.

☐ **STEP A8:** Clean the arm and its mounting(s).
➔ Check for cracks, corrosion and distortion in both these areas.
➔ Check the pivot bolt for wear.
➔ If it is scored or tapered, renew it.

13-A8

☐ **STEP A9:** Loosely refit the arm assembly, then tighten the bolts to their recommended torques. Start at the outboard end, and work inwards.

☐ **STEP A10:** Refit the wheel and lower the vehicle to the floor. Make sure the wheel bolts are tightened correctly.

Section B: Suspension arm bush renewal.

TOP TIP!

☐ **STEP B1:**
• Removing the old bushes is easier if you cut off one of the lips with a sharp craft knife as shown.

13-B1

• Lubricate the blade with soapy water – it makes a huge difference when cutting rubber!

☐ **STEP B2:** A suitably sized socket and vice can be used to press out the old bushes.

13-B2

STEP B3: The same goes for the large vertical rear bush – you need large-size sockets!

13-B3

STEP B4: Before pressing in the new bushes, coat them with a little silicon grease or soapy water – but don't use washing-up liquid because it may contain rust-inducing salt.

Section C: Balljoint renewal.

STEP C1: Original equipment balljoints are held by rivets. To remove these, the head has to be drilled off. ➔ Use a centre punch and accurately mark the centre of the rivet head.

STEP C2: Drill a pilot hole.

13-C2

STEP C3: Drill the head off the rivet.

13-C3

TOP TIP!

STEP C4: It's possible to renew the balljoint with the suspension arm in situ as shown. If you do this take care to keep the drill bit square to the job.

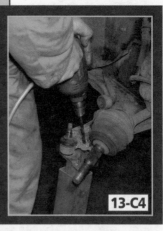

13-C4

STEP C5: This is the balljoint replacement kit, complete with bolts and locknuts.

13-C5

STEP C6: Take the opportunity to clean up the end of the suspension arm and apply paint to any bare metal.

13-C6

STEP C7: The balljoint flange fits into a sandwich between the main and upper ends of the suspension arm.

13-C7

STEP C8: Fit the bolts and locknuts.

13-C8

JOB 14: FRONT SUBFRAME - *replace.*

Section A: Cavalier and Calibra.

☐ STEP A1: Disconnect and remove the battery. See *Chapter 10, Electrical, Dash, Instruments, Fact File: Disconnecting the Battery* BEFORE doing so!

☐ STEP A2: Raise the front of the vehicle. If you are not using a vehicle hoist, secure the vehicle on axle stands. Remove the wheels.

☐ STEP A3: Remove the coolant expansion tank. See *Chapter 8, Cooling System*.

☐ STEP A4: Disconnect the oxygen sensor plug where fitted.

☐ STEP A5: The engine must be supported from above.
➜ This can be achieved using an engine crane.
➜ It is better to use an engine bar like this one from Sykes-Pickavant.

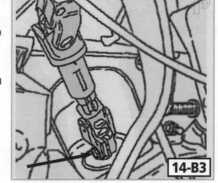
14-A5

☐ STEP A6: Remove the front section of the exhaust pipe - see *Chapter 9, Ignition, Fuel, Exhaust*.

☐ STEP A7: Split the lower suspension arm/hub balljoint – see *Job 13*.

☐ STEP A8: If oil cooler or refridgerant pipe brackets are attached to the subframe, remove them.

☐ STEP A9: Remove the transmission mount from the subframe – see *Chapter 7, Transmission, Clutch*.

☐ STEP A10: Place a trolley jack (**b**) under the subframe:
➜ Use a piece of strong wood or steel (**a**) to support both legs.
➜ Undo the nuts that hold the rear engine mounting to the subframe.

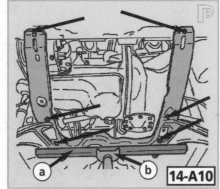
14-A10

➜ Undo the subframe mounting bolts (arrowed).

☐ STEP A11: Using the trolley jack, lower the subframe and suspension components, complete.

☐ STEP A12: Remove the anti-roll bar (**a**) and the lower arms (**b**) - see *Job 10*.

☐ STEP A13: Replacement is the reverse of removal. Use new nuts and bolts throughout, and tighten them to the specified torque settings.

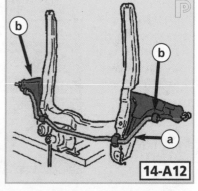
14-A12

Section B: Vectra.

☐ STEP B1: Disconnect and remove the battery. See *Chapter 10, Electrical, Dash, Instruments, Fact File: Disconnecting the Battery* BEFORE doing so!

☐ STEP B2: Turn the steering dead ahead, remove the ignition key and engage the steering lock.

☐ STEP B3: Undo the clamp bolt (arrowed) from the centre steering column shaft in the driver's footwell, and detach the shaft.

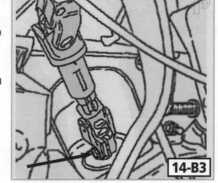
14-B3

☐ STEP B4: MANUAL TRANSMISSION: Slacken the gearshift rod clamp (**a**) and disconnect the shift rod. Remove the shift guide retaining clip (**b**) and pull the link assembly from the bracket.

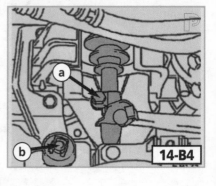
14-B4

☐ STEP B5: Disconnect the Lambda (oxygen) sensor wires – see *Chapter 9, Part A, Job 21*.

STEP B6: The engine/transmission assembly has to be supported from above.

→ This can be achieved using an engine crane though movement of the engine/transmission unit can make reassembly difficult because mounting bolts have to align with captive nuts.

→ It is far better to use an engine bar, though alignment problems can still arise.

→ Vauxhall supply an engine bridge with a range of adapters to suit different engines fitted to the Vectra. This is the recommended option.

STEP B7: Remove the radiator grille and tie the radiator to the air deflector panel – see *Chapter 8, Part B*.

STEP B8: Siphon as much fluid from the power steering reservoir as possible.

STEP B9: Raise the front of the vehicle, remove both front wheels, disconnect the steering track rod ends and the anti-roll bar drop links.

STEP B10: Disconnect the lower suspension arm to hub carrier balljoints.

STEP B11: Remove the rear engine damping bracket (bolts arrowed) from the transmission.

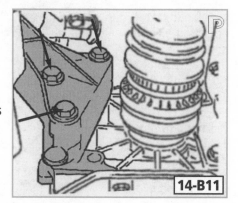
14-B11

STEP B12: Remove the front section of the exhaust pipe - see *Chapter 9, Ignition, Fuel, Exhaust*.

STEP B13: If oil cooler or refrigerant pipe brackets are attached to the subframe, remove them.

STEP B14: Remove the engine/transmission mounts - see *Chapter 6, Job 9*.

STEP B15: The subframe must be supported. A trolley jack and length of timber can be used.

STEP B16: Remove the subframe mounting bolts. Then lower the subframe sufficiently to detach the power steering pipes.

→ The subframe may now be lowered and removed.

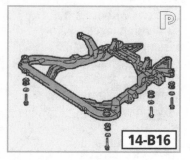
14-B16

STEP B17: Refit as the reverse of removal.

JOB 15: REAR DAMPERS - *replace*.

TOP TIP!

• Leaking dampers guarantee an annual-test failure!

• Moreover, tired suspension components can have a marked effect on a vehicle's handling and compromise its safety.

• Dampers and springs MUST be replaced in pairs, front and rear. If only one side is renewed, it will have a significant effect on the vehicle's handling, and make it extremely dangerous to drive!

• This job must be done one side at a time.

Section A: Cavalier and Calibra.

STEP A1: Raise the rear of the vehicle. If you are not using a vehicle hoist, secure the vehicle on axle stands. Remove the wheels. Position a trolley jack under the trailing arm you intend to work on first.

STEP A2: If the vehicle you are working on has its damper top mounting in the luggage bay area, you will have to remove a
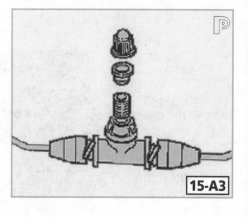
15-A2
rubber cover, or a flap in the luggage bay trim to access it.

STEP A3: If the vehicle you are working on has a suspension levelling system, release the pressure from the system by letting air out.

15-A3

→ There is a valve in the luggage bay for this.

→ Remove the pressure lines from the damper unit.

☐ STEP A4: ALL NON-ESTATE/VAN MODELS: Undo the damper top mounting bolt:
→ On some models, you will need to hold another spanner on the damper rod's hex head, to keep it from turning.
→ Remove the top washer and rubber buffer.

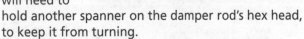

15-A4

☐ STEP A5: ESTATE AND VAN MODELS: The damper top mounting bolt is accessed through a hole in the chassis longitudinal member.

☐ STEP A6: Undo the fixing that holds the bottom of the damper to the trailing arm.

15-A6

☐ STEP A7: This is the estate/van model configuration, where the damper is mounted through the axle and secured with a nut.

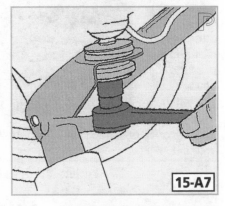

15-A7

☐ STEP A8: Compress the damper unit and remove the damper.
→ You may need to lever out the bottom end (arrowed), which can be difficult to remove from its mounting.

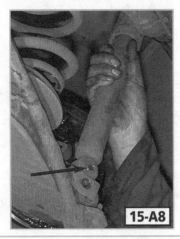

15-A8

TOP TIP!
• You can tell if a damper unit requires replacement simply by pushing the rod up and down a couple of times. If you feel any jerkiness to its movement, or hear a bubbling sound, or see fluid coming out of the piston, it should be replaced.
• New damper units need to be 'primed' before fitting. This is a simple process, involving depressing the damper rod and pulling it back out about five times. This draws oil into the reservoir and seals, so it can do its job.

☐ STEP A9: Refit as the reverse of removal.
→ Mounting bolts of the type that run sideways through the unit must be tightened to the specified torque settings.

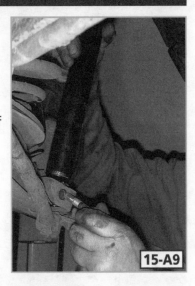

15-A9

☐ STEP A10A: However, mounting nuts, which fit the threaded parts of the damper shaft, either at the top (most passenger models) or bottom (estate and van models) of the unit, should be tightened using a different criteria.

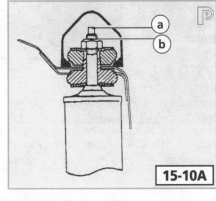

15-10A

→ The distance between the end of the shaft (**a**) and the shoulder of the nut (**b**), shown here by arrows, should be 8 mm (vehicles without a suspension levelling system) and 6 mm (vehicles with a suspension levelling system). This is the upper mounting…

☐ STEP A10B: …and this is the lower mounting. The same distances apply.

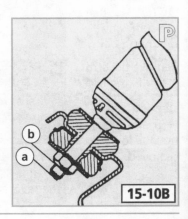

15-10B

STEP A11: If the vehicle you are working on has a suspension levelling system, re-pressurise it by adding air through the valve in the luggage compartment. The pressure should be 0.8 bar.

Section B: Vectra.

STEP B1: Remove the rear wheelarch liner.

15-B1

STEP B2: Remove the damper bracket bolts.

STEP B3: Remove the strut lower fixing bolt.

15-B3

STEP B4: Remove the strut/damper assembly complete.

15-B4

STEP B5: Refitting is the reverse of removal.

JOB 16: REAR SPRINGS - *replace*.

CAVALIER AND CALIBRA ONLY

Two configurations of rear suspension are fitted to the Cavalier and Calibra.
➔ Vehicles with SOHC engines have a semi-independent suspension.
➔ Vehicles with DOHC engines have a fully independent, semi-trailing arm suspension.

TOP TIP!

• Dampers and springs MUST be replaced in pairs, front and rear. If only one side is renewed, it will have a significant effect on the vehicle's handling, and make it extremely dangerous to drive!

IMPORTANT NOTE: On vehicles with torsion bar suspension, replace ONE spring at a time. Always leave one in place!

STEP 1: Raise the rear of the vehicle. If you are not using a vehicle hoist, secure the vehicle on axle stands. Remove the wheels. Position a trolley jack under the trailing arm you intend to work on first.

STEP 2: If the vehicle you are working on has a suspension levelling system, release the pressure from the system by letting air out. There is a valve in the luggage compartment for this – see illustration *15-A3*.

☐ **STEP 3A: FULLY INDEPENDENT REAR SUSPENSION:** Using the trolley jack placed under the hub assembly (arrowed) raise the semi-trailing arm (DOHC)...

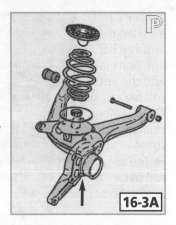

16-3A

☐ **STEP 3B: SEMI-INDEPENDENT REAR SUSPENSION:** ... or the trailing arm (arrowed)... ➜ ...in both cases by just enough to take spring pressure off the damper mounting bolts.

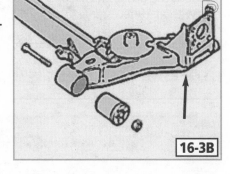

16-3B

☐ **STEP 4:** Undo the fixing that holds the bottom of the dampers to the trailing arm – see *Job 15, Steps 6 and 7*.

☐ **STEP 5:** Lever out the damper bottom ends from the mounting bracket – see *Job 15, Step 8*.

☐ **STEP 6:** If the vehicle you are working on has a brake pressure regulation valve, unhook its spring from the valve – see *Chapter 12, Job 12*.

☐ **STEP 7: DOHC MODELS:** Release the brake pipes from their clips on the axle.

☐ **STEP 8: DOHC MODELS:** Clamp the fuel line, disconnect it from the fuel filter and immediately plug the filter end. ➜ There will be spillage, so be ready to catch it.

☐ **STEP 9:** Use the trolley jack to lower the axle. ➜ Remove the springs, along with their rubbers. ➜ You may have to pull the axle further down to get them out. ➜ Be careful not to crush any brake pipes as you do so.

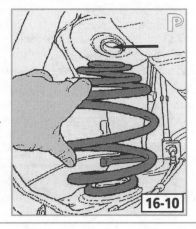

16-10

☐ **STEP 10:** Refit as the reverse of removal, making sure the springs are correctly positioned on their mounting cups (arrowed). Tighten the damper mounting bolts to the specified torque settings.

☐ **STEP 11:** If the vehicle you are working on has a suspension levelling system, re-pressurise it by adding air through the valve in the luggage bay. The pressure should be 0.8 bar.

JOB 17: REAR ANTI-ROLL BAR – *remove, refit.*

Section A: Semi-Independent rear suspension.

☐ **STEP A1:** These are the components of the anti-roll bar fitted to vehicles with SOHC engines.

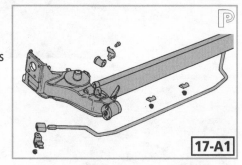

17-A1

☐ **STEP A2:** Raise the rear of the vehicle. ➜ Remove the wheels. If you are not using a vehicle hoist, secure the vehicle on axle stands. ➜ Position a trolley jack under the trailing arm you intend to work on first.

☐ **STEP A3:** If the vehicle you are working on has a suspension levelling system, release the pressure from the system by letting air out – see illustration *15-A3*.

☐ **STEP A4:** Remove the nuts and bolts fastening the anti-roll bar to the trailing arms.

17-A4

☐ **STEP A5:** Remove the anti-roll bar clamps to torsion beam fixings. ➜ The anti-roll bar can now be removed.

17-A5

STEP A6: Refitting is the reverse of removal.
→ Check the bushes for wear and condition, and renew them if necessary.

Section B: Fully independent semi-trailing arm suspension.

STEP B1: Raise the rear of the vehicle. If you are not using a vehicle hoist, secure the vehicle on axle stands. Remove the wheels.

STEP B2: If the vehicle you are working on has a suspension levelling system, release the pressure from the system by letting air out. There is a valve in the luggage compartment for this.

STEP B3: Release the brake pipes from their clips on the axle.

STEP B4: Clamp the fuel line, disconnect it from the fuel filter and immediately plug the filter end.
→ There will be spillage, be ready to catch this.

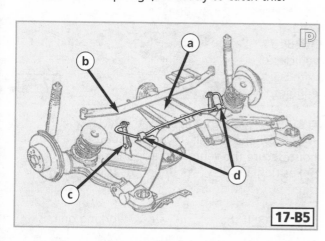

17-B5

STEP B5: Place a trolley jack under the rear plate (**a**).
→ Remove the bolts from the crossmember tube (**b**).
→ Carefully lower the axle assembly, checking that no wires, pipes or hoses are being strained.
→ Remove the nuts and bolts at the anti-roll bar drop link lower ends (**c**) and the clamp bracket bolts (**d**).
→ The anti-roll bar may now be removed.

STEP B6: Refitting is the reverse of removal.
→ Check the bushes for wear and condition, and renew them if necessary.

Section C: Vectra.

STEP C1: Raise the rear of the vehicle. If you are not using a vehicle hoist, secure the vehicle on axle stands. Remove the wheels.

STEP C2: Remove the anti-roll bar link bolts (**a**) and the anti-roll bar to crossmember clamp nuts (**b**).
→ The anti-roll bar assembly may now be withdrawn.

17-C2

STEP C3: Refitting is the reverse of removal.
→ Check the bushes for wear and condition, and renew them if necessary.

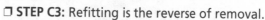

JOB 18: REAR AXLE – *remove, refit*.

Section A: Semi-Independent (SOHC – Cavalier and Calibra).

STEP A1: Raise the rear of the vehicle.
→ Remove the wheels. If you are not using a vehicle hoist, secure the vehicle on axle stands.
→ Position a trolley jack under the trailing arm you intend to work on first.

STEP A2: Place cling film over the brake master cylinder opening and refit the lid.
→ This creates a vacuum in the brake system and minimises the amount of brake fluid lost when you work on the brake pipes.

STEP A3: Using brake pipe clamps, clamp the rear flexible brake pipes, as close to their connection with the rigid pipes on the axle as possible.

STEP A4: Undo the connections between the rigid pipe and flexible hose.
→ Catch the fluid that escapes, then tape-up the pipe ends with masking tape to keep them clean.

STEP A5: Disconnect the parking brake cable from the parking brake and release the cable from the cable guides – see *Chapter 12, Job 19*.

STEP A6: Remove the springs – see *Job 16*.

STEP A7: Place a trolley jack under the centre of the axle crossmember to support it.

TOP TIP!
• If you are doubtful that the axle will stay on the jack, use a ratchet tie-down strap to tie the axle to the jack.

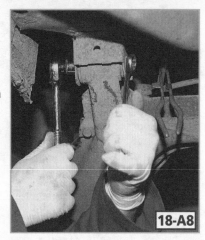

❐ **STEP A8:** With the axle secure on its support remove the nuts and bolts holding the axle to the chassis.

18-A8

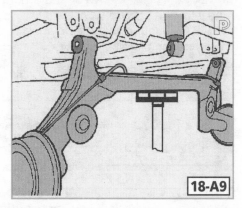

❐ **STEP A9:** Lower the axle on the jack, and pull it away from the vehicle.

18-A9

❐ **STEP A10:** Refit the axle as the reverse of removal:
→ Renew all nuts and bolts and tighten them to their specified torque settings.
→ The exception to this is the trailing arm to chassis mounting nuts and bolts, which should not be fully tightened at this stage.

❐ **STEP A11:** Check that the parking brake is correctly adjusted. See *Chapter 12, Jobs 12* and *18*.

❐ **STEP A12:** Bleed the brake system - see *Chapter 12, Job 16*.

Section B: Semi-Trailing arm suspension.

❐ **STEP B1:** Raise the rear of the vehicle.
→ Remove the wheels. If you are not using a vehicle hoist, secure the vehicle on axle stands.
→ Position a trolley jack under the trailing arm you intend to work on first.

❐ **STEP B2:** Remove the rear section of the exhaust system – see *Chapter 9, Part C*.
→ If the vehicle is fitted with a catalytic converter, remove the heat shield.

❐ **STEP B3:** Disconnect the parking brake cable from the parking brake and release the cable from the trailing arm brackets – see *Chapter 12, Job 19*.

❐ **STEP B4:** Lever off the brake line retaining clips from the trailing arms.

❐ **STEP B5:** Unscrew the wheel speed sensor bracket from the trailing arm and tie out of the way.

❐ **STEP B6:** Remove the brake calipers (and, if fitted, ABS sensor bracket) and tie out of the way – see *Chapter 12, Brakes*.

❐ **STEP B7:** Clamp the fuel line, disconnect it from the fuel filter and immediately plug the filter end.
→ There will be spillage, so be ready to catch it.

❐ **STEP B8:** Unbolt the anti-roll bar – see *Job 17, Section B*.

❐ **STEP B9:** Support each semi-trailing arm in turn on the trolley jack, undo the damper bottom bolt, then lower the semi-trailing arm.
→ Remove the spring.

❐ **STEP B10:** Place a trolley jack under the crossmember rear plate and remove the crossmember securing bolts.

❐ **STEP B11:** Carefully lower the jack, checking that no pipes or wires are strained.

SAFETY FIRST!

• The rear axle assembly is heavy and not well balanced on the trolley jack.
• Before attempting the following step, arrange for one or preferably two helpers to steady the assembly.

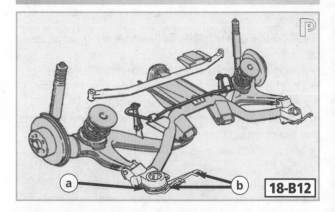

18-B12

❐ **STEP B12:** With the weight of the axle assembly taken by the jack and assistants, remove the two front crossmember mounting bolts (**a**) then the bolts securing the crossmember brace brackets (**b**).

❐ **STEP B13:** Lower the rear axle assembly to the ground.

❐ **STEP B14:** Replacement is the reverse of removal.

➜ Renew all rubber components.
➜ Preferably renew all nuts and bolts.

Section C: Vectra.

❒ **STEP C1:** Raise the rear of the vehicle. If you are not using a vehicle hoist, secure the vehicle on axle stands. Remove the wheels.

❒ **STEP C2:**
Remove the rear section of the exhaust system (**a**) and the rear silencer box (**b**) – see *Chapter 9, Part C*.

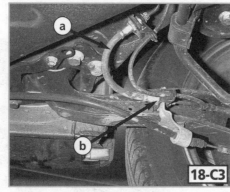

➜ This shows the location of the rear 'axle' on these models.

❒ **STEP C3:**
Using brake pipe clamps, clamp the rear flexible brake pipes (**a**), as close to their connection with the rigid pipes (**b**) as possible.

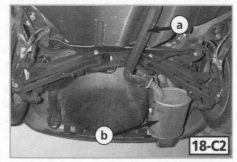

➜ Undo the connections between the rigid pipe and flexible hose at (**b**).
➜ Catch the fluid that escapes, then tape-up the pipe ends with masking tape to keep them clean.

❒ **STEP C4:** Pull the retaining spring clips and detach the hoses from their supports.

❒ **STEP C5:**
Disconnect the ABS wheel speed sensors' wires.

❒ **STEP C6:**
Disconnect the parking brake rear

cables from the secondary cables guides (position arrowed) – see *Chapter 12, Job 19*.

IMPORTANT NOTE:
• The rear wheel alignment on the Vectra is set by the position of the trailing arm mounting plates.
• Before removing the trailing arm mounting plates, their position must be very accurately marked.

❒ **STEP C7:**
Mark the positions of the trailing arm mounting plates on the chassis (arrowed).

❒ **STEP C8:** Support the trailing arms, then remove the mounting plate bolts.

❒ **STEP C9:** Take the weight of the axle assembly on a trolley jack.

❒ **STEP C10:** Remove the damper bottom end fastenings.

SAFETY FIRST!

• **The rear axle assembly is heavy and not well balanced on the trolley jack.**

• **Before attempting the following step, arrange for one or preferably two helpers to steady the assembly.**

❒ **STEP C11:**
Remove the four fastening bolts (positions of two, arrowed), then lower the assembly to the ground as your assistant(s) help steady it.

❒ **STEP C12:** Replacement is the reverse of removal.
➜ Check the condition of the four captive nuts into which the main bolts fit.
➜ If the suspension arms have been removed, do not fully tighten their inner bolts until the weight of the vehicle is back on the suspension.
➜ The rear wheel alignment should be checked at the earliest opportunity.

JOB 19: REAR TORSION BEAM AXLE BUSHES - *replace*.

JOB 20: REAR HUB - *remove, refit*.

It is easiest to change the chassis mounting bushes with the axle unit removed, as described in the last Job. However, it is perfectly feasible to drop the front end of the axle to do this.

→ Proceed as in *Job 18*, but leave the dampers connected and the coil springs *in situ*.

→ Remove the chassis mounting nuts and bolts, and lower the axle downwards on a trolley jack.

❒ STEP 1: Remove the bushes:
→ It helps to chisel the front edge of the bush free.

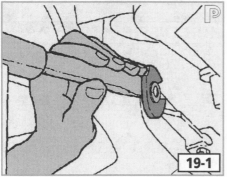
19-1

❒ STEP 2: Construct a suitable tool to remove the bushes similar to the Vauxhall tool shown here.
→ Normally, a combination of a large socket or suitable sized tubing, a long

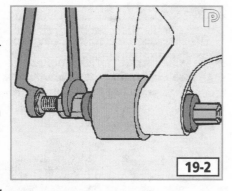
19-2

bolt and some suitably sized washers should enable you to pull the bush from its mounting.

❒ STEP 3: Lubricate the new bushes with the lubricant that comes with them, if supplied - or soapy water.

❒ STEP 4: Using your home-made tool in reverse, pull the replacement bushes into the correct position.
→ The angle (a) must be between 40 and 50 degrees.

19-4

❒ STEP 5: Refit the axle – see *Job 18*.

FACT FILE

FACT FILE: REAR HUBS.

• The rear hubs can be removed and refitted easily on Vectras and on Cavaliers and Calibras with semi-independent rear suspension (SOHC engines).

• The hub and bearing assembly fitted to the semi-trailing arm suspension on Cavaliers and Calibras is very difficult to work on without specialist pullers and a torque wrench capable of measuring the 300 lbs. ft. to which the hub nut has to be tightened.

Section A: Semi-Independent rear suspension with drum brakes (Cavalier and Calibra).

❒ STEP A1: Raise the rear of the vehicle. If you are not using a vehicle hoist, secure the vehicle on axle stands. Remove the wheels.

❒ STEP A2: Remove the brake drum – see *Chapter 12, Brakes*.

20-A2

❒ STEP A3: To remove the hub:
→ Lever the hub dust cap (a) off.
→ Remove the split pin (b) from the castellated hub nut (c).
→ Remove the hub nut and thrust washer (d).
→ Pull the hub (f) off the stub axle.
→ The outer wheel bearing (e) will probably drop off, so be ready to catch it if it is to be refitted.
→ Item (g) is the inner wheel bearing and (h) the seal.

20-A3

☐ **STEP A4:** Refitting is the reverse of removal.
➜ Apply fresh grease to the bearing and oil seal, and pack grease into the area between the two bearing tracks.
➜ Fit a new oil seal
➜ Adjust the parking brake.

☐ **STEP A5:**
Tighten the castellated nut to about 24 Nm torque. Undo the nut a few turns again, then tighten finger tight.

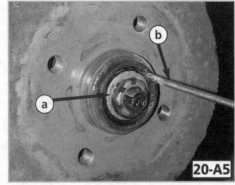
20-A5

You should just be able to move the thrust washer (**a**) by pushing (NOT levering) with a narrow, flat-bladed screwdriver (**b**). If not, turn the nut one flat at a time until the pressure on the washer is correct.
➜ Fit the split pin. You may have to turn the hub nut a fraction to align the holes – if so, recheck the thrust washer as already described.
➜ Check for play by fitting a wheel and rocking it from top to bottom. If it joggles at all, the bearing should be tightened.

Section B: Semi-Independent rear suspension with disc brakes (Cavalier and Calibra).

IMPORTANT NOTE: The hub is an integral part of the disc on these models.

☐ **STEP B1:** Raise the rear of the vehicle. If you are not using a vehicle hoist, secure the vehicle on axle stands. Remove the wheels.

☐ **STEP B2:** Remove the brake caliper, and tie it out of the way – see *Chapter 12, Brakes*.

☐ **STEP B3:** Lever the hub dust cap off.

☐ **STEP B4:** Remove the split pin from the castellated hub nut.

☐ **STEP B5:** Remove the hub nut and thrust washer.

☐ **STEP B6:** Remove the disc/hub assembly.
➜ It may be necessary to back the parking brake shoes off – see *Chapter 12, Brakes*.
➜ The outer wheel bearing will probably drop off, so be ready to catch it if it is to be refitted.

☐ **STEP B7:** Refitting is the reverse of removal.

➜ Apply fresh grease to the bearing and oil seal, and pack grease into the area between the two bearing tracks.
➜ Fit a new oil seal
➜ Adjust the parking brake.

Section C: Fully independent semi-trailing arm rear suspension (Cavalier and Calibra).

FACT FILE
HUB DIFFICULTIES
• The hub (wheel spindle) nuts on these vehicles are tightened to 300 Nm, they are fairly shallow and VERY difficult to undo!
• There is a Vauxhall specialised hub removal/refitting multi-part tool KM-610.
• This tool includes a socket, which passes through a hole in the main body of the tool (clamped to the trailing arm) to keep the socket from slipping when undoing the wheel spindle nut.
• The tool has accessories for pressing first the wheel spindle then the bearings out and also pulling them back in.
• We recommend that this job is carried out by a Vauxhall specialist.

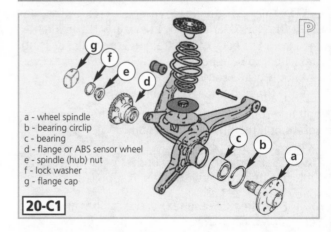

a - wheel spindle
b - bearing circlip
c - bearing
d - flange or ABS sensor wheel
e - spindle (hub) nut
f - lock washer
g - flange cap

20-C1

☐ **STEP C1:** These are the hub, spindle and wheel bearing components.
➜ Note how shallow that 300 Nm is!
➜ All of the illustration letters in this Section refer to this illustration.

☐ **STEP C2:** Raise the rear of the vehicle. If you are not using a vehicle hoist, secure the vehicle on axle stands. Remove the wheels.

☐ **STEP C3:** Remove the brake caliper and ABS wheel speed sensor bracket, and tie them out of the way.

☐ **STEP C4:** Back off the parking brake adjuster if

necessary, remove the brake disc securing screw and remove the brake disc – see *Chapter 12, Brakes*.

☐ **STEP C5:** Using a long 10 mm splined key through one of the plain holes in the hub flange, undo the four brake backplate bolts.
→ The top two are longer and must be replaced in the same positions.

☐ **STEP C6:** Lever the plastic flange cap (**g**) out of the flange (**d**), and lever the lock washer (**f**) from the spindle nut (**e**).

☐ **STEP C7:** Remove the spindle nut (**e**).
→ In the absence of special tool KN-620, try an hexagonal socket.
→ Prevent the wheel spindle from turning by fitting two wheel bolts into the flange and inserting a lever between them.

☐ **STEP C8:** Remove the ABS sensor flange (**d**).
→ A puller may prove necessary!

☐ **STEP C9:** Press the wheel spindle (**a**) out.

☐ **STEP C10:** Remove the hub bearing circlip (**b**).

☐ **STEP C11:** Press the bearing (**c**) out.
→ The inner bearing might have come out on the wheel spindle.
→ Apply pressure only against the outer bearing race.

☐ **STEP C12:** Refitting is the reverse of removal.
→ Fit the new bearing first, then support the inner race from the back of the trailing arm whilst drifting the wheel spindle home.
→ Refit the ABS sensor flange.
→ Refit the hub nut and locking washer.

Section D: Vectra.

☐ **STEP D1:** Remove the rear brake calliper and disc – see *Chapter 12, Brakes*.

☐ **STEP D2:** Disconnect the ABS sensor.

☐ **STEP D3:** Support the brake assembly and remove the hub assembly nuts from the rear of the trailing arm.

☐ **STEP D4:** The hub may now be removed.

☐ **STEP D5:** Reassembly is the reverse of removal.
→ New nuts MUST be used on the hub assembly bolts.

JOB 21: REAR WHEEL BEARINGS – *remove, replace, adjust.*

Section A: Cavalier and Calibra with semi-independent suspension (SOHC).

IMPORTANT NOTE:
• The hub and bearing assemblies fitted from 1993 are sealed units.
• This means they're maintenance-free, but also that if there is any problem with the bearings then the whole assembly must be renewed.

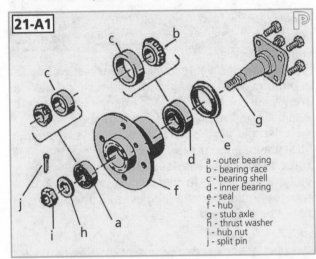

21-A1

a - outer bearing
b - bearing race
c - bearing shell
d - inner bearing
e - seal
f - hub
g - stub axle
h - thrust washer
i - hub nut
j - split pin

☐ **STEP A1:** This illustration shows the position of the components of a typical hub assembly.

☐ **STEP A2:** Remove the hub – see *Job 20*.

☐ **STEP A3:** Remove the outer bearing from the hub. If it is to be reused, keep it scrupulously clean!

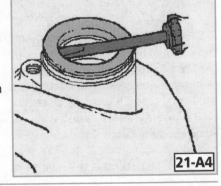

21-A3

☐ **STEP A4:** Lever the oil seal out using a broad bladed screwdriver.

21-A4

❏ **STEP A5:** Pull out the inner bearing.

❏ **STEP A6:** Use suitably sized tubes (**a**) and a vice or press, push the bearing tracks out of hub (**b**).

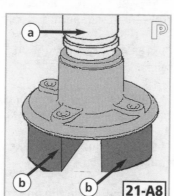

❏ **STEP A7:** Clean the hub and bearings thoroughly, grease the hub inside and the inner bearing race, then press the inner bearing race back into position.

21-A6

❏ **STEP A8:** Fit a new oil seal.
➜ This can be tapped into position or pressed (**a**) using a vice and suitable tube and supports (**b**) as shown here.

❏ **STEP A9:** Clean the stub axle and lightly grease it, then refit the hub.

21-A8

❏ **STEP A10:**
Grease the outer bearing and refit it.
➜ Position the washer (**a**) and castellated nut and tighten the nut finger tight.

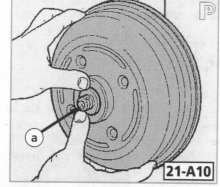

21-A10

❏ **STEP A11:**
Tighten the castellated nut to about 24 Nm torque. Undo the nut a few turns again, then tighten finger tight. You should just be able to move the thrust washer by pushing (NOT levering) with a narrow, flat-bladed screwdriver. If not, turn the nut one flat at a time until the pressure on the washer is correct.
➜ Fit the split pin. You may have to turn the hub nut a fraction to align the holes – if so, recheck the thrust washer as already described.
➜ Check for play by fitting a wheel and rocking it from top to bottom. If it joggles at all, the bearing should be tightened.

Section B: Cavalier and Calibra with semi-trailing arm suspension.

See *Job 20, Section C*.

Section C: Vectra.

IMPORTANT NOTE: A dial gauge is necessary to check for play in the rear wheel bearings.

❏ **STEP C1:** Raise the rear of the vehicle. If you are not using a vehicle hoist, secure the vehicle on axle stands. Remove the wheels.

❏ **STEP C2:** On the 1.6 litre SOHC model with rear drum brakes, remove the brake drum – see *Chapter 12, Brakes*.

❏ **STEP C3:** Set the dial gauge against the disc or, in the case of drum brakes, the hub.
➜ Lever the hub out and in and measure the bearing play.

❏ **STEP C4:** Check run-out by turning the disc or hub and reading the dial gauge.
➜ If the play or run-out are excessive, renew both the hub bearing and bracket.

> **JOB 22:** VECTRA REAR SUSPENSION ARMS – *remove, replace*.

Section A: Upper arm.

❏ **STEP A1:** Raise the rear of the vehicle. If you are not using a vehicle hoist, secure the vehicle on axle stands. Remove the relevant wheel.

❏ **STEP A2:**
Remove the upper arm to trailing arm bolt.
➜ Release the locking plate first.

22-A2

STEP A3: Remove the upper arm inner bolt.
➔ The upper arm may now be removed.

STEP A4: The inner bush can be pressed out and renewed using suitable sized tubes and a vice.

STEP A5: Replace as the reverse of removal.
➔ Do not fully tighten the bolts until the weight of the vehicle is on the suspension.

Section B: Lower arm.

STEP B1: Raise the rear of the vehicle. If you are not using a vehicle hoist, secure the vehicle on axle stands. Remove the relevant wheel.

STEP B2:
Remove the bolt holding the outer end of the lower arm to the trailing arm.
➔ Release the locking plate first.

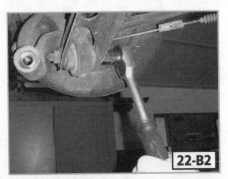

22-B2

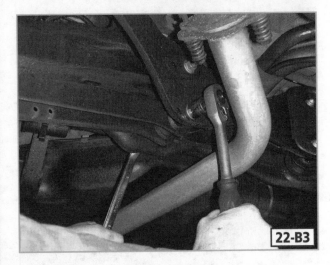

22-B3

STEP B3: Remove the inner end bolt.
➔ The arm may now be removed.

STEP B4: The inner bush can be pressed out and renewed using suitable sized tubes and a vice.

STEP B5: Replace as the reverse of removal.
➔ Do not fully tighten the bolts until the weight of the vehicle is on the suspension.

JOB 23: VECTRA SEMI-TRAILING ARMS – *remove, replace*.

The illustrations references here refer to illustration *1-5*.

STEP 1: Raise the rear of the vehicle. If you are not using a vehicle hoist, secure the vehicle on axle stands. Remove the wheels.

STEP 2: Disconnect the parking brake rear cables from the secondary cables guides – see *Chapter 12, Job 19*.

STEP 3: Remove the secondary parking brake cable from the trailing arm support.

STEP 4: Disconnect the ABS wheel speed sensors' wires.

STEP 5: Remove the hub, bracket and brake backplate – see *Job 20, Section D.*

STEP 6: Remove the brake line support from the trailing arm.

STEP 7: Remove the anti-roll bar link bolt (**e**).

STEP 8: Remove the bolts holding the upper and lower suspension arms to the trailing arm – see *Job 21*.

STEP 9: Remove the suspension strut lower bolt – see *Job 15, Section B*.

IMPORTANT NOTE:
• The rear wheel alignment on the Vectra is set by the position of the trailing arm mounting plates.
• Before removing the trailing arm mounting plates, their position must be very accurately marked.

STEP 10: Mark the positions of the trailing arm mounting plates.
➔ Typists' correction fluid is not accurate enough and may be accidentally wiped away.
➔ It is better to make marks using a centre punch.

STEP 11: Remove the front mounting bolts.
➔ The trailing arm (**a**) can now be lifted away.

STEP 12: The rubber bushes may be pressed out and in using a press or vice and suitable tubing.

STEP 13: If the front bracket is to be removed, use a centre punch to make accurate alignment marks beforehand – this must be precisely positioned.

STEP 14: Reassembly is the reverse of removal.
➔ Have the rear wheel tracking checked at the earliest opportunity.

CHAPTER 12: BRAKES

*Please read **Chapter 2 Safety First** before carrying out any work on your car.*

CONTENTS

FACT FILE

SYSTEMS EXPLAINED

• All the models covered in this manual have front disc brakes, operated by a single piston, floating caliper. Some models have drum brakes at the rear, while others have disc brakes. All of these are self adjusting. The parking brake operates, via a cable, on the rear wheels, operating the drum brakes, or on separate drums within the rear discs, when fitted.

• There are two basic designs of caliper.

TYPE ONE: Fitted to earlier models. The pads can be fitted and removed without removing the caliper.

TYPE TWO: Fitted to later models. The caliper (**a**) must be removed from it's bracket (**b**) for the pads to be removed.

• If the vehicle has ABS (anti-lock brakes), this is in two circuits, split front and rear. If not, it is in two circuits split diagonally. This ensures that, in the event of a leaking hydraulic pipe, the vehicle retains some braking ability.

• The system has an inherent bias towards the front wheels, to increase stability under heavy braking manoeuvres. On most models, there is a pressure regulation valve for each rear brake. On some models, pressure is variable, regulated by a single valve, which changes the pressure according to the payload.

• Directly connected to the brake pedal mechanism is the servo, which is powered by the inlet manifold vacuum or, in the case of diesels, a separate vacuum pump driven by the alternator or camshaft. This loads the whole system via the master cylinder.

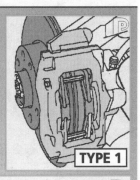

TYPE 1

• Some models are equipped with ABS. This uses wheel speed sensors to tell a central control unit whether a wheel is starting to lock under heavy braking manoeuvres. The control unit then pulses the braking pressure being supplied to that wheel to stop it from skidding.

TYPE 2

JOB 1: FRONT BRAKE PADS - *check, replace*

SAFETY FIRST!

• Raise the front of the vehicle off the ground, after carefully reading the information in *Chapter 2, Safety First!* on lifting and supporting the vehicle.

• Obviously, a vehicle's brakes are among its most important safety related items. Do NOT dismantle or attempt to perform any work on the braking system unless you are fully competent to do so. If you have not been trained in this work, but wish to carry it out, we strongly recommend that you have a garage or qualified mechanic check your work before using the vehicle on the road. See also the section on *BRAKES AND ASBESTOS in Chapter 2, Safety First!* for further information.

• Always start by washing the brakes with a proprietary brand of brake cleaner - brake drums removed where appropriate. Never use compressed air to clean off brake dust.

• ALWAYS replace disc pads in sets of four across the vehicle: that is, both front wheels or both rear wheels at the same time. Failure to do this will result in unequal and dangerous braking, due to the imbalance in efficiency.

• After fitting new brake shoes or pads, the driver should avoid heavy braking - except in an emergency - for the first 250 to 300 km (150 to 200 miles).

• Wear gloves when working on the braking system!

• Brake dust gets in cut and skin cracks and pores and can take ages to 'wear' out! It could also cause skin complaints.

➜ You can check the pad thickness without removing the caliper. The maker's recommended minimum thickness for the friction material-plus-backing plate is 7 mm for Cavalier and Calibra and 7.5 mm for Vectra.
➜ You may want to renew earlier than this because you won't be looking at the pads again for another 6,000 miles.
➜ It is normal for one pad to wear slightly more than the other but, if it appears that only one pad is doing the work and the other has hardly worn at all, it is a sign that the caliper is sticking. See *Job 2*.

Section A: Type One – Calibra, Cavalier.

The design of these calipers allows the pads to be fitted and removed from outside, without removing the caliper itself.

TOP TIP!

• A special pin-punch will be required to drift the pad retaining pins from the calipers.
• Because the pins are hollow, the punch requires a small 'step' to be included at the 'business' end that locates inside the pin, thereby keeping the punch central.
• Suitable punches are available from most good accessory shops and are worth purchasing - do not try to use a conventional 'straight' punch as there is a risk it will damage the retaining pin or become jammed in the pin.

FACT FILE
PREPARATION

• In preparing to re-fit or renew the pads, wash any oil or grease from your hands so that you don't get it onto the friction materials.
• Also, wipe the disc surface with spirit wipe to remove any traces of oil, grease or dirt. In order to make room for new (obviously much thicker) disc pads, the caliper piston must be pushed back into the cylinder bore.
• This is one good reason why you have to be sure a piston's exposed dirt doesn't get pushed back into the cylinder bore along with the piston.
• Before reassembly, lightly smear a little special brake grease or 'copper grease' (NOT ordinary grease) around the top and bottom edges of the steel backing plate of the pads.
• Be extra careful not to allow the grease to get onto friction material - use just a small amount of grease on each edge.

❑ **STEP A1:** Drive out the upper pad retaining pin, using the special punch, from the inside of the caliper as shown.
➜ The pins may well be corroded and stiff to move - if so, it may help to apply a little penetrating oil to the pins.

1-A1

→ If the pads are to be used again, make absolutely sure no oil or grease gets on to the friction material.
→ As the upper pin is moved across, the two anti-rattle springs will be freed!
→ Take note of the position of the springs in relation to the pads and pins, then remove the lower pin.

☐ **STEP A2:** By levering the outer pad away from the disc slightly, the pads can be removed from the caliper with pliers, along with their respective shims...

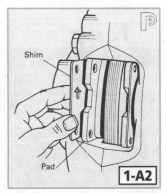

☐ **STEP A3:** ...which must be retained for re-fitting. Scrape away any rusty scale from the pad housings using an old screwdriver, brushing away any loose debris afterwards with

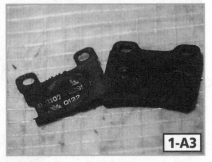

a small brush, but remember - DO NOT BREATH IN ANY DUST as it will contain potentially dangerous irritants.

TOP TIP!

• As the pistons are pushed back into their bores the brake fluid level in the reservoir will rise accordingly, often to the point of over-flow!
• Be prepared for this by removing a quantity of fluid first.
• The easiest way is to draw-off the fluid using an old battery hydrometer, to 'suck' the fluid a little at a time from the reservoir.

☐ **STEP A4:** On reassembly:
→ Position the pads and shims (dark arrows) in the caliper housing and fit the bottom pin first, followed by the anti-rattle springs, then the top pin (light arrows).
→ The springs will need to be tensioned as the top pin is tapped home.

→ Remember to fit the pins from the outside of the caliper i.e. tap them home TOWARDS the engine side of the caliper, to finish flush with the caliper sides, and with the openings in the pins facing each other.

Section B: Type 2 – Cavalier, Calibra and Vectra.

FACT FILE
VEHICLES EQUIPPED WITH PAD WEAR INDICATORS

• When buying new disc pads, note that pads bought from Vauxhall agents or dealers may not come equipped with new wear sensors – these will be a separate 'extra'.
• However, pads bought from independent suppliers may well come with the sensor wires already fitted.
• It may be that you are replacing the pads before the wear sensors have contacted the disc and become damaged, in which case there is strictly no need to replace the sensors but as the cost of new ones is relatively small, you might want to replace them regardless.

☐ **STEP B1:** Lever the pad retaining spring away from the caliper (also see *Step B3* for alternative type) using a screwdriver, being careful not to bend it if it sticks.
→ A little penetrating oil and careful levering will release a stubborn spring.
→ Remove the wear indicator wire, if fitted, taking care not to pull the wire from the connector on the end of the wire.

☐ **STEP B2:** The caliper securing/guide bolts are protected from the elements by a rubber sleeve topped by a plastic cap which is easily pulled off. Insert the Allen key fully and apply tension.

• Caliper guide bolts are notoriously tight, but they also have a habit of suddenly 'giving' as pressure on the tool mounts.

• Be extra careful not to apply pressure to the tool in a way that the vehicle becomes unstable.

• Direct effort downwards and be ready for the point of 'give' when the bolt may suddenly release and offer little or no resistance without warning.

☐ **STEP B3:** With the bolts removed, it will probably be necessary to retract the piston sufficiently to allow the pads to clear any wear ridge on the outer edge of the disc.

➔ Use a large pair of 'grips' or lever the inside pad away from the disc slightly using a flat lever.

➔ DO NOT apply excessive pressure to the outer edge of the disc, which can distort and be ruined.

➔ ALWAYS leave the 'other' pad in place when levering the caliper piston so as not to risk cracking the disc/rotor.

☐ **STEP B4:** Lift-off the caliper:

➔ The inner pad is fixed to the piston by way of a spring clip and will come away with it, while the 'free' pad may remain with the caliper – simply lift it out.

1-B4

☐ **STEP B5:** This shows the pads and the spring clip which secures the inner pad to the (hollow) piston.

1-B5

☐ **STEP B6:** Spray brake cleaner over the carrier bracket, paying particular attention to the pad seatings and those areas not accessible when the pads are fitted.

➔ Wire brush the pad seatings on the bracket, removing any hard scale or rust by scraping with an old screwdriver or chisel.

1-B6

☐ **STEP B7:** Retracting the piston is necessary when fitting new pads.

➔ This can be done earlier in the sequence when the caliper is still attached to the carrier, or now before cleaning.

➔ See TOP TIP! earlier in this Job, regarding fluid level in the master cylinder reservoir.

1-B7

☐ **STEP B8:** Check the piston rubber dust excluder for splits, chafing or perishing. If any sign of brake fluid is present a leak is indicated – replace the caliper – see *Job 2*.

1-B8

☐ **STEP B9:** Reassemble the new pads to their seats and ensure the two raised steps on the piston face are positioned at the 12 o'clock and 6 o'clock positions - the pistons can be turned using the square shaft of a screwdriver, or a large pair of grips.

1-B9

• The dust caps MUST be refitted to the slide bolts.

• If they are missing, replace them.

• If road dirt gets into the slide mechanism, the caliper will seize up and cease to work.

☐ **STEP B10:** Reassembly follows the guidelines above, but note:

➔ The tightening torque of the guide bolts is 30 Nm (22ft/lb) – nothing like as tight as the struggle of removal would suggest!

➔ The use of a thread-locking compound is advisable, just a 'spot' on the threads of each bolt being sufficient.

➔ Fit the 'inboard' pad (with the spring clip) to the piston and the 'free' pad to the caliper, before replacing the latter.

➔ Don't forget to fit the wear indicators if necessary.

JOB 2: FRONT BRAKE CALIPER - *remove, replace.*

There are two different approaches to brake caliper removal:

→ If the caliper needs to be removed in order to take off the disc/hub, you don't need to open up the hydraulic system.

→ If the caliper itself is to be replaced, the hydraulics do need to be broken in to and will need bleeding afterwards - see *Job 16*.

REMOVE CALIPER FROM VEHICLE

❑ **STEP 1:** Clamp the flexible fluid hose as close to the caliper connection as possible with a suitable hose clamp – see the Sykes-Pickavant clamp (illustration *5-2*).

2-1

→ The flexible hose is unbolted from the back of the caliper, as shown.

→ After fitting the new caliper, you will need to bleed the hydraulic system. See *Job 16* but only bleed this 'corner' of the brake system if the rest of it is okay.

❑ **STEP 2:** To remove the two caliper bolts:

→ Take off the two rubber covers (arrowed).

→ Use a suitable driver extension (varies between models) and remove both bolts.

2-2

❑ **STEP 3:** With both bolts removed, the caliper can now be lifted free.

→ If there is any wear lip on the disc, the pads will have to be pushed

2-3

back in to the caliper so that they clear the wear lip.

→ You should always fit new caliper retaining bolts when replacing the caliper. They are designed to stretch when tightened to the correct torque.

CALIPER NOT BEING REMOVED FROM VEHICLE

❑ **STEP 4:** Use a cable tie to support the caliper on the road spring.

→ Do not leave the caliper hanging under the

2-4

weight of its flexible hose and do not put strain on the flexible hose or connections.

JOB 3: FRONT BRAKE DISC - *remove, replace.*

IMPORTANT NOTE:

• Discs should be renewed if they have heat discoloration, usually visible as blue, which means they have overheated and may have softened.

• Discs should be renewed if the faces are scored.

• Discs should be renewed if they run out of true – see *Step 2*.

❑ **STEP 1:** Disconnect the caliper assembly from the hub carrier, and suspend it from the strut using a strong cable tie or wire – see *Jobs 1* and *2*.

❑ **STEP 2:** To check a disc, fit a roadwheel nut opposite to the small disc securing screw. Rotate it slowly and check it visually for scoring, then measure the disc 'run out'. The ideal tool for this is a dial gauge, but it is possible to use feeler gauges.

→ If using a dial gauge, set the tip against the disc 10 mm from the outer edge, turn the disc and note the reading.

→ In the absence of a dial gauge, set a fixed metal pointer 2 mm from the disc, 10 mm from its outer edge, turn the disc and insert feeler gauges at several points to measure any run out.

→ If more than 0.1 mm is recorded, either the disc should be replaced or the wheel bearings need adjustment.

❑ **STEP 3:** Remove the small screw that holds the disc onto the hub.

→ The screw can be very tight, in

3-3

which case, use an impact screwdriver.

STEP 4: Clean the friction surface of the new unit with brake cleaner or industrial degreaser, as it may be supplied covered in a protective layer.

STEP 5: The disc is now removed from the hub.

3-5

SAFETY FIRST!

• Before refitting the caliper, replace the pads. Old pads must never be used on a new disc.

• Brake discs should always be replaced in pairs.

STEP 6: Replace the disc as the reverse of removal.

→ It is most important, if necessary, to clean the mating

3-6

surface of the hub, using abrasive paper. This ensures the new disc fits flush and runs true.

→ Dress the screw thread with copper grease before refitting.

JOB 4:	REAR BRAKE SHOES AND DRUMS - *check, replace.*

FACT FILE

REAR AND PARKING BRAKES.

• Some models of Cavalier, Calibra and the 1.6 SOHC Vectra are fitted with rear disc brakes. These have integral drums in which conventional parking brake shoes operate. For details of these see *Job 9*.

• Early Cavalier rear brake shoes are manually adjusted via two cams on the brake backplate.

• Later Cavalier, Calibra and the 1.6 SOHC Vectra were fitted with rear/parking brake assemblies that were self-adjusting. Models from 1992 had modified shoe anchorage points and different shoes – check to make sure to obtain the correct type before starting work!

• Although there are detail differences between the rear drum brakes of Cavalier/Calibra and Vectra, the principles of operation are identical, and so they are all grouped together in **Section A**.

TOP TIP!

• You can quickly check the thickness of the brake shoe lining material through the inspection hole in the backplate.

• If you remove the blanking plug and shine a light through you should be able to see the side of the brake shoe.

• Unless you are experienced it may be difficult to be sure of the lining thickness.

• To be certain (and to check for fluid leaks) remove the drum.

Section A: Cavalier from 1984 and Calibra.

STEP A1: Remove the brake drum – see *Chapter 11, Job 22*. It is highly likely that, as you slide the drum off the stub

4-A1

axle, it will be obstructed by the shoes due to a 'wear ridge' on the drum.

→ Place a screwdriver in the hole provided on the backplate and push the handle towards the front of the vehicle.

→ This applies pressure to the parking brake lever inside the drum and causes the shoes to contract slightly, allowing the drum to pass over them.

STEP A2: Spray the brake components with brake cleaner to 'wash' away all traces of brake dust.

→ NEVER use a brush or air line

4-A2

to blow the dust away.

→ Check the thickness of the lining material. The minimum thickness is specified as 1.5 mm above the rivet heads. If the shoes are the bonded type, the thickness is 2 mm above the backing plate (arrowed).

→ As the shoes may not be examined again for some time, they should be renewed well before this point.

STEP A3:
After 'washing' the drum with brake cleaner, check the interior surface for excessive scoring, corrosion and deep 'wear ridges' (arrowed).

→ If these are present, the drum must be replaced.

STEP A4:
Carefully pull back the rubber gaiters at each end of the wheel cylinder and check there are no signs of brake fluid seeping past the piston.

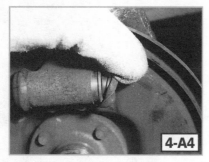

→ If there is, replace the cylinder – see *Job 5*.

TOP TIP!

Before dismantling the brake shoe assembly, but after the brake drum has been removed:
• Make a careful record of where everything goes, especially brake shoe return springs.
• Only work on one side at a time, so that you've always got the other one to refer to. However, note that some components are 'handed' and each side of the vehicle will be arranged as a mirror image of the other.

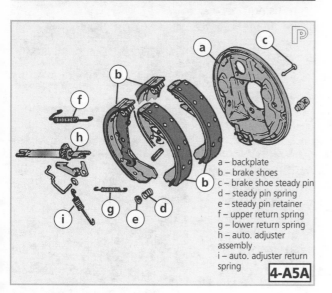

a – backplate
b – brake shoes
c – brake shoe steady pin
d – steady pin spring
e – steady pin retainer
f – upper return spring
g – lower return spring
h – auto. adjuster assembly
i – auto. adjuster return spring

STEP A5A: These are typical self-adjusting brake components…

STEP A5B:
…and these are the brake components *in situ*.

STEP A6:
There is no 'set' order for removing the brake components.
→ The method described is easier than removing components individually, and allows you to see the components assembled, which makes reassembly easier!

STEP A7:
Release the spring clips on the shoe steady pins:
→ Hold the pin at the rear of the brake backplate

and push against the spring with pliers.
→ Turn it through 90 degrees and allow it to pass over the holding pin head.
→ These springs can be strong, so it is advisable to wear eye protection.

STEP A8: Lever the bottom of one shoe away from the anchor plate, which will relieve the spring tension and allow the other shoe to be released.

STEP A9:
Unhook the parking brake cable from the operating arm on the rearmost shoe (arrowed). It will now be possible to lift the shoes away from the backplate in one piece for cleaning and reassembly.

STEP A10: Place the assembly flat on the ground and build up the new shoes a step at a time, transferring springs and adjuster in sequence.
→ In this way, you will avoid any confusion as to what part goes where, and can you clean each part as it is dealt with.

❏ **STEP A11:** This picture shows the arrangement of the automatic adjuster mechanism on the front shoe of the left-hand brake assembly.

4-A11

TOP TIP!

• It is permissible to apply a little copper grease to the auto-adjuster rod and click wheel, but make absolutely sure none gets onto the lining surfaces.

❏ **STEP A12:** Clean the backplate with brake cleaner and a rag before fitting the new shoes.

4-A12

❏ **STEP A13:** Also smear a little copper grease where the backplate and the shoes make contact (arrowed).

4-A13

❏ **STEP A14:** Remember to attach the parking brake cable to the respective shoe before assembling the components. Otherwise the installation operation is the reverse of removal.
➜ Take care not to allow oil or grease to contaminate the new shoes!

4-A14

❏ **STEP A15:** Fit the rear shoe first, and secure it with the keeper pin and spring.

4-A15

❏ **STEP A16:** Assemble the adjuster lever onto the front shoe, and assemble into position with the adjuster mechanism.

4-A16

❏ **STEP A17:** Fit the adjuster spring.

4-A17

❏ **STEP A18:** Fit the lower return spring.

4-A18

❏ **STEP A19:** Clean the inside of the drum using brake cleaner which degreases it.

4-A19

❏ **STEP A20:** Back off the adjuster mechanism to reduce the strut length and allow the shoe tops to move inwards.

4-A20

☐ **STEP A21:** Refit the drum.

4-A21

☐ **STEP A22:** Make sure the bearing is correctly adjusted – see *Chapter 11, Job 23*.
→ If the slots in the castellated nut are not aligned with the hole in the stub axle, loosen (DO NOT tighten) the nut a fraction of a turn, until the split pin can be fitted.
→ Lower the vehicle to the ground, then apply the foot brake and parking brake several times, to centralise the shoes and allow the automatic adjuster to take up any slack in the assembly.

IMPORTANT NOTE: Check for excess bearing play after reassembly, and again after about 150 km (100 miles) of use. If there is excess play you can't get rid of, excess stiffness in spinning the wheel (which can be very dangerous!) or excess noise, reset the clearance – see *Chapter 11, Job 23*.

Section B: Cavalier up to 1984.

☐ **STEP B1:**
Pre-1984 vehicles have a basic rear brake layout, with non-automatic brake shoe adjustment. There is a fixed slider bar (**a**)

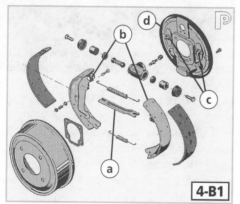

4-B1

between the shoes (**b**), and two cams (**c**) on the backplate (**d**), which are adjusted by turning their hex heads on the rear of the backplate, one adjuster for each shoe.

☐ **STEP B2:**
Depending on model, the brake shoe retaining clips (arrowed)

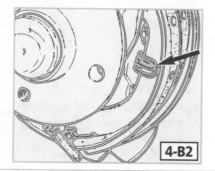

4-B2

may differ from the spring clips seen in *Section A*.
→ They may be the type seen here, which you squeeze together and remove.
→ They might be a curved spring from under which the shoe has to be slid, the spring staying in place on the backplate.

☐ **STEP B3:** Adjust each wheel as follows:
→ Turn one of the adjusters in whichever direction is required to lock the wheel, then slacken it so the wheel is just free to turn without 'binding' on the shoes.
→ Press the brake pedal to centre the shoes, then see whether further adjustment is possible.
→ Repeat this operation for the second adjuster, and then repeat on the other wheel.

JOB 5: **REAR BRAKE WHEEL CYLINDER -** *check, replace.*

TOP TIP!

• If the wheel cylinder leaks, fluid will inevitably contaminate the brake lining material. Replacing the shoes will then be necessary.

☐ **STEP 1:** Remove the brake drum and shoes – see *Job 4*.

☐ **STEP 2:** Using a pipe clamp, clamp the flexible brake fluid hose (arrowed).

5-2

☐ **STEP 3:** Remove the bleed nipple cover.

5-3

STEP 4:
Disconnect the fluid pipe from the back of the cylinder.

➔ This union could well be seized, so penetrating fluid may well be useful to help free it.

➔ Cap the pipe, using a suitable rubber or plastic cap.

➔ The spare bleed nipple cover is just right for this.

5-4

STEP 5: Undo the wheel cylinder mounting bolts.

5-5

STEP 6: Lever the wheel cylinder off the backplate.

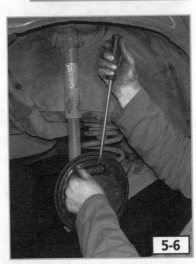

5-6

STEP 7:
Replace the cylinder as the reverse of removal, remembering to bleed the brakes once the drums are fitted.

5-7

TOP TIP!

• Although it is possible to overhaul leaking brake cylinders, replacements are inexpensive enough to make this unnecessary.

JOB 6: REAR BRAKE PADS - *check, remove, replace.*

REAR DISC BRAKE MODELS ONLY

IMPORTANT NOTE: This procedure is similar to that covered in *Job 1* for the front brake pads. However, note that a 3 mm punch will be necessary to remove the pad retaining pins.

STEP 1:
Raise the rear of the vehicle and secure it on axle stands under the axle beam, or raise with a vehicle hoist.

6-1

➔ Examine the disc for any signs of scoring, pitting or corrosion. Check also for wear ridges at the inner and outer limits of the pad contact area.

STEP 2:
The pads are clearly visible through the 'window'

6-2

in the caliper. Check that both pads display a similar degree of wear.

➔ If one pad is more worn than the other, the corresponding piston may be sticking. If this is the case, the caliper must be replaced – see *Job 7*.

➔ Check that the pad thickness is at least the recommended 7 mm (including the backplate).

➔ If the pads do not require replacing, place a rag or newspaper under the assembly and spray brake cleaner liberally around the caliper to remove dust and debris.

➔ Wire brush any corrosion from the pads and retaining spring.

TOP TIP!

• Wear gloves – or at least use barrier cream – because brake dust is very hard to shift from nails and cracks in the skin.

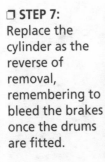

STEP 3: If the pads do need replacing, drive out the upper retaining pin with a 3 mm punch. Drift it inwards.

STEP 4: As you withdraw the pin and drift, the anti-rattle spring can be removed…
➜ …followed by the lower pin

STEP 5: As the caliper is fixed, the pistons will usually need to be retracted to get the pads past the inevitable 'wear ridge'.
➜ Use a pair of grips, as shown here.
➜ Pushing the pistons raises the fluid level in the master cylinder. Wrap rag around the reservoir to catch any spillage.

STEP 6: Pull the pads from the caliper with pliers. Also remove the anti-squeal shims that are sandwiched between the pad backplate and the piston.

STEP 7: Clean any brake dust and road dirt from the caliper using a wire brush, an old screwdriver to scrape any stubborn deposits, and liberal amounts of brake cleaning spray…

STEP 8: …then wire brush the anti-rattle spring clean.

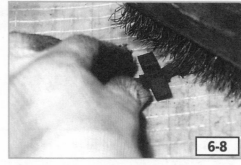

STEP 9: Before refitting the pads, make sure the recessed part of the piston face is aligned as shown here.
➜ You can buy an alignment tool like the one shown (arrowed), from Vauxhall or your local auto accessory store.
➜ Alternatively, make a template, so that the angle shown is 23 degrees.

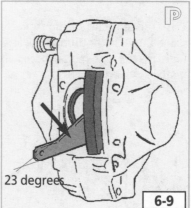

23 degrees

STEP 10: Smear a thin layer of copper grease on the backplates of the pads, and where they slide in the caliper.
➜ Don't overdo this, however, to avoid getting grease on the pad or disc surfaces.

STEP 11: On reassembly, replace the pads and fit the bottom pin first, followed by the anti-rattle spring which has to be tensioned 'inwards' so that the top pin can be fitted.

STEP 12: Tap the pins fully home, so that they are flush with the caliper.

STEP 13: With the wheels on and the vehicle back on the ground, top up the brake master cylinder, replace the cap and press the brake pedal hard several times to settle the new pads into position.

JOB 7: REAR BRAKE CALIPER - *remove, replace.*

REAR DISC BRAKE MODELS ONLY

SAFETY FIRST!

• Although it is possible to fit new seals to brake calipers, we do not recommend doing so – new seals will only be a short-term fix at best.

• We recommend fitting replacement or 'exchange', reconditioned calipers if the old ones leak, if seals are split, if bores are corroded or pistons seized.

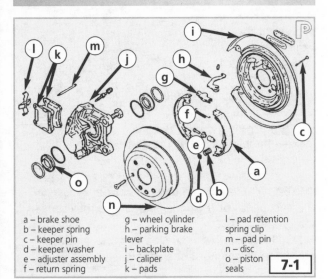

a – brake shoe
b – keeper spring
c – keeper pin
d – keeper washer
e – adjuster assembly
f – return spring
g – wheel cylinder
h – parking brake lever
i – backplate
j – caliper
k – pads
l – pad retention spring clip
m – pad pin
n – disc
o – piston seals

7-1

❒ **STEP 1:** These are the components of the combined rear disc brake and parking brake assembly.

❒ **STEP 2:** Remove the brake pads – see *Job 6*.

❒ **STEP 3:** Clamp the flexible fluid hose, as close to the caliper connection as possible, with a suitable hose clamp – see illustration *5-2*.

❒ **STEP 4:** Disconnect the brake line from the caliper. Have a suitable glass container ready to catch brake fluid.

❒ **STEP 5:** Remove the mounting bolts, and draw the caliper away from the hub.

7-5

❒ **STEP 6:** Refit the caliper as the reverse of removal.

❒ **STEP 7:** Bleed the brake system to remove any air in the flexible hoses – see *Job 16*.

JOB 8: REAR BRAKE DISC - *remove, replace.*

REAR DISC BRAKE MODELS ONLY

❒ **STEP 1:** Remove the brake pads and calipers – see *Jobs 6* and *7*.
➔ Do not disconnect the caliper from the brake fluid pipe.
➔ Instead, use a cable-tie (arrowed) to support it out of harm's way, but without straining the flexible hose.

8-1

❒ **STEP 2:** Undo the small screw that holds the disc onto the hub.
➔ If this screw is very tight, use an impact screwdriver.

8-2

❒ **STEP 3:** Remove the disc/hub unit.

8-3

TOP TIP!

❒ **STEP 4:** • The disc could well be quite difficult to remove, as the parking brake's shoes operate on the inside of it, as with a drum brake.
• Try tapping a few times with a soft headed mallet (NOT a steel one!), in case the brake shoes are binding.

8-4

TOP TIP!

❏ **STEP 5:** • If this doesn't work, insert a flat-bladed screwdriver through the inspection hole in the front of the disc and lever it upwards to undo the brake self adjuster and help release the disc.

8-5

❏ **STEP 6:** Replace the disc as the reverse of removal. It is a good idea to clean the mating surface of the hub first, using abrasive paper. This ensures the new disc fits flush.

➜ If the retaining screw has become distorted, replace it.

➜ Dress the screw with copper grease before refitting.

SAFETY FIRST!

• Before refitting the caliper, replace the pads. Old pads must never be used on a new disc.

• Brake discs should always be replaced in pairs.

| JOB 9: | PARKING BRAKE SHOES – *remove, replace.* |

REAR DISC BRAKE MODELS ONLY

❏ **STEP 1:** In the rear disc brake assembly the parking brake mechanism and shoes fit within the disc hub, making it in effect a drum brake inside a disc brake assembly.

9-1

➜ You can check the thickness of the friction material visually, through the inspection hole provided in the front of the disc.

➜ Shine a torch through the hole, and slowly turn the disc so that the entire edge of both shoes can be checked.

❏ **STEP 2:** Disconnect the parking brake cable (**a**) and its return spring (**b**) from the operating lever, at the back of the backplate.

9-2

❏ **STEP 3:** Remove the brake caliper – see *Job 7* and *Job 8*.

❏ **STEP 4:** Remove the screws holding the disc/drum assembly (**a**) to the hub (**b**) and remove the disc/drum assembly.

9-4

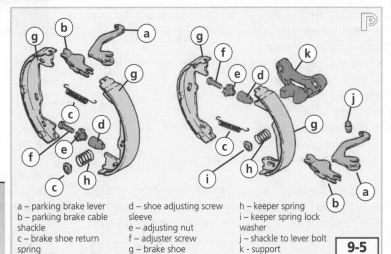

a – parking brake lever
b – parking brake cable shackle
c – brake shoe return spring

d – shoe adjusting screw sleeve
e – adjusting nut
f – adjuster screw
g – brake shoe

h – keeper spring
i – keeper spring lock washer
j – shackle to lever bolt
k - support

9-5

❏ **STEP 5:** These are the components of the parking brake shoe mechanism used with rear disc brake vehicles.

➜ The positions of the parking brake lever and adjusting mechanism vary according to the model and engine, but the process of removing components is broadly the same.

➜ The process of removing and replacing the parking brake shoes is very similar to that involved in removing the drum brake shoe assembly – see *Job 4*, and refer to it throughout.

❏ **STEP 6:** Spray brake cleaner liberally around the assembly to remove dust and debris.

❏ **STEP 7:** Remove the retaining spring clips (**a**), by compressing the spring with a pair of strong pliers and turning them by 90 degrees.

➜ This spring (**b**) is very strong.

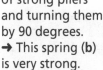

9-7

Wear eye protection just in case it slips whilst compressed.

➜ Pull the assembly from the backplate entire.

➜ Lay it out on a piece of newspaper and disassemble it.

→ Clean any remaining dirt from the springs, adjuster and operating lever mechanism.

❒ **STEP 8:** Apply a thin film of copper grease to the adjuster threads, then screw it closed.

9-8

→ Apply a thin film of copper grease to the points of the backplate that make contact with the shoes.

→ Fit the frontward shoe, securing it with its retaining spring clip.

→ Locate the operating lever in the shoe and its hole in the backplate.

→ Fit the rearward shoe, securing it with its retaining spring clip.

→ Fit the upper spring (**a**), then the adjuster (**b**), and finally the lower spring.

→ Fit the brake disc – see *Job 8*.

❒ **STEP 9:** Repeat the process on the other side of the vehicle, then adjust the parking brake – see *Job 18*.

→ Fit the wheels, and lower the vehicle off its jacks.

JOB 10: BRAKE MASTER CYLINDER - *remove, replace.*

SAFETY FIRST!

• Although it is possible to overhaul some (but not all!) master cylinder types, we do not recommend doing so.

• We recommend fitting a replacement or 'exchange', reconditioned master cylinder if the old one is faulty in any way.

❒ **STEP 1:** Before dismantling, disconnect the battery negative (-) earth/ground terminal. See *Chapter 10, Electrical, Dash, Instruments, Fact File: Disconnecting the Battery* BEFORE doing so!

❒ **STEP 2:** Press the brake pedal a few times until all pressure left in the servo has dissipated.

TOP TIP!

• Cover the bodywork with paper or cloths.
• Spilt brake fluid will harm paintwork unless thoroughly and immediately washed away with warm, soapy water.

❒ **STEP 3:** Disconnect the wiring connection from the brake fluid level indicator sensor on the master cylinder cap. Remove cap and float.

❒ **STEP 4:** Drain the master cylinder reservoir with an old battery electrolyte tester or, as shown here, with a special syphoning kit.

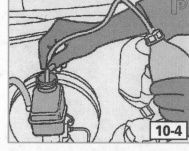

10-4

→ After removing as much fluid as possible, place a plastic bag under the reservoir cap and close the cap, to create a seal and minimise fluid loss as you remove the master cylinder unit.

❒ **STEP 5:** To remove the master cylinder:

→ Loosen the unions of the brake pipes on the master cylinder.

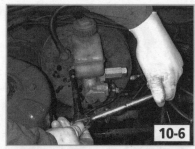

10-5

→ Return to the brake pipe unions and undo them fully.

→ Have a suitable container ready to catch any brake fluid that escapes.

→ Cap the pipe ends with suitably sized rubber caps, such as bleed nipple covers.

❒ **STEP 6:** Loosen the two mounting nuts which hold the master cylinder to the servo body.

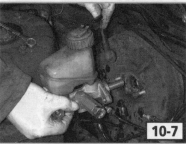

10-6

❒ **STEP 7:** Undo the master cylinder mounting nuts fully and remove it.

→ Take great care not to let any brake fluid get onto paintwork.

10-7

❒ **STEP 8:** Lever the tank from the master cylinder body. Have a cloth handy, to mop up any brake fluid.

10-8

STEP 9: Reassemble and refit as the reverse of removal. The entire brake system will have to be refilled and bled – see *Job 16*.

JOB 11: BRAKE SERVO - *check, replace.*

FACT FILE

CHECK SERVO OPERATION

• Press the brake pedal several times until all pressure left in the servo has dissipated.
• Start the engine, with the brake pedal pressed.
• As the vacuum increases in the servo, the pedal should lighten in feel. It will not move, but the resistance against your foot pressing it will decrease.
• After two to three minutes, switch the engine off, release the pedal, then press it again.
• Listen for a hissing sound from the servo.
• Press and release the pedal a few more times.
• The hissing should lessen each time, and the pedal feel increasingly firmer.
• If there are no such changes throughout this sequence, the servo is malfunctioning and should be replaced.

STEP 1A: VERY EARLY RIGHT-HAND DRIVE VEHICLES: Pre-1985 Cavaliers had the left-hand drive servo, connected to the brake pedal by a shaft running across the rear of the engine bay.

STEP 1B: ALL OTHER RIGHT-HAND DRIVE VEHICLES: These are the right-hand drive brake servo fittings.

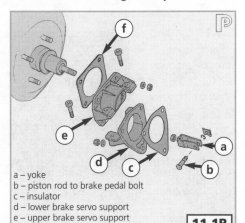

a – yoke
b – piston rod to brake pedal bolt
c – insulator
d – lower brake servo support
e – upper brake servo support
f – gasket

11-1B

STEP 1C: LEFT-HAND DRIVE: These are the left-hand drive brake servo fittings.

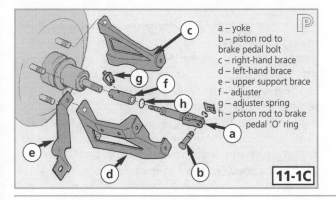

a – yoke
b – piston rod to brake pedal bolt
c – right-hand brace
d – left-hand brace
e – upper support brace
f – adjuster
g – adjuster spring
h – piston rod to brake pedal 'O' ring

11-1C

STEP 2: Before dismantling, disconnect the battery negative (-) earth/ground terminal. See *Chapter 10, Electrical, Dash, Instruments, Fact File: Disconnecting the Battery* BEFORE doing so!

STEP 3: If the model you are working on has a trim panel under the dashboard, remove the right-hand section - see *Chapter 14, Interior, Trim*.

STEP 4: If necessary, disconnect the brake pedal (depending on model):
→ Disconnect the wiring connection to the brake light switch (arrowed) – located on the brake pedal mounting bracket - then unscrew the switch unit.

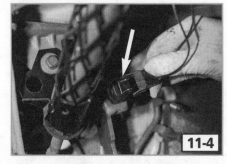

11-4

STEP 5A: Remove the spring clip (**a**) from the pin (**b**) that joins the servo rod to the brake pedal mechanism.

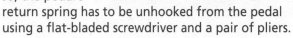

11-5A

STEP 5B: For clarity, this is the servo unit removed.
→ The pin (*11-5A, item b*) goes through the clevis (**item b**).
→ Tap the pin out.
→ Before doing so, the pedal's return spring has to be unhooked from the pedal using a flat-bladed screwdriver and a pair of pliers.

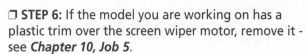

11-5B

STEP 6: If the model you are working on has a plastic trim over the screen wiper motor, remove it - see *Chapter 10, Job 5*.

STEP 7: Remove the wiper motor - see *Chapter 10, Job 5*.

STEP 8: If it is necessary in order to access the servo, remove the air induction trunking.

STEP 9: Remove the nuts securing the brake master cylinder to the brake servo body.
→ Pull the master cylinder away, and lay it to one side.

STEP 10: Remove the securing clip, and lever the vacuum pipe from the brake servo body.

STEP 11: Working in the recess where the wiper motor sits, remove the nuts that hold the servo and/or its mounting to the bulkhead (depending on version). See illustration *11-5B, item a* for stud positions.

11-11

→ On some models, these lie below plugs, which must be levered out first.

STEP 12: Pull the servo from the bulkhead.
→ There is sealant on the joint between the body and the servo's mounting bracket.
→ It may require

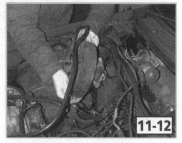

11-12

levering free with a broad-bladed screwdriver.
→ Clean all old sealant off the mounting point on the vehicle's body.

STEP 13: Undo the bolts holding the servo body to its mounting bracket and remove the bracket. Again, this joint has sealant on it, so a broad-bladed screwdriver may be needed to lever it free.
→ Clean all traces of old sealant off the mounting bracket.

STEP 14: Refitting is the reverse of removal. However, remember to apply a suitable sealant to the two joints which require it.

STEP 15: When fitting the yoke, the distance should be set to 141.5 mm.

STEP 16: Once the servo is fitted, test it - see *Step 1*.

JOB 12: BRAKE PRESSURE REGULATION VALVE - *replace, adjust*.

FACT FILE

REGULATION VALVE TYPES

• Among the models covered by this manual, there are two basic types of brake pressure regulation system – hydraulic and electronic.
• Hydraulic systems comprise modulators which limit the brake fluid pressure to the rear brakes to prevent them seizing under extreme braking.
• Electronic systems are programmed into the ECU.
• Some electronic systems contain a switch on the rear suspension that measures load on the rear wheels and sends a signal to the ECU, which makes the necessary adjustments.

Section A: Hydraulic systems.

STEP A1: The modulators are usually situated either adjacent to the master cylinder (**1**) or on the vehicle's underside, one in each brake line (**2**).

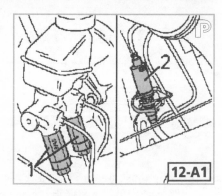

12-A1

STEP A2: If a fault is suspected, renew both as a pair.

STEP A3: To minimise spillage of brake fluid, remove the reservoir cap, place plastic sheet over it and replace the cap.
→ In the case of modulators mounted under the vehicle, it is possible to clamp the brake hose instead.

Section B: Electronic systems.

STEP B1: These systems are controlled entirely by the ECU.

STEP B2: The vehicle load sensor (arrowed) is situated on a bracket attached to the rear crossmember and upper suspension arm.
→ Diagnostic equipment is needed to test it.

12-B2

JOB 13: ABS SENSOR - *remove, clean, replace*.

Problems with the ABS system are indicated when the dashboard warning light comes on. If you have use of diagnostic equipment, you can isolate the affected sensor.
→ The most common fault is failure of the ABS sensors. This can either be due to electrical malfunction or just simply an encrustation of road dirt.
→ Without diagnostic equipment, it is impossible to tell which sensor is affected, so work round the vehicle, cleaning or replacing each sensor.

FACT FILE
SENSOR TYPES

• On all models covered by this manual, the front sensors are essentially similar in form.
• The rear sensors, however, vary between models.
• Some have a sensor attached to the brake drum back plate.
• Models with rear disc brakes have a sensor that is integral to the hub assembly.
• This Job covers removal of the former two types. If a rear sensor fault develops on a model with rear disc brakes, the whole rear hub will have to be replaced – see *Chapter 11, Job 22*.

❑ **STEP 1:** If you are using diagnostic equipment, test the function of the sensors.

❑ **STEP 2:** Before dismantling, disconnect the battery negative (-) earth/ground terminal. See *Chapter 10, Electrical, Dash, Instruments, Fact File: Disconnecting the Battery* BEFORE doing so!

❑ **STEP 3: FRONT SENSOR AND REAR SENSOR EXCEPT VECTRA:** There is an electrical connector in the electrical cable.
➜ Unclip the connector from its mounting, and separate the two halves.
➜ Clean the connector and, if necessary, use a small screwdriver to lever the clips open and apart.

13-3

❑ **STEP 4:** Undo the bolt holding the sensor to the front hub or rear brake backplate.
➜ Remove the sensor, unclipping its wiring from any mounting points as you do so.
➜ Carefully clean the sensor with brake cleaner.

13-4

❑ **STEP 5:** Refit the sensor, as the reverse of removal.

❑ **STEP 6: REAR SENSOR VECTRA:** Unclip the lead from the terminal at the rear of the hub.

❑ **STEP 7:** To remove the sensor, remove the hub – see *Chapter 11, Job 20*.

JOB 14: FLEXIBLE BRAKE HOSES - *replace*.

❑ **STEP 1:** Undo the brake pipe unions. Have a suitable container ready to catch any brake fluid that escapes.
➜ Cap the pipe ends with suitably sized rubber caps, such as bleed nipple covers.

14-1

❑ **STEP 2:** Remove the clip (arrowed) that holds the flexible hose to its inboard mounting.
➜ Brush dirt off with a wire brush and apply penetrating oil, before levering the clip loose with a screwdriver.

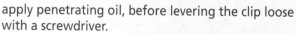
14-2

❑ **STEP 3:** Use pliers to pull the clip free.
➜ Tap back in place with a hammer on reassembly.
➜ To disconnect rigid pipes – see *Job 15*.

14-3

❑ **STEP 4:** Replace the hose as the reversal of removal.

❑ **STEP 5:** Bleed the brake system - see *Job 16*.

JOB 15: RIGID BRAKE PIPES - *replace*.

Although it is possible to make up appropriate piping runs using proprietary brake piping and an end forming tool, ready made pipes from your Vauxhall dealer or other specialist supplier offer the most time-efficient way of carrying out this job.

TOP TIP!

• Before starting work, remove the brake master cylinder reservoir cap, then refit it with polythene under the reservoir cap.
• This will minimise fluid loss as you remove the hose.

□ STEP 1: Undo the brake pipe unions.
→ Lock the larger nut while undoing the smaller one.
→ Have a suitable container ready to catch any brake fluid that escapes.
→ Cap the pipe ends with suitably sized rubber caps, such as bleed nipple covers.

□ STEP 2: Free the pipe from its securing clips, working progressively along its length.
→ There are several clip types: Some with screws; others with fold-back tabs; all are obvious.

□ STEP 3: Refit as the reverse of removal.

□ STEP 4: Bleed the brake system - see *Job 16*.

JOB 16: BRAKE BLEEDING.

This job is best done with the aid of an assistant. If this option is not available, there are two main options: the one-way valve bleeding kit, and the pressure bleeding kit.
→ The important point, no matter what technique is employed, is to never let the master cylinder reservoir run out of fluid.
→ If you do so, the system will ingest air, the element you are trying to displace. Then you'll have to start all over again!

TOP TIP!

• This process is the most efficient one for removing air from all of the pipe runs.
• Always start with the (rear brakes') bleed nipple furthest away from the master cylinder.
• Next, bleed the other rear brake.
• Third, bleed the front brake on the opposite side to the master cylinder.
• Finish with the front, master cylinder-side brake.

Section A: Both types.

□ STEP A1: Start by topping up the brake fluid reservoir, unless you are first draining the system, of course – see *Job 16*.
→ This automatic 'feeder' is left sitting on top of the reservoir and ensures that the fluid is kept topped up.
→ Otherwise, carry out regular visual checks.

Section B: With a one-way bleeding valve.

□ STEP B1: Remove the bleed nipple cap and keep it safe – it helps prevent the valve from seizing.

□ STEP B2: Connect the one-way tube to the relevant bleed nipple.
→ Undo the nipple half a turn.
→ Push the brake pedal smoothly and steadily to the floor and release it slowly.
→ Top up the fluid level in the master cylinder reservoir and repeat several times, until the fluid flows with

no air bubbles in it.
→ Tighten the bleed nipple and remove the tube.

□ STEP B3: Repeat this on each of the four brakes. If the pedal still feels 'spongy', repeat the process until the pedal feels firm.

Section C: With a pressure bleeding kit.

☐ **STEP C1:**
Attach the
pressure kit to
the brake
master cylinder
reservoir.

☐ **STEP C2:**
Follow the
instructions in
*Section B, Step
B2-on*.

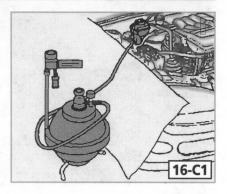
16-C1

JOB 17: BRAKE FLUID – *check, change.*

Although most models covered by this manual have
a dashboard-mounted warning light to tell you
when the brake fluid level is getting low, it's wise to
regularly check the level by inspection. The master
cylinder reservoir has a minimum level marking on
the side. If the level of fluid in the tank is below this,
top up the fluid.

TOP TIP!

☐ **STEP 1:** • As the reservoir needs to be translucent
for the fluid level to be checked, keep it clean. Road
dirt, oil and even overspray from paintwork repairs
can cover it, making spot checks impossible.

☐ **STEP 2:** Renew the fluid by bleeding the brake
system, referring to the last Job.
➜ However, start by syphoning the existing fluid
out of the reservoir, and flush the fluid through until
the old fluid in the pipes has been expelled, from all
bleed nipples.

JOB 18: PARKING BRAKE CABLE – *adjust.*

The approach to this job varies, according to
whether the vehicle has self-adjusting rear brakes.
➜ **SELF-ADJUSTING REAR BRAKES:** Excessive parking
brake travel indicates either worn brake shoes or
stretching of the brake cable. Check the brake shoe
lining thickness – see *Job 4* - and replace the shoes if
necessary before adjusting the cable.

➜ **NOT SELF-ADJUSTING REAR BRAKES:** Whether
they're disc or drum (disc brakes incorporate drum
parking brakes), adjusting the parking brake entails
adjusting first the shoes then the cable.

Section A: Models with drum brakes.

☐ **STEP A1:** Fully release the parking brake.
➜ If the model you are working on has an exhaust
heat shield, you will need to remove its mounting
bolts and take it off.

ADJUSTABLE DRUM BRAKES

☐ **STEP A2:** Pre-1985 Cavaliers have two cams on
each brake backplate. These can be turned via their
hexagonal heads to move the brake shoes. With the
parking brake off, turn each adjuster until the wheel
locks, then turn them back until the wheel turns
freely.

ALL DRUM BRAKES

☐ **STEP A3:**
These are
the two
cable and
adjuster
systems.
➜ One type
(**A**) has an
adjustment
nut on the
parking brake lever rod.
➜ The other type (**B**) has an adjuster on the rear axle.

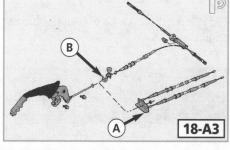
18-A3

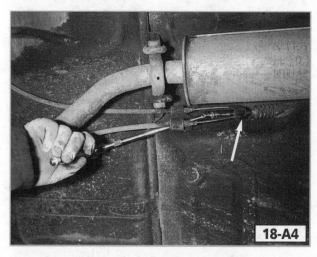
18-A4

☐ **STEP A4: TYPE A:** Adjust the cable adjuster
(arrowed) under the mid-section of the vehicle until
the brakes start to bind as you or an assistant turn
the wheel.

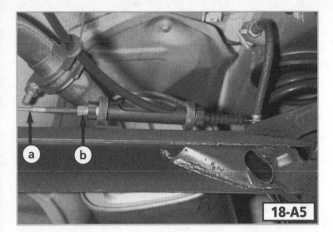

18-A5

☐ **STEP A5: TYPE B:** Hold the square end (**a**) of the adjuster with a spanner and adjust the nut (**b**) until the brakes start to bind as you or an assistant turn the road wheel.

☐ **STEP A6: BOTH TYPES:** Back the adjuster off until the wheels are JUST free to turn without the brakes binding.
➔ Now check that the brakes begin to bind when the parking brake is on the second 'click' of its ratchet.
➔ Put copper grease on the adjuster threads to stop them from seizing.

☐ **STEP A7:** If applicable, replace the exhaust heat shield.

Section B: Models with adjustable disc/drum brakes.

☐ **STEP B1:** With the parking brake off and using a screwdriver through the inspection hole on the front of the

18-B1

disc, turn the brake adjustment wheel until the brake shoes can be heard to contact the parking brake drum when you turn the disc.
➔ Back off the adjuster until the disc turns freely.
➔ Do this on both sides of the vehicle.

☐ **STEP B2:** Apply the parking brake and footbrake to centre the shoes, then check whether further adjustment is needed.

☐ **STEP B3:** Pull the parking brake to the second 'click' of its ratchet.

☐ **STEP B4:** If the model you are working on has an exhaust heat shield, you will need to remove its mounting bolts and take it off.

☐ **STEP B5:** Tighten up the cable adjuster (see *Steps 18-A3, A4* and *A5*) until the shoes can be heard to rub when the drum is turned.

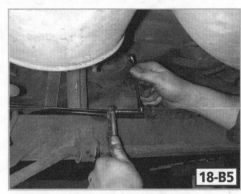

18-B5

☐ **STEP B6:** Back the adjuster off until the drums turn freely.

☐ **STEP B7:** If applicable, refit the exhaust heat shield.

☐ **STEP B8:** Replace the wheels, and lower the vehicle off its jacks.

JOB 19: PARKING BRAKE CABLE - *renew*.

☐ **STEP 1:** Fully release the parking brake.

☐ **STEP 2:** If the vehicle you are working on has an exhaust heat shield, remove its mounting bolts and take it off.

☐ **STEP 3:** Disconnect the parking brake cable from the parking brake mechanism, by removing the adjustment nut (see *Job 18, Step A4*)...

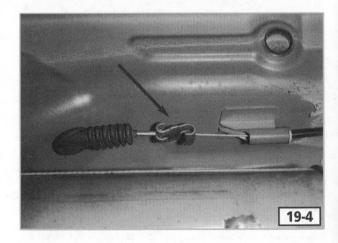

19-4

☐ **STEP 4:** ...or, if the adjuster is on the rear axle (see *Job 18, Step A5*), unclip the 'S' connection (arrowed).

STEP 5: If the model you are working on has rear drum brakes, remove the drums and brake shoes - see *Job 4*.

STEP 6: Remove the clips holding the parking brake cable in the brake drum backplate.

19-6

STEP 7: Pull the parking brake cable out from the brake back plate.

19-7

STEP 8: Remove the cable from its various guides and mounts.
➜ Some of the clips may need to be carefully bent open to allow the cable out.
➜ If the model you are working on has rear disc brakes, disconnect the cable ends from the operating levers on the brake backplates.

19-8

STEP 9: Carefully refit the cable (a) through the hole (b) in the brake backplate...

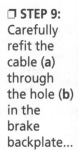
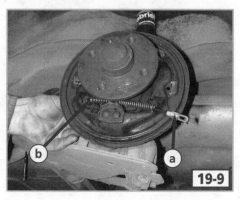

b   a

19-9

STEP 10: ... and fit a new retaining clip.

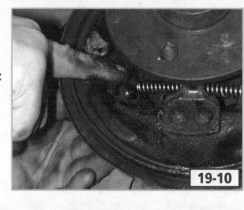

19-10

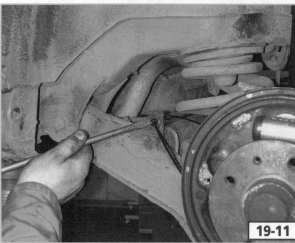

19-11

STEP 11: Replace the cable in each of its various mounting points.

STEP 12: Renewing the operating cable of the disc/drum parking brake is generally similar to the same job on drum brakes.
➜ The cable is in three sections – a long centre section and short sections at either end.

19-13

STEP 13: The rear cable is attached to the centre section by a connector.
➜ With the parking brake lever off, the rear cable end fitting can be levered out of the connector.

STEP 14: The front section runs from the parking brake lever to the compensator plate.

19-14

→ Remove the parking brake lever gaiter and remove the adjustment nut.

→ Under the vehicle, remove the rear exhaust section and heatshield to allow access to the compensator plate.

→ Turn the end fitting of the cable through a quarter of a turn and disengage it from the compensator plate.

STEP 15: To remove the centre section of the parking brake cable, disconnect the end cables as described.

→ Release the cable from the various guides.

ALL

STEP 16: Refitting is the reverse of removal.

→ Adjust the parking brake cable and brake shoes as appropriate.

JOB 20: **PARKING BRAKE LEVER -** r*emove, replace.*

STEP 1: The parking brake lever is situated in the centre console on many models, and on the side of the centre console on others.

→ It is usually necessary to remove the driver's seat in order to access the parking brake lever bolts – see *Chapter 14, Interior, Trim.*

→ In the case of the Vectra, it is necessary to remove the exhaust system and heatshield to gain access.

STEP 2: Remove the right-hand seat and drift out the sliding seat rail mountings – see *Chapter 14, Interior, Trim.*

STEP 3: If the parking brake lever is situated in the centre console, remove this – see *Chapter 14, Interior, Trim.*

STEP 4: Pull back the carpet and unscrew the parking brake assembly securing bolts.

→ The bolts are sometimes accessible through slits in the carpet.

→ In some cases, the plastic gaiter must be removed.

CAVALIER AND CALIBRA

STEP 5: Slacken the cable adjuster – see illustration *19-14.*

STEP 6:SOHC: Disconnect the parking brake cable front end from the 'S' connector – see *Job 19.*

→ **DOHC:** Disconnect the parking brake rod from the compensator.

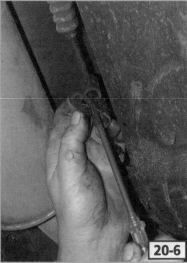

20-6

STEP 7: Working inside the vehicle, remove the rear section of the centre console – see *Chapter 14, Job 7*

20-7

– parking brake lever gaiter, cable adjusting nut and parking brake lever securing bolts.

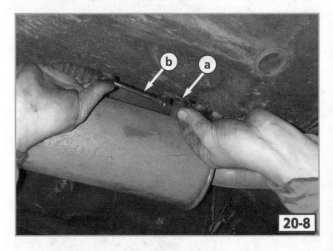

20-8

STEP 8: SOHC: Remove the 'S' connector (**a**) from the parking lever operating rod (**b**).

STEP 9: Slide the parking brake lever operating rod rubber boot.

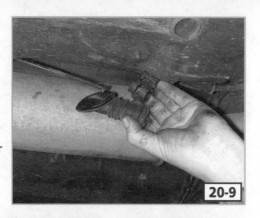

20-9

20-10

STEP 10: You can now lift the parking brake lever out from inside the vehicle.

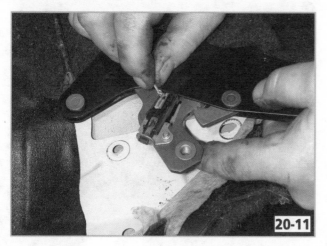

20-11

STEP 11: Disconnect the parking brake warning light wire before you pull the lever away.

VECTRA

STEP 12: Remove the exhaust and heat shield – see *Chapter 9, Part C*.
→ How much of the exhaust system you have to remove varies according to the year and engine.
→ Before removing the entire system, check whether you can at least leave the front (down) pipe in position!

STEP 13: Disconnect the cable.
→ Push the compensator fully forward and twist it to achieve this.

STEP 14: Inside the vehicle, remove the cable end adjustment nut and disconnect the parking brake warning lamp wire.

STEP 15: Undo the four fixing parking brake lever mounting nuts.
→ The lever may now be withdrawn.

JOB 21:	BRAKE SERVO VACUUM PUMP (DIESEL ONLY) - *remove, replace*.

FACT FILE
VACUUM PUMPS

• Diesel engines lack an inlet manifold Venturi. To provide a vacuum for the brake servo, a vacuum pump is used.
• On some engines, the vacuum pump is driven by the camshaft and mounted at the left hand end of the cylinder head.
• On other engines, this is mounted on and driven by the alternator.

Section A: Camshaft driven.

21-A1

STEP A1: Remove the servo vacuum hose. Hold nut (a) still and turn pipe nut (b).
→ If fitted, remove the EGR vacuum hose.
→ Undo the Allen-headed fixing screws, as shown, and withdraw the pump.
→ There may be some small amount of oil spillage.
→ Retrieve the small oil pipe and driving dog.

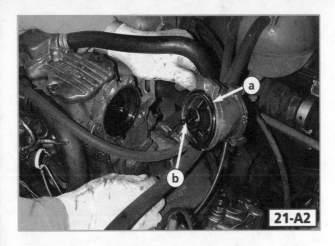

21-A2

☐ **STEP A2:** Reassembly is the reverse of stripping.
➜ Use new 'O' ring oil seals (**a**).
➜ Fit the central oil pipe and drive dog (**b**) before offering pump body into position.

Section B: Alternator mounted pumps.

☐ **STEP B1:** Raise the vehicle and remove the exhaust heat shield to improve access to the vacuum pump. Alternatively, to remove the alternator and pump complete – see *Chapter 10, Electrical, Instruments*.

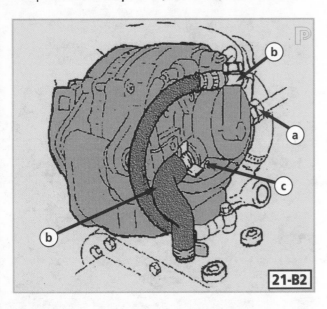

21-B2

☐ **STEP B2:** Disconnect the servo vacuum pipe (**a**) and the oil feed/return pipes (**b**).
➜ Remove the securing screws (**c**) and pull the pump off.

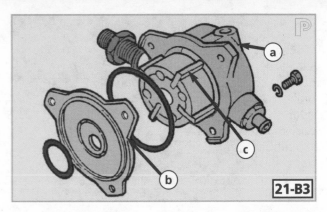

21-B3

☐ **STEP B3:** To check the pump, remove the impeller cover (**a**), which is a bayonet fitting.
➜ Hold the pump body securely upside down and drift the cover (**b**) counter-clockwise.
➜ If the vanes (**c**) are less than 13 mm they should be renewed.
➜ If the pump body internal diameter is less than 57.1 mm the pump should be replaced.
➜ If the pump or vanes have to be renewed, also renew the valve.

☐ **STEP B4:** Refitting is the opposite of removal.
➜ Pour a small quantity of fresh engine oil into the oil feed aperture before reconnecting the pipe.
➜ Check that the alternator turns freely – if not, remove the pump and reassemble.

CHAPTER 13: BODYWORK

*Please read **Chapter 2 Safety First** before carrying out any work on your car.*

CONTENTS

Jobs featured in this chapter concern the parts of the bodywork that can be 'unbolted and replaced'. Certain specific areas that may seem on first consideration to be within the remit of this chapter – such as lighting – are covered in other sections.

Some of the models covered by this manual feature glass panels 'bonded' in place, rather like bonded front screens. Replacing these requires the help of your Vauxhall dealer or a specialist such as Autoglass. This is explained further in *Job 1, Section B*, where we show an Autoglass operative at work.

TOP TIP!

• Lay an old quilt across the front bodywork to offer protection if the windscreen slips while being lifted away.
• Consider fitting a new windscreen rubber at the same time as the glass (early vehicles).

JOB 1:   SCREEN GLASS – *remove, replace.*

Section A: Screen held in with rubbers.

On some of the early vehicles covered by this manual it is possible to fit a new windscreen yourself, because these vehicles have a screen held in place with windscreen rubbers. However, we don't recommend it! A company such as Autoglass will be able to supply and fit a screen for little more than the cost of buying one. The risk of breaking a screen when you fit it yourself is quite high.

STEP A1: Mel from Autoglass removed the old screen from the vehicle and cleaned off any adhesive from the aperture.

1-A1

→ Allow time for de-rusting and painting any affected areas.
→ The new rubber has been fitted to the glass.
→ Fitting a string into the groove on the rubber which will fit over the lip in the windscreen aperture.
→ The loose ends seen here have simply been tucked into the upper edge of the rubber so that they are out of the way as the glass is being fitted.
→ Soapy lubricant is applied to the rubber to help it slip in to place.

STEP A2: If you are doing this yourself, it's best to have two people offer up the new glass with rubber. The rubber must be seated evenly in to the aperture all the way around.

1-A2

STEP A3: As the fitting string is pulled out of the rubber, the inner lip is eased over the windscreen aperture. Particularly on an older, brittle rubber, you have to take great care not to pull the string through the rubber.

1-A3

STEP A4: There is always a little tidy up of rubber flanges to carry out, to make sure that they are all sitting evenly. If the screen is pushed in too far, the rubbers will start to recess in to the

1-A4

seating around the windscreen aperture. The outer lip should lie more or less flush with the bodywork.

Section B: Bonded glass.

It is best to have a windscreen specialist replace a bonded windscreen because of the equipment required. Once again, Autoglass demonstrated how it replaced a typical type of bonded windscreen.

TOP TIP!

STEP B1: • Mel from Autoglass recommends putting masking tape over the screen vent outlets to prevent debris from falling into the vents.

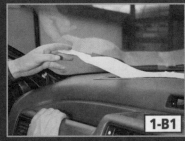

1-B1

• If debris – or worse still, shards of broken glass – fall into the vent, it could be thrown into the eyes of the vehicle's occupants next time the screen vent fan is turned on.

STEP B2: There will inevitably be interior trim to remove from around the screen. In this case, the trim is partly screwed and partly clipped into place.
→ Use an upholstery lever for freeing the trim clips.
→ Take special care where, as in this case, an alarm sensor is fitted on the trim. Be sure not to disturb the wiring.

1-B2

STEP B3: In cases where a trim cover is fitted at the base of the screen, remove the wipers and unscrew

1-B3

or clip off the trim cover as necessary.

STEP B4: Mel wears protective goggles and gloves for this stage of the work.
→ The bonded glass has to be cut from the screen aperture.
→ This special tool has a blade which reaches behind the glass and cuts through the sealant as Mel pulls it around the perimeter of the screen.

1-B4

STEP B5: The old glass can now be lifted away.

1-B5

STEP B6: Mel spends quite a considerable amount of time cleaning up the screen aperture before applying the adhesive hardener to the aperture frame.

1-B6

→ On any vehicle more than a few years old, there will almost inevitably be rust found behind the glass or glass trim.
→ In worst cases, some welding may be necessary; in almost all cases it will be necessary to clean off the rust, apply a product such as Wurth Rust Killer and then prime and paint the metal.

STEP B7: As with the screen frame, Mel uses a panel wipe to clean all traces off the surface of the glass (where it will touch the

1-B7

screen frame) before wiping hardener on to the surface of the glass.

STEP B8: The glass seals are now fitted in place around the edge of the glass – only applicable to certain types of screen.

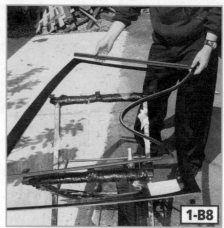

1-B8

STEP B9: A very thick bead of bonding filler is now applied by Mel all the way around the screen frame.

1-B9

STEP B10: Note the special suction lifting pads that Mel uses to grip the glass so that he can lower it accurately in to position on the screen frame. Note also the tabs of masking tape fitted ready in place on top of the screen...

1-B10

1-B11

STEP B11: ...so that they can be used to hold the screen at the correct height while the bonding filler goes off.

❑ **STEP B12:** Where the rear view mirror is fitted to the glass, use special double-sided mirror fixing tape, but be

1-B12

sure to clean both the mounting pad on the mirror and the area of the glass to which it is fitted with panel wipe so that there are no traces of grease on either.

TOP TIP!

• The screen that Mel took from this vehicle was cracked right down the middle, starting from the rear view mirror mounting position.
• The crack was caused by someone having previously fitted the mirror using Superglue or epoxy resin.
• Either of them will cause differential expansion to take place in the glass and will, in every case, cause it to crack.

1-B13

❑ **STEP B13:** A properly fitted screen will be free from leaks and also from crack-inducing stresses.

TOP TIP!

• Be sure to clean the screen wiper blades and, if necessary, replace them so that the new screen is not instantly marked with disfiguring scratches.

JOB 2:	**BONNET** - *remove, replace, adjust.*

❑ **STEP 1:** Open the bonnet and rest it on its stay. Draw round the hinges with a marker pen, so they can be replaced in the same positions.

2-1

❑ **STEP 2:** On the Vectra, remove the windscreen washer hose.
➜ If there is a underbonnet light situated in the bonnet, disconnect the wiring.

2-2

❑ **STEP 3:** Some models have struts - fitted to the hinges - to hold the bonnet open. Others have support stays.
➜ In either case, one assistant either side of the vehicle will be needed to support while the hinge bolts are undone.

2-3

TOP TIP!

• Lay an old quilt across the windscreen and another across the front bodywork to offer protection if the bonnet slips as it is being lifted away.

❑ **STEP 4:** Undo the four bolts (arrowed – two each side) holding the hinges to the bonnet.

2-4

STEP 5: Lift the bonnet carefully away.

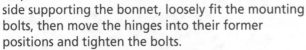

2-5

STEP 6: Refit as the reverse of removal.
➜ With a helper either side supporting the bonnet, loosely fit the mounting bolts, then move the hinges into their former positions and tighten the bolts.
➜ If you have fitted a new bonnet, transfer any rubber buffers and/or liner material.

STEP 7: Close the bonnet.
➜ Make sure the panel gaps between it and the wings are both equal and parallel.
➜ If not, loosen the hinge mounting bolts and retighten in the desired position.
➜ This can be a process of trial and error.

STEP 8: If you have to adjust the hinges, the catch striker may now be out of alignment. If so, loosen it with an open spanner, then move it to the desired position and tighten it. Make sure it is also set at the correct height should be 40-45 mm.

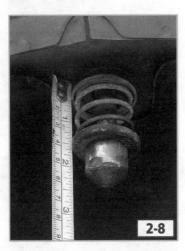

2-8

JOB 3: BONNET RELEASE CABLE - *remove, replace.*

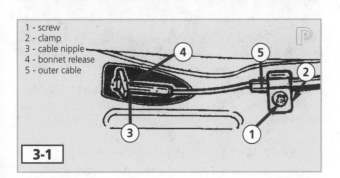

1 - screw
2 - clamp
3 - cable nipple
4 - bonnet release
5 - outer cable

3-1

STEP 1: These are the basic components of the bonnet lock.

STEP 2: Undo the clamp holding the release cable to the front panel.

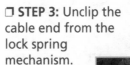

3-2

STEP 3: Unclip the cable end from the lock spring mechanism.
➜ It is usually easier (on the Vectra it's necessary) to remove the grille – see *Chapter 14, Interior, Trim*.

3-3

STEP 4: Release all the clips and fixings along the cable (a) and make a note of how it is routed.

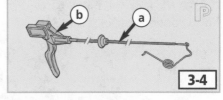

3-4

➜ Finally, undo the bonnet release lever (b) in the driver's footwell.

TOP TIP!

• It is a good idea to tie string to the end of the cable. That way, you can pull the new cable back into position.

STEP 5: The cable end simply clips into the release handle.

STEP 6: Pull the cable into the footwell.
➜ Remove

3-5

the cable's rubber grommet from the bulkhead, to make refitting easier.

STEP 7: Route the replacement cable through the bulkhead and replace its rubber grommet.

❑ **STEP 8:** Secure the release lever.

❑ **STEP 9:** Secure the cable in all its correct fixings, and connect it to the locking spring mechanism.

❑ **STEP 10:** Slide the outer cable in its clamp, until there is no free play in the inner cable. Then tighten the clamp on the slam panel.

JOB 4: TAILGATE AND STRUT - *remove, replace, adjust.*

SAFETY FIRST!

• **The tailgate is surprisingly heavy!**

• **Have one or preferably two people on hand to assist when removing or replacing it**

Section A: Strut – replacement.

❑ **STEP A1:** Prop the tailgate open or have someone hold it open for you.

❑ **STEP A2:** Lift off the clips on the ball joints with a flat-bladed screwdriver, then pull the ball joint socket off the pin.

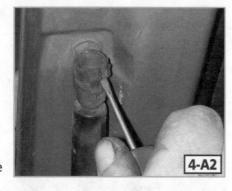

4-A2

❑ **STEP A3:** Replace as the reverse of removal.

Section B: Tailgate.

❑ **STEP B1:** If the vehicle you are working on has an interior trim panel on the tailgate, remove it first.

❑ **STEP B2:** Disconnect the tailgate wiper and heated rear screen connectors.

4-B2

❑ **STEP B3:** Attach a strong fish wire to the connectors.

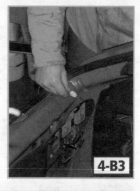

4-B3

❑ **STEP B4:** Pull the wires and fish wire up and out of the rear body.

❑ **STEP B5:** Pull the rear wash jet out, disconnect the pipe, attach a fish wire and pull this through.

4-B4

❑ **STEP B6:** The tailgate is held by pins and circlips.
➔ Lever off the circlips (a).
➔ Support the tailgate (it's very heavy) and use a narrow punch to drift out the pins (b) in the direction of the arrow.
➔ This is a three-person job.

4-B5

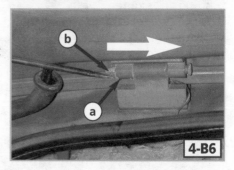

4-B6

Section C: Luggage bay lid.

❑ **STEP C1:** Draw round the hinges with a marker to indicate their original position, then unbolt the hinges.

❑ **STEP C2:** With your helper, lift the lid away from the vehicle. Store it carefully if it is to be re-used.

❑ **STEP C3:** Refit as the reverse of removal. At first, loosely fit the hinge mounting bolts, then move the hinges into their former positions and tighten the bolts.

STEP C4: The lid should now be in its former position.

→ To check this, carefully close it, and make sure the panel gaps between it and the wings are both equal and straight.

→ If not, loosen the hinge mounting bolts and retighten in the desired position.

→ This can be a process of trial and error.

STEP C5: If you have to adjust the hinges, the catch striker may now be out of alignment.

→ If so, loosen its retaining bolt, then move the striker to the desired position and tighten the bolt.

JOB 5:	**TAILGATE HANDLE, LOCK, LOCK MOTOR AND STRIKE PLATE** - *remove, replace, adjust.*

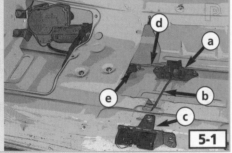

a - lock cylinder and handle assembly
b - handle to lock connecting rod
c - lock
d - handle to central locking servo
e - central locking servo

5-1

STEP 1: This is a typical tailgate locking mechanism.

→ Working inside the tailgate panel, remove any trim panels necessary to access the locking mechanism.

FACT FILE

LOCKS AND STRIKERS.

There are a number of strikers and locks fitted.

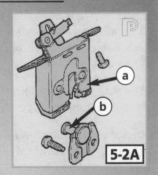

STEP 2A: • This is the Cavalier/Vectra tailgate lock (a) with striker pin (b).

STEP 2B: • This is the Calibra tailgate lock (a) with fabricated striker plate (b).

5-2B

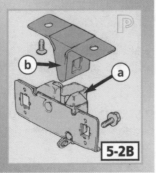

FACT FILE

continued...

STEP 2C: • This is the Vectra estate tailgate lock.

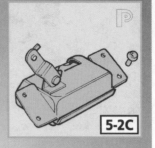

5-2C

STEP 2D: • This is the Vectra hatchback luggage bay lock (a) with 'bent wire' striker (b).

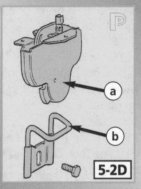

5-2D

STEP 3: Undo and remove the nuts (a) holding the handle/cylinder assembly.

→ This, depending on the model

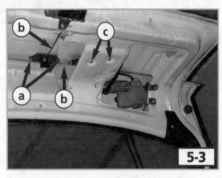

5-3

you are working on, could be incorporated into the rear panel.

→ It may be necessary to remove the tailgate wiper motor (see *Chapter 10, Job 6*), and possibly licence plate lights, to access them.

→ Remove the handle.

→ Now pull the connector rodor rods (b) out of their location(s) on the lock barrel.

→ Depending on the model you are working on, removing the handle's mounting nuts may well have freed the lock assembly. If not, remove its mounting bolts or nuts and pull the lock barrel from its location.

→ **MODELS WITH CENTRAL LOCKING:** If the vehicle you are working on has central locking, disconnect the wiring connector from the lock barrel assembly. Undo the lock motor's mounting bolts (c) and remove the motor, unhooking the connector rod from the lock barrel as you do so.

❐ **STEP 4:** Pull the connector rod(s) out of their location(s) on the lock catch.

5-4

➔ Draw around the lock catch with a marker pen, to help you replace it in the right position.
➔ Remove its mounting bolts (arrowed), and take it from its location.

❐ **STEP 5:** Draw around the lock strike plate with a marker pen, to help you regain its correct position.
➔ Remove its mounting bolts (arrowed), and remove the plate.

❐ **STEP 6:** Refit as the reverse of removal.
➔ Fit the lock catch and strike plate in their original positions, as governed by the marks.
➔ Shut the boot/tailgate and note if the lock catch engages cleanly. If not, adjust the catch plate until it does.

5-6

JOB 6: TAILGATE SPOLIER - *remove, replace.*

FACT FILE
SPOILERS

• Cavalier SRi and CD were fitted with a tailgate rubber spoiler, estates with an air deflector. These are simply bolted through the tailgate.
• Vauxhall supplied tailgate spoilers for the Cavalier and Calibra. These are also bolted through the tailgate, but some were also bonded.
• Certain Vectra models have been supplied with tailgate spoilers.

❐ **STEP 1:** Working inside the luggage bay, remove the lower trim panel - see *Chapter 14, Job 1*.

❐ **STEP 2:** Identify and unfasten the spoiler fixings.
➔ These will vary in number and position according to the model,

6-2

year and design of spoiler. This is typical of those fitted to the Calibra.

❐ **STEP 3:** If the spoiler is bonded, the application of heat may be needed to remove it.
➔ Beware that both the application of heat and the removal of bonded spoilers may damage paintwork.

❐ **STEP 4:** Refitting non-bonded spoilers is the reverse of removal.
➔ Clean and degrease the contact areas before fitting.
➔ Renew any seals or gaskets and/or apply sealant, as necessary.

❐ **STEP 5:** Refitting bonded spoilers entails removing any material that remains on the contact area of the paintwork and degreasing it.
➔ New spoilers should have self-adhesive patches; pull the covering strips off these before fitting.

JOB 7: DOOR - *remove, replace, adjust.*

SAFETY FIRST!
VEHICLES WITH SIDE AIRBAGS
• Side airbags are located in the front seats of some vehicles and are activated by control units situated in the doors.

• The side airbag system must be de-activated before starting work on the doors.

• Disconnect the battery earth terminal and allow one minute or preferably longer for the capacitor to dischage. See *Chapter 10, Electrical, Dash, Instruments, Fact File: Disconnecting the Battery* BEFORE doing so!

• See *Chapter 14, Job 11, Section E* for more information on side air bags.

❐ **STEP 1:** Open the door and place a suitable, padded support under it.

STEP 2: Disconnect any wiring that feeds into the door.

→ Some models have a connector **(a)** for this purpose. Simply twist its collar anti-clockwise to release it and pull apart the connector.

→ In some cases it will be necessary to remove the door trim to access wiring connectors – see *Chapter 14, Interior, Trim*.

7-2

STEP 3: Remove the check rod fitting.

→ In some cases, this is a roll pin that has to be drifted out.

→ In some cases, this is a nut and bolt (see illustration *7-2, item b*).

> ## TOP TIP!
>
> • The hinges are welded in place, so if they are worn, they can't be replaced. One solution is to replace the door pins with slightly fatter items from larger vehicles in the Vauxhall range. Your local Vauxhall dealer should be able to help you source suitable pins.

STEP 4: While a helper takes the weight of the door, remove the door pins.

→ Remove the end caps.

→ Drift the pins out with a suitably sized punch.

→ Better still, use a door pin remover as illustrated.

→ Lift the door carefully away from the vehicle.

7-4

STEP 5: Refit as the reverse of removal.

→ Check that the door shuts cleanly, with the outer panel flush to the surrounding panels.

→ If it does not, loosen the door striking pin and move it to the correct position.

→ This can be a process of trial and error.

7-5

STEP 6: Remember to lubricate the catch mechanism lightly with white silicone spray grease.

JOB 8:	DOOR HANDLE, DOOR LOCK – *remove, replace*.

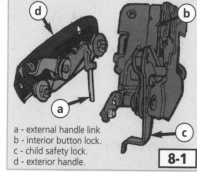

STEP 1: Remove the door trim panel – see *Chapter 14, Job 2*.

a - external handle link
b - interior button lock.
c - child safety lock.
d - exterior handle.

8-1

STEP 2: Undo the two nuts holding the door handle to the door skin or lock mechanism plate, depending on model.

→ To access them, you may have to remove the rear window guide channel.

→ Remove the mounting bolts and take it out of the door – see *Job 9*.

8-2

STEP 3: Remove the handle mechanism from the door.

→ You may have to disconnect the central locking electrical connector before doing so.

8-3

STEP 4: If the model you are working on has its lock mechanism separate from the handle:

→ Lever off the clip holding the control rod in place.

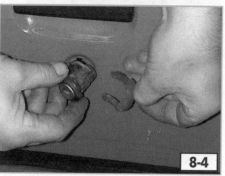

8-4

→ Pull the rod out of the lock arm.
→ Remove the clip holding the lock in place.
→ Remove the lock mechanism from the door.

STEP 5: If the model you are working on has its lock mechanism integral with the handle:
→ Lever off the clip holding the control rod(s) in place.
→ Pull them out of the lock arm.

8-5

→ Remove the bolts holding the lock mechanism and its bracket to the rearwards panel of the door.
→ Withdraw the mechanism.

STEP 6: CENTRAL LOCKING: If the vehicle you are working on has central locking, undo the mounting bolts, and remove the central locking motor and microswitch.

8-6

STEP 7: Reassemble and refit the mechanisms as the reverse of removal.

JOB 9:	DOOR GLASS AND MECHANISMS - *remove, replace*.

SAFETY FIRST!

VEHICLES WITH SIDE AIRBAGS

• Side airbags are located in the front seats of some vehicles and are activated by control units situated in the doors.

• The side airbag system must be de-activated before starting work on the doors.

• Disconnect the battery earth terminal and allow one minute or preferably longer for the capacitor to dischage. See *Chapter 10, Electrical, Dash, Instruments, Fact File: Disconnecting the Battery* BEFORE doing so!

• See *Chapter 14, Job 11, Section E* for more information on side air bags.

The approach to this Job differs slightly depending on whether you are dismantling a front or rear door.

Section A: Front doors.

STEP A1: Remove the door trim panel and, if present, the polythene damp protection sheet – see *Chapter 14, Job 2*.

9-A1

STEP A2: Make sure the window glass is fully up, then remove the rear window guide channel, by removing its mounting bolts and taking it carefully out of the door.
→ Depending on model, the bolts could be in the glass aperture, on the inside face of the door, or on the rear edge.

STEP A3: Wind the window half way down. Ask someone to hold the glass steady for you throughout the rest of the Job.
→ If the vehicle has electric windows, temporarily connect the battery and lower the window until the guide channel can be accessed through the aperture in the door frame.

9-A2

STEP A4: The regulator guide has two retaining bolts, which also provide a degree of adjustment. Mark round them with a marker pen, so you can replace them in the same position. Then undo them.

9-A4

STEP A5: If the the winder/ regulator mechanism is to be removed, drill out the securing rivets using an 8.5 mm drill...

9-A5

STEP A6: ...and remove the mechanism securing bolts (arrowed).

9-A6

STEP A7: If the model you are working on has one, push the plastic end stop from the guide channel with a flat-bladed screwdriver.

STEP A8: Manoeuvre the mechanism out of the door.

9-A8

STEP A9: Items (1) and (2) indicate the securing points. The slide (A) may be removed – early models.

9-A9

STEP A10: VECTRA WITH ELECTRIC WINDOWS: If the model you are working on has electric windows, remove the wiring connector from the motor as you do so.
→ Undo the electric motor's mounting screws, and remove it with the mechanism.
→ **ALL MODELS:** If the glass is not to be removed, you can secure it in position with tape as seen here.

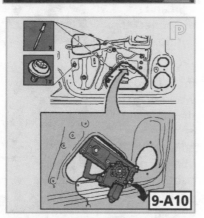

9-A10

STEP A11: Remove the inner and outer sealing strips from the door top.
→ You may have to remove the door mirror to extract the outer rubber - see *Job 11*.

9-A11

STEP A12: Remove the glass upwards and outwards through its aperture.

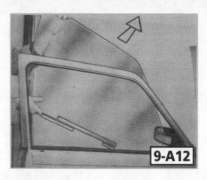

9-A12

STEP A13: Replace as the reverse of removal.

TOP TIP!

• Use double-sided tape to stick the plastic damp shield back in place inside the door trim.

Section B: Rear doors.

STEP B1: Remove the door trim panel and, if present, the polythene damp shield – see *Chapter 14, Job 2*.

STEP B2: Remove the inner and outer sealing strips from the door top.

STEP B3: Make sure the window is fully down, then remove the rear window guide channel.
→ Remove the guide chanel mounting bolts and take it out of the door.

9-B3

STEP B4: From here, you will be able to manoeuvre the glass off its guide and out through the door aperture.

9-B4

STEP B5: Remove the bolts or drill out the rivets holding the window winder mechanism to the door skin, and manoeuvre the mechanism out of the door.
→ If the model you are working on has electric windows, remove the wiring connector from the motor as you do so.

STEP B6: If an electric window motor needs replacing, undo the mounting screws from the mechanism, and remove it.

STEP B7: Replace as the reverse of removal.

JOB 10: QUARTER LIGHTS AND WINDOWS - *remove, replace.*

Section A: Quarter lights in rear doors.

❒ **STEP A1:** Remove the door glass as described in *Job 9*.

❒ **STEP A2:** Remove the screws holding the rear guide channel.

❒ **STEP A3:** Pull the quarter light and its surrounding rubber away from the frame and out of the window aperture.

❒ **STEP A4:** Remove the rubber from the glass.

❒ **STEP A5:** Refit as the reverse of removal.

Section B: Rear quarter glass.

In some models, these windows are bonded in place, and require the attentions of a Vauxhall dealer or vehicle glass specialist. In earlier models, however, they are more simply held in with plastic nuts.

❒ **STEP B1:** Remove the rear seat belt upper guide – see *Chapter 14, Interior, Trim*.

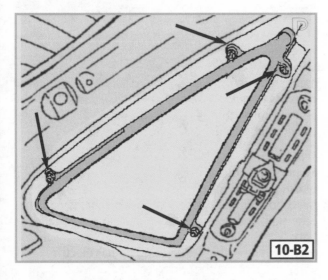

10-B2

❒ **STEP B2:** Undo the nuts (arrowed).
➜ The window may be removed.

❒ **STEP B3:** Reassembly is the reverse of removal.
➜ Check the cemented weatherstrip and renew if damaged.

JOB 11: DOOR MIRRORS - *remove, replace.*

SAFETY FIRST!

VEHICLES WITH SIDE AIRBAGS

• Side airbags are located in the front seats of some vehicles and are activated by control units situated in the doors.

• The side airbag system must be de-activated before starting work on the doors.

• Disconnect the battery earth terminal and allow one minute or preferably longer for the capacitor to discharge. See *Chapter 10, Electrical, Dash, Instruments, Fact File: Disconnecting the Battery* BEFORE doing so!

• See *Chapter 14, Job 11, Section E* for more information on side air bags.

Section A: Complete mirrors.

❒ **STEP A1:** If the model you are working on has electrically heated and adjusted mirrors, remove the door trim and disconnect the wiring.

11-A2

❒ **STEP A2:** If the model you are working on has manually remote-adjustable mirrors, pull the handle off the adjuster lever and carefully lever the inner trim from the mirror mounting.

11-A3

❒ **STEP A3:** Undo the fixing screws or nuts and remove the mirror.

11-A4

❒ **STEP A4:** Refitting is the reverse of removal.

Section B: Glass replacement.

❏ **STEP B1:** If you need to renew the mirror glass.
➜ Carefully lever the glass out of the mirror housing with a blunt lever, such as a wooden spatula.
➜ It is located on plastic clips or ball joints.

TOP TIP!

❏ **STEP B2:** • On older vehicles, the plastic may have become brittle with age, and could well snap as you do this. Replacement will then be the only viable option.
• Renewing the mirror glass can be done without removing the mirror body from the vehicle.

❏ **STEP B3:** If the vehicle you are working on has heated mirrors, disconnect the wiring connection to the glass.

Section C: Mirror motor.

❏ **STEP C1:** If you need to replace a mirror motor;
➜ Remove the mirror glass.
➜ Remove the motor fixing screws, disconnect the wiring connector and remove the motor.

JOB 12: SUNROOF - *remove, replace, adjust.*

Among the models covered by this manual, various types of sunroof are featured. These all require different approaches to repair procedures, and allow different jobs to be done. If you can avoid dismantling the sunroof, do so! The refitting and adjustment of the most types of sunroof is invariably complex and difficult. The following information is supplied for reference.

TOP TIP!

• If removal or adjustment of the sunroof involves the removal of headlining, have a trimmer carry out the work.
• Most Vauxhall sunroofs are designed to be fitted as complete 'cassette' assemblies. A severely damaged sunroof may be best replaced with a new or good second-hand one.

Section A: Sliding roof.

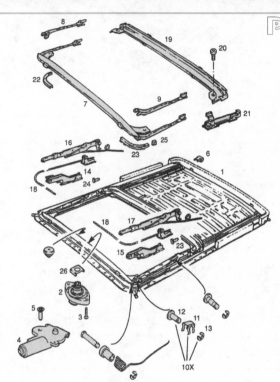

1 - operating unit
2 - crankdrive - not used with electric sun roof
3 - screw, m4 x 16, crankdrive to sun roof frame. note: not used with electric sun roof
4 - sun roof motor - used with electric sun roof
5 - screw, m5 x 25, sun roof motor to frame. Note: used with electric sun roof
6 - slide cover, rear end
7 - wind deflector
8 - wind deflector

operating rod, rh
9 - wind deflector operating rod, lh
10x - adjusting set for operating rod, rh
10x - adjusting set for operating rod, lh
11 - operating rod clip
12 - operating rod bolt
13 - operating rod safety washer
14 - front lever guide, with catch, rh
15 - front lever guide, with catch, lh
16 - side guide, rh
17 - side guide, lh

18 - control cable, with rear guide
19 - water deflector
20 - screw, m5 x 10, water deflector
21 - catch lever and guide, rh and lh
22 - inner cable guide, rh
23 - inner cable guide, lh
24 - frame insert
25 - rubber bumper for cable guide
26 - spring plate

12-A1A

❏ **STEP A1A:** This is the Rockwell Golde sunroof fitted as an optional extra to Cavalier and Vectra models.
➜ The main job that you can expect to do on this is removing the panel.
➜ If there are problems with the opening mechanism, refer them to your local Vauxhall dealer or specialist, as they are very difficult to work on.

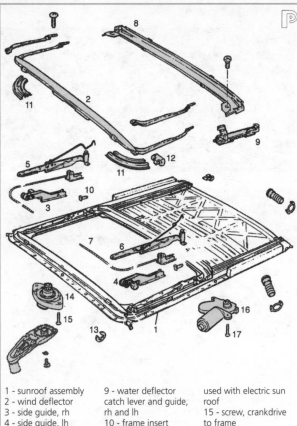

1 - sunroof assembly
2 - wind deflector
3 - side guide, rh
4 - side guide, lh
5 - front lever guide, with catch, rh
6 - front lever guide, with catch, lh
7 - control cable, with rear guide
8 - water deflector
9 - water deflector catch lever and guide, rh and lh
10 - frame insert
11 - clamp retaining plate
12 - rubber bumper
13 - spring plate plug - not used with electric sun roof
14 - crankdrive - not used with electric sun roof
15 - screw, crankdrive to frame
16 - electric motor
17 - screw, m5 x 32, motor to operating unit

12-A1B

❏ **STEP A1B:** This, for reference, is the Tudor Webaster sunroof fitted to some Vectra models.

IMPORTANT NOTE: If the motor should fail on an electrically operated example of this type of sunroof, check the relevant fuse (see *Chapter 15, Wiring Diagrams*). If this does not fix the problem, refer it to your Vauxhall dealer or specialist.
→ If the roof becomes stuck partly open, lever off the small trim panel at the rear edge of the sunroof and close the sunroof with an Allen key.
→ There should be one stored in the glove compartment for this very purpose.

❏ **STEP A2:**
The commonest reason for a leaking sunroof is blocked or damaged drain tubes.
→ If blocked, try cleaning with a long piece of flexible wire.

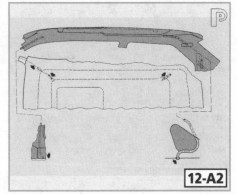

12-A2

→ Tubes deteriorate and leak over time. Although tricky to replace, it may be the only solution.

Section B: Calibra.

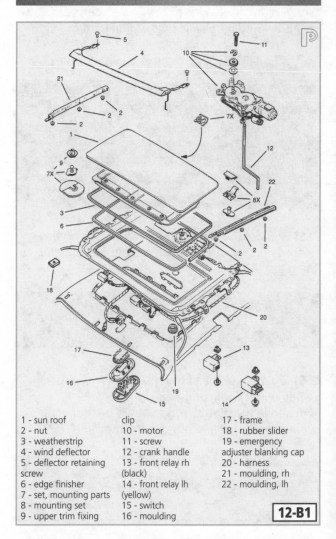

1 - sun roof
2 - nut
3 - weatherstrip
4 - wind deflector
5 - deflector retaining screw
6 - edge finisher
7 - set, mounting parts
8 - mounting set
9 - upper trim fixing
clip
10 - motor
11 - screw
12 - crank handle
13 - front relay rh (black)
14 - front relay lh (yellow)
15 - switch
16 - moulding
17 - frame
18 - rubber slider
19 - emergency adjuster blanking cap
20 - harness
21 - moulding, rh
22 - moulding, lh

12-B1

❏ **STEP B1:** This was fitted as standard to the Calibra. The main jobs that you can expect to do on this are: removing the panel and replacing its headlining trim panel. If there are problems with the opening mechanism, refer them to your local Vauxhall dealer or specialist, as they are excessively difficult to work on. It differs from Vauxhall's other sliding roof in that the panel slides up and out of the roof, rather than retreating into a space between the headlining and roof skin.

IMPORTANT NOTE: If the motor should fail on an electrically operated example of this type of sunroof, check the relevant fuse (see *Chapter 15, Wiring Diagrams*). If this does not fix the problem, refer it to your Vauxhall dealer or specialist. If the roof becomes stuck partially open, lever off the small trim panel at the rear edge of the sunroof and close the sunroof with an Allen key. There should be one stored in the glove compartment for this very purpose.

JOB 13: SPOILERS, SIDE SKIRTS, WHEELARCH TRIMS - *remove, replace*.

Rear tailgate or luggage bay lid spoilers are covered separately in *Job 6*.

Some of the models covered by this manual have body kits, featuring front and/or rear spoilers plus, and in some cases, side skirts and wheelarch trims. These are not difficult to remove and replace, but require slightly different approaches. Therefore, we will deal with them in several sections.

Section A: Front spoiler.

☐ **STEP A1:** If necessary, raise the front of the vehicle and support it off the ground.

☐ **STEP A2:** If there is a panel covering the spoiler fixings as on the Vectra seen here, remove it.

13-A2

☐ **STEP A3:** Remove the clips or screws holding the spoiler to the front valance. If these are expanding type clips with a central pin:

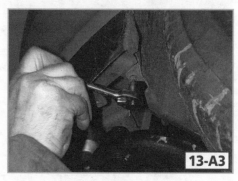
13-A3

➔ Drive the pin backwards with a suitable punch.
➔ Use a trim clip removing tool to lever out the round type of clip.
➔ These clips become brittle with age, and can break easily when removed with a screwdriver.

☐ **STEP A4:** Remove the spoiler.

☐ **STEP A5:** Refit as the reverse of removal.

Section B: Side skirts and/or wheelarch trims.

☐ **STEP B1:** Raise the vehicle, secure it on axle stands or a vehicle hoist and remove the appropriate wheel.

☐ **STEP B2:** If you are working on a wheelarch trim, disconnect the outer edge of the plastic wheelarch liner. If this doesn't give enough access, unclip the whole liner – see *Job 14*.

13-B3

☐ **STEP B3:** Remove the nuts and bolts securing the trim, then pull it away from the vehicle. This may reveal further clips, which should be carefully undone.

13-B4

☐ **STEP B4:** Remove the trim.

☐ **STEP B5:** Refit as the reverse of removal.

13-B5

JOB 14: WHEELARCH LINERS AND UNDERTRAY - *remove, replace*.

Section A: Wheelarch liner.

☐ **STEP A1:** Raise the front of the vehicle, secure it on axle stands or a vehicle hoist and remove the appropriate wheel.

☐ **STEP A2:** Remove the clips and self-tapping screws holding the wheelarch liner in place. A trim clip removing tool (which looks rather like a broad-bladed screwdriver, forked in the middle) is useful here – see illustration *13-A2*.

☐ **STEP A3:** Withdraw the liner from the wheelarch.

☐ **STEP A4:** Refit as the reverse of removal.

14-A3

Section B: Undertray.

☐ **STEP B1:** Raise the front of the vehicle, secure it on axle stands or a vehicle hoist.

☐ **STEP B2:** Remove the clips and self-tapping screws holding the engine undertray in place. A trim clip removing tool (which looks rather like a broad-bladed screwdriver, forked in the middle) is useful here – see illustration *13-A2*.

14-B2

☐ **STEP B3:** Withdraw the undertray.

☐ **STEP B4:** Refit as the reverse of removal.

JOB 15: RADIATOR GRILLE, FRONT AND REAR BUMPER - *remove, replace.*

FACT FILE

BUMPERS

• A range of different bumpers was fitted to the various models covered by this manual.

• Some of the models covered in this manual have their radiator grille separate from the front bumper, while in the majority of cases it is integral to the front bumper and valance moulding.

• Some models have a plastic, 'slimline' rear bumper, while in others the bumper is integral to the rear valance moulding.

FACT FILE

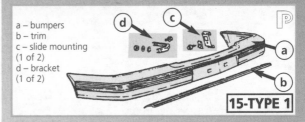

a – bumpers
b – trim
c – slide mounting (1 of 2)
d – bracket (1 of 2)

15-TYPE 1

☐ **TYPE ONE:** These are the front bumper and mountings as fitted to Cavaliers up to 1988.

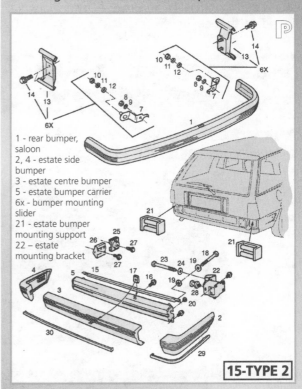

1 - rear bumper, saloon
2, 4 - estate side bumper
3 - estate centre bumper
5 - estate bumper carrier
6x - bumper mounting slider
21 - estate bumper mounting support
22 – estate mounting bracket

15-TYPE 2

☐ **TYPE TWO:** These are the rear bumper and mountings used on Cavaliers up to 1988.

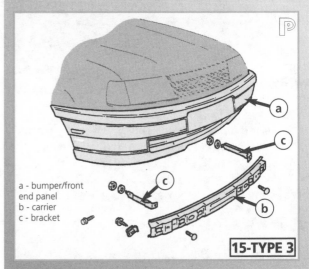

a - bumper/front end panel
b - carrier
c - bracket

15-TYPE 3

☐ **TYPE THREE:** These are the front bumper and fixings fitted to later Cavaliers from 1988 to 1995.

FACT FILE

1 – bumper mounting panel
2 – bumper
3 – bumper moulding
4 – trim
5 – towing bracket flap
6 – catch
7 – washer
8 – nut
9 – spacer, bracket to body
11 – washer
12 – nut
13,14 – bracket
15 – slider bracket, rh
16 – slider bracket, lh

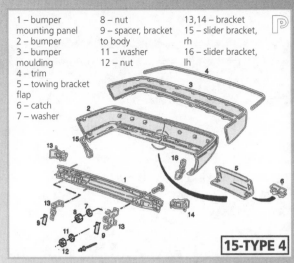

15-TYPE 4

❏ **TYPE FOUR:** These are the rear bumper fittings fitted to Cavaliers from 1988 to 1995.

a - front bumper
b - side support
c - side brackets

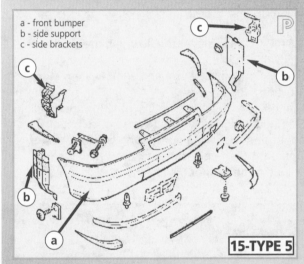

15-TYPE 5

❏ **TYPE FIVE:** These are the front bumper and fixings fitted to Calibras.

a - rear bumper carrier
b - rear bumper
c - side supports

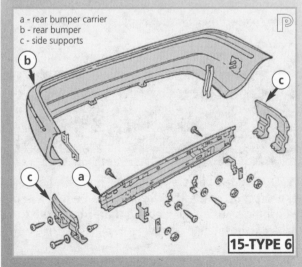

15-TYPE 6

❏ **TYPE SIX:** These are the Calibra's rear bumper and fixings.

FACT FILE

a – bumper
b – grille
c – spoiler
d – finisher
e – supports
f – clip – bumper to body
g – spring clamp cover

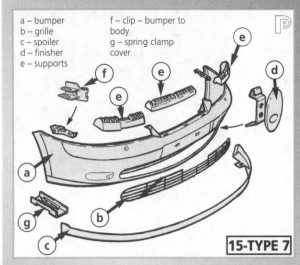

15-TYPE 7

❏ **TYPE SEVEN:** The Vectra's front bumpered mountings appear like this.

a - bumpers
b - impact absorber
c - bumper extension
d -support

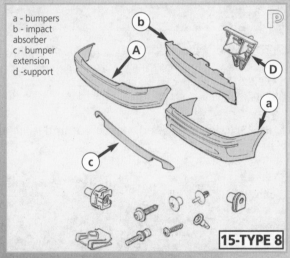

15-TYPE 8

❏ **TYPE EIGHT:** This is the rear bumper and fixings of Vectras.

Section A: Separate grille.

❏ **STEP A1:** Lift the bonnet and support it on its stay.

❏ **STEP A2:** To remove the grille:
➔ Release the clips holding the top edge of the grille to the front panel (arrowed).
➔ Pull the grille forwards and up to remove it from its

15-A2

locators on the bottom edge.
→ This is an early Cavalier.

STEP A3: This is the grille of a Vectra.
→ Remove the screws at either end of the top, raise the grille and disconnect the clips (arrowed).

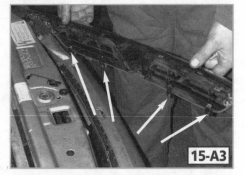

15-A3

STEP A4: Refit as the reverse of removal lowering the bottom locators into place first.

Section B: Front bumper/valance moulding, either with or without grille.

STEP B1: Lift the bonnet and support it on its stay.

STEP B2: The grille can be removed at this stage if desired.

STEP B3: Raise the front of the vehicle and remove the front wheels.

STEP B4: To remove the bumper:
→ Using the illustrations above, locate the main fixings and remove them,

15-B4

leaving the main topmost fixing till last.
→ Have an assistant take the weight of the bumper while the last fixing is removed.

STEP B5: Pull the bumper assembly forwards and away from the vehicle.

15-B5

Before the sides of the bumper fully disengage from their guides, disconnect the front fog lights as the bumper is lifted away, if applicable.

STEP B6: Replace as the reverse of removal. Before tightening the top mounting bolts, make sure the bumper sits level, as these fixings have a small degree of adjustment in them.

Section C: Rear bumper, either with or without integral valance moulding.

Refer to illustrations *TYPE TWO, TYPE SIX and TYPE EIGHT*.

STEP C1: Working inside the rear luggage compartment, remove any trim panels necessary to access the two bumper mounting bolts. Remove them.

STEP C2: If the model you are working on has its rear valance integral with the bumper, remove the mounting bolts on the top and bottom edges of the assembly.

STEP C3: Disconnect the license plate light units.

STEP C4: Pull the bumper assembly away from the vehicle.

STEP C5: Replace as the reverse of removal. Before tightening the two mounting bolts, make sure the moulding sits level, as these fixings have a small degree of adjustment in them.

JOB 16: BODY PROTECTION STRIPS - *remove, replace.*

Most of the models covered by this manual have bodyside protection strips secured by a proprietary type of double-sided tape which comes ready fitted to new, replacement body strips.

TOP TIP!

• If a self-adhesive body moulding is removed, it cannot successfully be replaced.
• As the old moulding is pulled away, the aluminium backing on the moulding stretches and distorts.

STEP 1: If, the vehicle you are working on has 'clip-on' protection strips, the clips have to be depressed by pushing a suitable tool behind the strip from the top. A plastic kitchen spatula is perfect for this.

STEP 2: Lift the trim strip away.

STEP 3: Replace as the reverse of removal.

JOB 17: COWLING - *remove, replace.*

Not every model covered by this manual has a cowling at the base of the front screen. However, if fitted, you will need to remove it to access the wiper motor, brake servo mounting bolts and ventilation pollen filter.

❏ **STEP 1:** Remove the screen wiper arms.
➜ Remove the large nuts from the wiper arm spindles and pull the arms of the splines.

❏ **STEP 2:** Remove the screws and nuts that secure the cowling in place.
➜ In some cases, the screws can be found behind small blanking plugs.

❏ **STEP 3:** If fitted, remove the edge sealing trim by pulling it off the lip.

❏ **STEP 4:** Lift the cowling panel off but do so with care – it is relatively large and may be easily damaged.

❏ **STEP 5:** Replace as the reverse of removal fitting the edge trim last of all.

JOB 18: LUGGAGE RACK CHANNEL TRIM - *remove, replace.*

❏ **STEP 1:** Using a well padded lever:
➜ Lever the trim from the inside edge of the roof rack channel.
➜ Be extremely careful not to damage the surrounding paintwork!
➜ Replace as the reverse of removal.

JOB 19: FRONT WING - *remove, replace.*

❏ **STEP 1:** Remove the aerial, if fitted to the wing.

❏ **STEP 2:** Remove the splash guard OR wheelarch liners – see *Job 14*.
➜ The wheel splash guard is fixed by a single bolt.

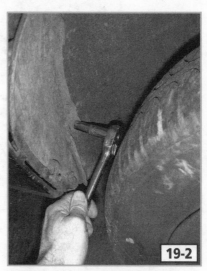

❏ **STEP 3:** Remove the front bumper – see *Job 15*.

❏ **STEP 4:** Remove the side indicator if fitted.

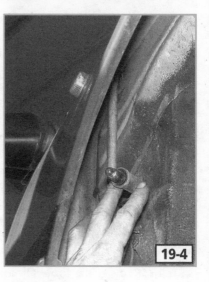

STEP 5: Undo the nuts and bolts attaching the wing front end to the front panel or valance.

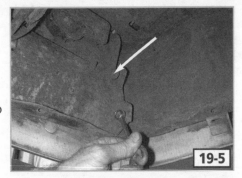

19-5

STEP 6: Remove the bolts in the A-pillars (arrowed).
→ Depending on which model you are working on, these can sometimes be reached by half-opening the door and pushing a socket extension through the gap.

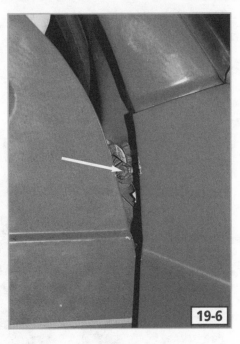

19-6

→ Protect the door's leading edge with padding if you do this or by removing the interior kick panel.

19-7

STEP 7: Remove the footwell side trim panel, and remove the wing fixing bolt underneath.

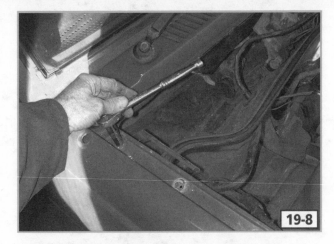

19-8

STEP 8: Remove the row of bolts in the wing top guttering – leaving a central one lose but still attached.
→ Check that nothing else sill connects the wing to the vehicle.

19-9

STEP 9: Remove the last wing guttering bolt and lift the wing away from the vehicle.

STEP 10: Refit as the reverse of removal.

IMPORTANT NOTE: Make sure the wing is correctly aligned against the shutlines of the bonnet and door, before fully tightening its mounting bolts.

CHAPTER 14: INTERIOR, TRIM

*Please read **Chapter 2 Safety First** before carrying out any work on your car.*

CONTENTS

SAFETY FIRST!

ALL INTERIOR TRIM AREAS:

• Before carrying out any work on any part of the interior, read *Job 9: AIRBAGS AND PRE-TENSIONERS – ESSENTIAL SAFETY FIRST!*

• There are very important safety hazards attached to working on, or *IN THE VICINITY OF* these components.

JOB 1: TRIM FIXINGS.

INTERIOR STRIPOUT

At first sight, it may look as though interior trim was never designed to be removed! But with a little thought and the certain knowledge that all the fixings *can* be found once you know where to start looking for clues, you'll find yourself able to work through the whole proceedings, step-by-step.

➡ Manufacturers usually hide trim fixings for safety reasons as well as for appearance, though their type and position is fairly predictable.

➡ Start with the outer fixings first. Work patiently and methodically.

➡ Pull carefully on plastic trim - you'll be able to see where it flexes and this often tells you where fixings are located and whether panels are separate, with a concealed join, or all-in-one.

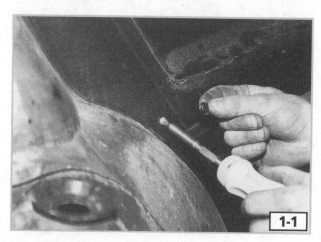

1-1

❑ **STEP 1:** In some cases, self-tapping screws are used to hold trim in place. Whenever you see a cap or other type of finisher, you can assume that it's there to cover something up. In this case, the cap is levered off, exposing the screwhead beneath.

☐ **STEP 2:** Similar at first sight, but in fact much flatter, is this type of plastic clip. To remove it, all you need to do is carefully lever under the head of the clip.

☐ **STEP 3:** Some similar looking clips (1) have a peg under the head, rather like a plastic nail, which pushes into a plastic socket mounted in the bodywork. The peg is pulled out first and then the socket (sometimes still attached

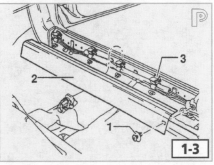

to the peg; sometimes not) pulls out after it. On these later-type door aperture trims (2), the cover is also clipped down to spring clips (3) on the body.

☐ **STEP 4:** Door seals are simply pushed on to the edge of the door and tread plates are either clipped or screwed down. Replacement door seals are remarkably expensive so, if yours are in good condition, do your best to save them.

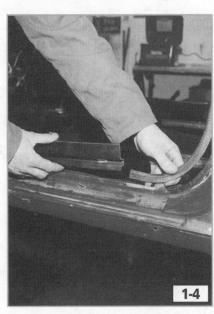

☐ **STEP 5:** On some models, the tread plates may be held in place with metal clips which stay on

the body seam after the trim has been removed and need to be levered off separately.

☐ **STEP 6:** Door trim and rear side trim is often fitted with concealed spring clips which push into the bodywork. If there is no evidence of any other type of clip, try easing the trim

carefully back and see if it springs away.

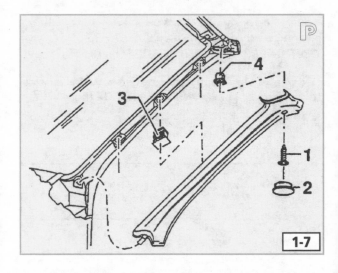

☐ **STEP 7:** Two fixing types shown here:
→ Another version of the hidden clip (3) may pull out of the trim panel, or might stay in the body. Whichever - before refitting, make sure it's first properly clipped into the trim panel.
→ The threaded screw (1) is covered by a cap (2) which has to be carefully levered out with a flat-bladed screwdriver. The screw goes into an expanding plastic plug (4) which is pushed into a

square hole in the body. On some versions, the cap is a softer material and is flush-fitting to the trim.

☐ **STEP 8:** Some sections of plastic trim and, more often, carpet are glued down. Take the very greatest care when pulling them off because the carpet is moulded to shape and will be impossible to replicate from a flat piece.

☐ **STEP 9:** Some fixing screws are cunningly concealed behind a cover plate which first has to be carefully levered off.

TOP TIP!

• Some plastics become brittle at low temperatures and more malleable in warmer conditions. If the workshop is cold, warming the car interior can make trim removal easier.

JOB 2:	DOOR AND SIDE TRIM PANELS - *remove, refit*.

Section A: Door panels.

Most door trim panels are held by a row of three self-tapping screws along the bottom edge and screws hidden by trim. They are usually overlapped by the door mirror interior trim panel, which has to be removed first.

☐ **STEP A1:** Pull the door mirror handle from the stalk and remove the trim, which is held by three expanding plugs.

2-A1

☐ **STEP A2:** Lever out the interior door pull filler and remove the interior door pull handle.

2-A2

☐ **STEP A3:** Lever off the interior door handle trim.

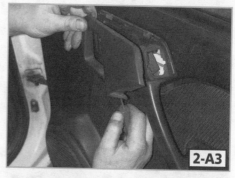

2-A3

☐ **STEP A4:** The trim varies between years and models.

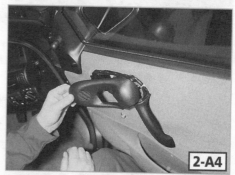

2-A4

☐ **STEP A5:** On some models there are self tapping screws underneath the trim.

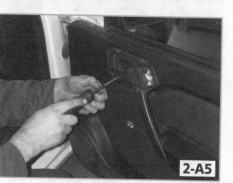

2-A5

☐ **STEP A6:** The window winder is held by a spring clip, which can be levered out using the special tool (arrowed).

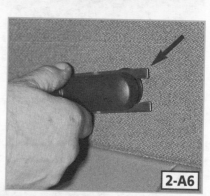

2-A6

☐ **STEP A7:** The spring clip can be removed by using a piece of cloth as shown.

2-A7

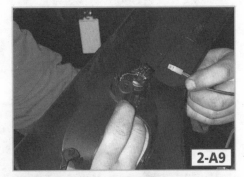

2-A8

❑ **STEP A8:** Later models have electrically operated window winders and mirror controls. It is usually possible to lever off the switch plate...

❑ **STEP A9:** ...and to disconnect the wiring from behind the plate.

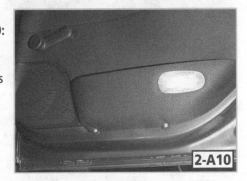

2-A9

❑ **STEP A10:** Some models have screws along the lower trim edge - remove them.

2-A10

❑ **STEP A11:** The door trim can now be removed.

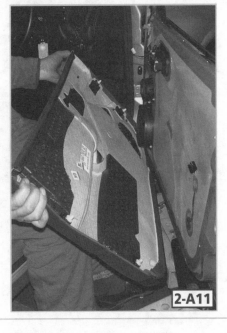

2-A11

Section B: Side trim panels.

There are many differences in the type and fastening methods of interior trim in the models covered by this manual, and the order in which they are removed.
→ In earlier models, the seat belt mounts are on top of the trim and have to be removed first.
→ In later models, the seat belt mountings are underneath the trim and the trim has to be removed first.

❑ **STEP B1:** Side trim looks complicated to remove because it is tucked under adjacent panels, so the first job is to remove them.

❑ **STEP B2:** This is the inner sill cover and, like most of the trim that overlaps the side trim, it simply clips into position.

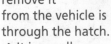

2-B2

❑ **STEP B3:** Depending on the model, the seatbelts might run through the side trim – see *Job 3*.

❑ **STEP B4:** On three door hatchbacks, the side trim is a large panel. The only way to remove it from the vehicle is through the hatch.
→ It is usually necessary to remove the rear seat.

2-B4

❑ **STEP B5:** The upper side trim is often held under the upper seatbelt guide.

2-B5

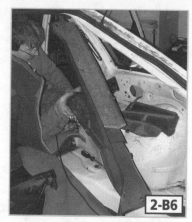

STEP B6: The upper trim is usually clipped into position.
➔ In some instances there will be wiring underneath to be unclipped as the trim is lifted away.

2-B6

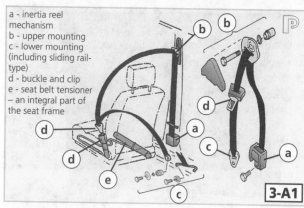

a - inertia reel mechanism
b - upper mounting
c - lower mounting (including sliding rail-type)
d - buckle and clip
e - seat belt tensioner – an integral part of the seat frame

3-A1

JOB 3: FRONT AND REAR SEAT BELTS - *check, remove, refit.*

STEP A1: Seat belts vary with model and year, and these are typical layouts.

SEAT BELT CHECKS

Visually check the seat belt webbing for fraying and cuts. In the UK, even a small cut or slight abrasion is sufficient to cause an MOT fail, and then the whole seatbelt/inertia mechanism must be renewed.
➔ Test the inertia mechanism by snatching the webbing. Check that the belt mounting points are secure in the same manner.
➔ Fasten and release the connector, checking for correct operation.

STEP A2: If fitted, disable the tensioning mechanism as described in *Job 4, Step A1*.

STEP A3: Remove the plastic cover from the upper mounting (runner) and remove the bolt, noting the positions of the washer and spacer which allow the component to rotate freely.

3-A3

Section A: Front seat belt remove/replace.

STEP A4: The bolt heads are generally shallow, so take care not to let the socket slip off!
➔ On some vehicles, the top mount is under the B-post trim. See *Job 2, Section B*.

3-A4

SAFETY FIRST!

• Read *Job 9* BEFORE working on seats fitted with seat belt pre-tensioners.

• The front seat belts on all the vehicles covered in this manual – all models except early Cavaliers - were fitted with tensioning mechanisms which, if they deploy while being handled, can cause injury.

• If you are unsure whether the car you are working on has them, look under the front seats - the mechanism is attached to the seat base.

• Before removing the seat belt or seat, take the plastic safety clip from the rear end of the tensioning unit and fit it as described in *Job 4, Steps A1 to A3*.

STEP A5: Remove the sill trim panel to reveal the lower web mounting.
➔ On some models, this also reveals the inertia mechanism.

3-A5

STEP A6: When removing the bolt holding the mechanism in place (position arrowed), note the positions of the washers and spacer.

3-A6

STEP A7: On some models, the lower seat belt is on a rail. Remove the rail front cover and the mounting bolt.
→ The rail can then be lifted away.

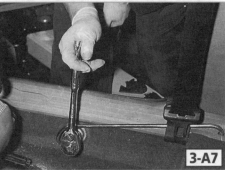

3-A7

STEP A8: On early vehicles the inboard belt end simply unbolts.
→ On later vehicles fitted with seat belt tensioners, the inboard end of the seat belt (the seat belt stalk and catch) is an integral part of the seat and cannot be removed on its own.
→ Because of the difficulties and dangers of working on the seat belt tensioning system, it is recommended that the seat unit complete is taken to a Vauxhall dealer should any work prove necessary.

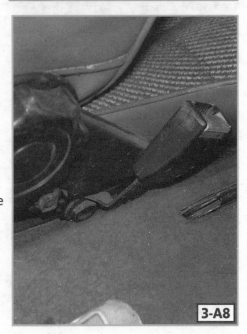

3-A8

STEP A9: Replace as the reverse of removal.

Section B: Rear seat belt remove/replace.

FACT FILE

• Inertia rear seat belts were fitted as standard from 1987. Earlier cars may be fitted with after-market belts.
• Slightly different rear seat belts were fitted to the Cavalier, Calibra and Vectra.

TYPE 1: These are the rear seat belt components fitted to the Cavalier.

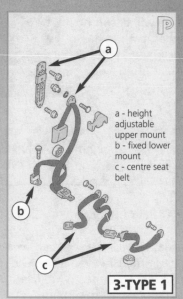

a - height adjustable upper mount
b - fixed lower mount
c - centre seat belt

3-TYPE 1

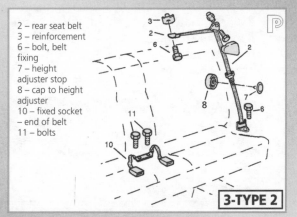

2 – rear seat belt
3 – reinforcement
6 – bolt, belt fixing
7 – height adjuster stop
8 – cap to height adjuster
10 – fixed socket – end of belt
11 – bolts

3-TYPE 2

TYPE 2: This is the simple arrangement of rear seat belt components fitted to the Calibra.

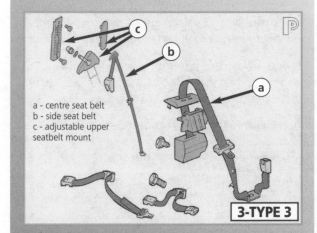

a - centre seat belt
b - side seat belt
c - adjustable upper seatbelt mount

3-TYPE 3

TYPE 3: The Vectra rear seat belt includes a centre inertia reel mechanism and belt.

STEP B1: Remove the seat base or, if the vehicle has a split folding rear seat, tilt the seat and undo the lap belt fixing bolts.

3-B1

STEP B2: Lever off the trim from the upper seat belt mountings and undo the fixing bolts.
➜ Note the correct positions of the washers and spacers.

3-B2

STEP B3: Remove the rear quarter trim panels. Unbolt the inertia mechanism.

3-B3

STEP B4: Replacement as the reverse of removal.

JOB 4: SEATS - *remove, refit.*

Section A: Seat belt tensioners – make safe.

SAFETY FIRST!

• Read *Job 9* BEFORE working on seats fitted with seat belt pre-tensioners.

• The front seat belts on all the vehicles covered in this manual expect early Cavaliers were fitted with tensioning mechanisms which, if they deploy while being handled, can cause injury.

• If unsure whether the car you are working on has them, look under the front seats - the mechanism is attached to the seat base.

• Before removing the seat belt or seat, take the plastic safety clip from the rear end of the tensioning unit and fit it as shown in *Steps A1 to A3*.

4-A1

STEP A1: If seat belt tensioners are fitted, remove the plastic trim as shown.

4-A2

4-A3

STEP A2: You can now de-activate the tensioners by placing the plastic safety clip in each tensioner body as shown.

STEP A3: Ensure that the clip is fully home and that the small 'keeper' pin is engaged before proceeding further.
➜ While working on the front seats, take care not to snag the safety clip tether.

Section B: Front Seats – *remove, replace.*

☐ STEP B1: There may be plastic trim covering the seat runner sides. If so, remove it.

4-B1

☐ STEP B2: Slide the seat fully forward. If there are plastic mouldings on the runner ends, remove them.

4-B2

☐ STEP B3: Remove the head restraint.

☐ STEP B4: Unbolt the rear seat runners.

4-B4

☐ STEP B5: Later vehicles have Torx headed bolts.

4-B5

☐ STEP B6: Unbolt the front seat runner bolts.

4-B6

☐ STEP B7: Tilt and lift the seat out.

☐ STEP B8: Replace as the reverse of removal.
➜ All work on front seats fitted with seat belt tensioners should be entrusted to a Vauxhall dealer.

4-B7

Section C: Rear seats – *remove, replace.*

The design of rear seats varies between models, but the principles of removing them are common to hatchbacks and estates.

☐ STEP C1: Lift the seat base and pull it forward to reveal the hinges.
➜ These usually incorporate a pin and circlip.
➜ Remove the circlip.

4-C1

☐ STEP C2: Pull out the hinge pins.

☐ STEP C3: Lift out the seat base.

4-C2

☐ STEP C4: To remove the seat backrest, release the top catches and pull the seat backrest forward.

4-C4

Remove any trim that covers the hinges.

STEP C5: The seat backrest can then be unbolted and removed.
→ On some cars, the backrest will be secured by screws rather than bolts.

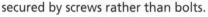

STEP C6: On saloons, pull the handles on the seat base and lift the base out.
→ The seat backrest is generally held by folding tabs at the base. Open these and the backrest can be lifted off its mounting hooks.

JOB 5: DASHBOARD - *remove, refit.*

See *Chapter 10, Job 3*.

JOB 6: HEADLINING - *remove, refit.*

The headlining fitted to early cars comprises material suspended from rods and glued along the edges. Fitting this type of headlining is a job for professional upholsterers.
→ Later cars have a moulded rigid headlining that springs into position and is held by the various fittings that pass through it. With care, this type can be removed and replaced.

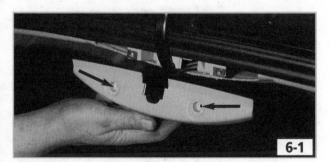

STEP 1: Where fitted, remove the tailgate wiring plastic cover from the rear centre of the roof.

STEP 2: Remove the grab handles and the coat hooks.

STEP 3: Remove the sunroof surround.
→ On some models, it will be necessary to remove the sunroof assembly – see *Chapter 13, Job 12*.

STEP 4: Remove the sun visors and the sun visor brackets.

STEP 5: Disconnect the sunroof operating switch plug or winder handle and disconnect the interior light plug.

STEP 6: Remove any trim items that overlap the edge of the headlining.
→ Depending on model, this may include the upper B-post trim, C-post trim, and the door and tailgate weather seals.

STEP 7: The headlining is held to the roof by clips.
→ Remove the headlining by pulling it downward.

JOB 7: CENTRE CONSOLE – *remove, refit*.

The centre console is secured by self-tapping screws and/or bolts. The number and position of these varies according to the model.

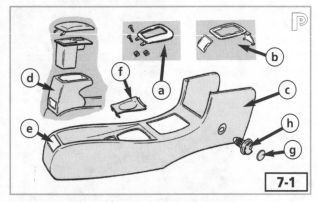

7-1

☐ **STEP 1:** Release the gear lever (**a**) or auto-transmission selector (**b**) and parking brake covers and gaiters from the centre console.
→ Some models have a built-in arm-rest (**d**).
→ Remove any screw-head covers (**g**) to expose screws (**h**) beneath.

☐ **STEP 2:** Locate the remaining fixing screws.
→ On some models they are under the ash tray (**e**) and centre 'oddment' trays (**f**) – see illustration **7-1**.

7-2

☐ **STEP 3:** Disconnect wiring plugs if appropriate.

☐ **STEP 4:** The console can then be lifted out.
→ On some models, the console must first be moved backwards to release the front end.

JOB 8: INTERIOR MIRROR - *remove, refit*.

The interior rear view mirror base mount is bonded to the windscreen and can not be removed. The mirror arm and mirror can be removed.

☐ **STEP 1:** Undo the fixing screw using a 2 mm hexagonal key.

8-1

☐ **STEP 2:** If a new windscreen is to be fitted, the interior mirror mount must be bonded using adhesive available from Vauxhall dealers.
→ One of several different styles of mirror may be fitted:
→ The mirror base is bonded to the screen and is not normally removed, except when the screen is replaced.

☐ **STEP 3:** If the base comes adrift from the screen:
→ Mark the outside of the screen with a felt pen to show where the mirror is to be relocated.

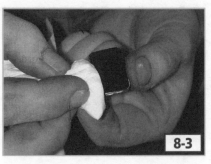

8-3

→ Use ONLY a proprietary mirror base adhesive for relocating the base.
→ Other types of adhesive may 'work', but will probably cause the screen to crack.
→ Clean off all traces of old adhesive from screen and base with a sharp blade.
→ Wipe the base and screen with degreaser and allow to dry.

☐ **STEP 4:** The Wurth mirror adhesive we used comes in two parts:
→ Break the phial of green adhesive and use the integral applicator to

8-4

wipe adhesive onto the base pad.
→ Some types of mirror adhesive involve special double-sided tape – the principles are the same. Follow instructions on the pack.

STEP 5: Use the markings on the outside of the screen. Apply hardener to the area of the screen where the base is to be fixed.

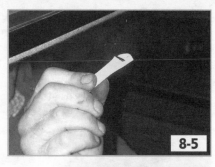

8-5

STEP 6: Hold the base into position. It will stick into place very quickly, but will need to be allowed to cure before re-attaching the mirror - see instructions on the pack.

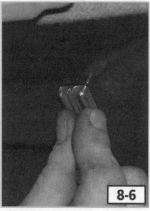

8-6

JOB 9:	AIRBAGS AND PRE-TENSIONERS - *essential Safety First!*

SAFETY FIRST!

• **THE WHOLE OF THIS JOB IS '*SAFETY FIRST!*'**

• **Read ALL of the information in this Job before working ON OR NEAR air bags and seat belt pre-tensioners.**

Section A: Essential safety notes.

A seat belt pre-tensioner is a device which is designed to pull the seat belt tight in the event of a frontal crash.

➜ Air bags and seat belt pre-tensioners all normally contain a pyrotechnic (explosive) charge.

➜ It is MOST IMPORTANT that the whole of this Job is read and understood before you consider how to deal with air bags and seat belt pre-tensioners and the components surrounding them.

➜ The risks of harm from these items are not extreme, but they are real and must be acted upon in an appropriate way.

STEP A1: Note the location of all airbags and seat belt pre-tensioners before starting work.

➜ **AIR BAGS:** These may be fitted, for example, in the steering wheel, in front of the passenger seat, in the door or in the side of the seat. There is normally an embossed label at the position of an air bag.

➜ **SEAT BELT PRE-TENSIONERS:** These may not easily be visible until the seat is partly dismantled. See the vehicle's handbook and also see *Job 3* and *Job 4* in this Chapter.

➜ If you are not sure whether or not a vehicle is fitted with air bags or pre-tensioners, or which type you are dealing with, or you are not fully competent to carry out any of the work described here, consult with, or take the vehicle to the vehicle manufacturer's dealership.

STEP A2: All manufacturers recommend that air bags and seat belt pre-tensioners are worked on only by their own, trained personnel.

➜ IF FOR ANY REASON, YOU ARE UNABLE TO COMPLY EXACTLY WITH THE INSTRUCTIONS AND WARNINGS GIVEN HERE, DO NOT HANDLE OR WORK ON THE AIR BAG OR SEAT BELT PRE-TENSIONER COMPONENTS - leave it to the vehicle manufacturer's dealership.

STEP A3: You may choose to have a dealer carry out the removal and replacement of air bag and/or pre-tensioner components so that you can carry out other work on the vehicle.

➜ If you choose to drive the vehicle to or from a main dealer with any of the pyrotechnic safety devices removed or disarmed, you will be taking a risk. In the event of a crash, you and/or your passengers will not be able to benefit from the normal function of the safety devices.

➜ Take care not to commit an offence or illegal act. For instance, in the UK, the seat belts must be functional and used by those in the vehicle. Do not contravene any laws or regulations by using the vehicle with safety devices disabled.

STEP A4: If a tensioner device or airbag needs to be renewed:

➜ Remember that levering an airbag can cause it to be triggered.

➜ If an airbag or seatbelt tensioner receives a voltage across its terminals it can deploy. Almost any voltage is enough, even the very low current from a test meter or one inducted from an electric welder. Disconnect the battery and the airbags and controller – see *Job 10* - before welding the body shell. Never use a test meter on any airbag circuit.

➜ Make sure the new unit is supplied in its correct safety packaging. It must be kept in its packaging at all times until fitted to the vehicle.

➜ Place the old unit in the safety packaging after removing and fitting the new one.

➜ Immediately take the old unit to the vehicle's main dealer and have them dispose of it

➜ Make sure that you carry out the work when you have access to the main dealer's premises - do not store new or old pyrotechnic safety devices

➜ If air bags or pre-tensioners have to be moved in

a motor vehicle, the tensioner/s should be placed in the (closed) luggage compartment and EMPHATICALLY NOT in the passenger compartment.

❏ **STEP A5:** It must not be assumed that a replaced air bag or seat belt pre-tensioner will work properly.
→ Take the vehicle to the vehicle's main dealer to have the unit checked and, if necessary, re-programmed.
→ Take note of the warning in *Step A3*.

❏ **STEP A6: IMPORTANT GENERAL NOTES:** The following notes apply to all seat belt pre-tensioners and air bags, where relevant.
→ Both air bags and pre-tensioners are referred to as 'safety devices' in the following notes.
→ Where a safety device has a safety locking mechanism, the safety device should not be removed from a vehicle or handled unless the locking mechanism is used, activated or fitted, as described in the relevant part of this manual.
→ A belt tensioner which has been dropped on the floor, or dented or damaged in any way must not be used.
→ On some vehicles, a seat belt on which the tensioner has been 'fired' is not safe to use. On those vehicles, the seat belt cannot be checked for locking once 'fired'.
→ Do not subject safety devices to blows, drilling, mechanical working or heating.
→ DO NOT drop safety devices or subject them to impacts. If one is accidentally dropped, it should not be used but returned to the main dealership.
→ If the safety devices has been activated, ALWAYS wait for at least 30 minutes after the activation before carrying out any operations to it - it may be hot enough to burn skin.
→ If a safety devices which has been activated has to be handled, use protective gloves and goggles.
→ Wash your hands with soap and water after handling a safety device.

❏ **STEP A7: WHEN WORKING ON THE VEHICLE:**
→ Both air bags and pre-tensioners are referred to as 'safety devices' in the following notes.
→ Do not subject an area surrounding the safety device to strong impacts. When, during bodywork repairs, for example, the use of a hammer is necessary, remove the complete unit.
→ If it is necessary to heat the area surrounding any of the safety devices or to carry out welding or brazing, then the complete safety devices in that area MUST be removed.
→ Safety devices have been designed to be fitted only on the type and model of vehicle for which they are intended. They cannot be adapted, reused or fitted on other vehicles, but only on those for which they were designed and produced. Any attempts to reuse, adapt or fit safety devices on different types of vehicles could cause serious or

fatal injuries to the occupants of the vehicle, either in the case of an accident or in normal usage.

❏ **STEP A8: IN THE DEPLOYMENT AREA OF THE AIRBAG...**
→ NEVER fit accessories or store objects
→ Use ONLY seat covers approved by the vehicle's manufacturer.

Section B: Airbags and Seat Belt Pre-tensioners - handling and storage.

This Section is taken from general information on how to handle and store airbags and seat belt pre-tensioners aimed originally at garages and workshops which handle and store only limited numbers, i.e. up to three or four, at any one time.
→ *Anyone working in domestic premises is strongly advised to follow the guidelines laid down here.*
→ This information is reproduced from the leaflet INDG280 10/98 C400 produced by the British Health & Safety Executive (HSE), and contains notes on good practice which are not at the time of publication compulsory in the UK but which you may find helpful in considering what you need to do.
→ Users in other territories must consider laws and regulations which may apply to them.
→ This information is current at August 1998 and is reproduced with thanks to HSE.

More and more vehicles are being fitted with a range of airbags and seat belt pre-tensioners. There is therefore an increasing likelihood that you will come across these devices at work.
Even though these devices are designed to save lives, there is the possibility of:
   • *physical injury; and*
   • *poisoning;*
if they are not handled correctly. While the likelihood of an accident involving an airbag or seat belt pre-tensioner is low, a few simple precautions can be taken to reduce the risks further.

❏ **STEP B1: WHAT TO DO:**
→ Find out from your supplier the UN hazard classification of the airbags and seatbelt pre-tensioners that you may handle.
→ **If any are classed as UN Hazard Class 1 (the explosives class) and you want to keep them on the premises,** you will need to register for a **Mode B Registered Premises** with your local authority under the Explosives Act 1875. The department dealing with registration varies from region to region, but it is usually the:
   • fire brigade;
   • trading standards; or
   • environmental health

☐ **STEP B2: REGISTRATION:** The HSE recommend that, as a garage or workshop, you should register even if you don't plan to keep these devices, as delays in fitting them to the vehicle may mean they need to be kept on the premises, overnight, for example.

TOP TIP!

• For airbags or seat belt pre-tensioners which are classed as **UN Hazard Class 2** or **UN Hazard Class 9,** the HSE recommend that you keep them under similar conditions to those required for Mode B registration.

☐ **STEP B3: STORING AIRBAGS AND SEAT BELT PRE-TENSIONERS:**

→ You can buy cabinets or containers which meet the requirements for Mode B registration. In general terms, these requirements are for a substantial container which:
  • has no exposed steel;
  • is easy to keep clean; and
  • can be closed and locked.
→ You should keep the container away from:
  • oils, paints and other flammable material;
  • areas where hot work, such as welding or brazing, is taking place; and
  • electricity cables, sockets, distribution boards etc.
→ Also make sure the container is:
  • secured to the wall or floor if possible; and
  • kept dry at all times.

☐ **STEP B4: HANDLING AIRBAGS AND SEAT BELT PRE-TENSIONERS:**

→ **It is essential that the manufacturer's or supplier's information is checked before starting work on vehicles containing airbags, as procedural differences will occur from make to make.**
→ Never place your head or body close to the front of an un-deployed airbag, especially when fitting it, or removing it from a vehicle.
→ Always carry the airbag module with the trim cover facing away from you.
→ Never place an airbag module, or steering wheel assembly fitted with an airbag, face (trim side) down or with the trim against a hard surface
→ Never attempt to repair or modify airbag modules.
→ Never expose airbag modules to excessive heat (over 90 ^{0}c), impact, electrical current (including static electricity) or radio transmitters.
→ Always use new components. Return any modules which are damaged or appear suspect to your supplier, **except** where the damage has resulted in the contents of the inflator cartridge being exposed or spilt, in which case obtain specialist advice from your supplier.

→ Return undeployed airbags to your supplier using the packaging the replacement device is supplied in. If for any reason this packaging is not available, contact your supplier and ask them to provide you with it.
→ Airbags should only be deployed by appropriately trained personnel working to the manufacturer's procedures.
→ Seek the advice of your supplier before disposing of any deployed airbags and seat belt pre-tensioners. Some manufacturers advise that their deployed airbags or seat belt pre-tensioners can be disposed of, or recycled, as normal waste; others recommend that they are treated as hazardous waste.
→ It is illegal to dispose of explosives as normal waste and domestic/commercial waste bins **must not** be used for disposing of **undeployed** airbags or seat belt pre-tensioners in Class 1.

☐ **STEP B5: TO SEEK MORE COMPREHENSIVE INFORMATION:**

→ Comprehensive guidance for those handling, storing or transporting larger numbers of these devices is provided in the HSE publication: *The handling, storage and transport of airbags and seat belt pre-tensioners* **HSG184 HSE Books 1998 ISBN 0 7176 1598 7.**
→ HSE priced and free publications are available by mail order from: HSE Books, PO Box 1999, Sudbury, Suffolk CO10 6FS. Tel: 01787 881165 Fax: 01787 313995.
→ HSE priced publications (i.e. those for which a charge is made) are also available from good booksellers.
For other enquiries, ring HSE's InfoLine on 0541 545500, or write to HSE's Information Centre, Broad Lane, Sheffield S3 7HQ.

JOB 10:	AIRBAGS AND SEAT BELT PRETENSIONERS – *deactivate, reconnect.*

DEACTIVATING THE AIRBAG SYSTEM

☐ **STEP 1:** Turn off the ignition, disconnect the battery and remove it from the vehicle so that there is no possibility of the battery leads accidentally contacting the terminals.

☐ **STEP 2:** Leave the vehicle for 30 minutes to ensure that any stored electrical energy in the airbag/pre-tensioner has been dispersed.

☐ **STEP 3:** If the vehicle body shell is to be welded, disconnect all airbag, airbag control unit, pyrotechnic seat belt pre-tensioner and control unit wiring harness leads.

RECONNECTING THE BATTERY

❑ **STEP 4:** As the battery is reconnected, there is a small risk that an airbag may deploy.

❑ **STEP 5:** Before reconnecting the battery, close the vehicle's doors and leave a side window open by a small amount.

❑ **STEP 6:** Park the vehicle out of doors and make sure that no-one is standing closer than 10 metres to the vehicle. Reconnect the battery.

JOB 11: AIRBAGS - *remove, refit*.

SAFETY FIRST!

• All models covered by this manual except early Cavaliers are fitted with air bags.

• These are easily found as their covers, on the steering wheel centre and passenger side of the dashboard, are embossed with the word 'AIRBAG'.

• We strongly advise that, if it is necessary to disconnect an air bag, unless you have the necessary understanding, expertise and safety equipment, the help of a Vauxhall dealer or specialist is sought.

• These systems are activated by sensitive triggers and contain pyrotechnic devices. They can cause serious injury if triggered by mistake!

FACT FILE

• Airbags are triggered by a frontal impact of around 20mph. They inflate at very high speed, and so no accessories should ever be mounted in the airbag inflation area.
• Some passenger airbags inflate through the top of the dashboard, and nothing should ever be placed on that part of the dashboard.

Section A: Systems explained.

11-A1

❑ **STEP A1:** The airbag system comprises the airbags themselves, a central control unit and the triggering mechanism.
➜ The driver's airbag is situated in the steering wheel boss.
➜ The presence of airbags should be indicated by the word 'AIRBAG' embossed in the trim (arrowed).

❑ **STEP A2:** The airbags are triggered by capacitor discharge. Before working on or near them, disconnect the battery earth terminal and allow one minute or preferably longer for the capacitor to discharge.
➜ Before disconnecting the battery, make sure you have the radio security code.

Section B: Driver's airbag.

❑ **STEP B1:** Remove the airbag screws (arrowed).
➜ On some models it will be necessary to turn the steering wheel 90 degrees each side of the straight ahead position to access the screws

11-B1

STEP B2: Carefully lift the airbag unit away and disconnect the feed wire. Handle the airbag with care, carry it with the padded face upward.

11-B2

STEP B3: Store the airbag with the padded face upward, preferably in some sort of container, so that it cannot accidentally be knocked.

→ **DO NOT** drop the airbag, subject it to abnormal pressures or allow it to become wet.

STEP B4: Refit as the reverse of removal. The airbag bolts must be tightened to the appropriate torque.

Section C: Passenger airbag.

It is recommended that the airbag is not disturbed unless it has triggered and has to be replaced.

→ Before working on or near airbags, disconnect the battery earth terminal and allow one minute or preferably longer for the capacitor to discharge. See *Chapter 10, Electrical, Dash, Instruments, Fact File: Disconnecting the Battery* BEFORE doing so!

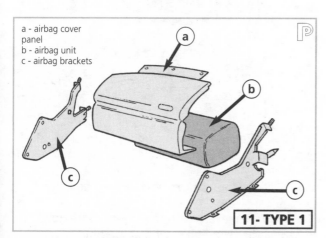

a - airbag cover panel
b - airbag unit
c - airbag brackets

11- TYPE 1

TYPE 1: This is the passenger airbag fitted to the Calibra and Cavalier.

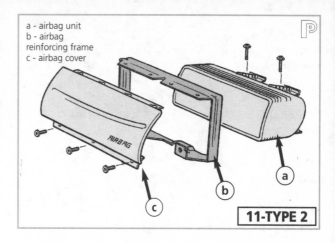

a - airbag unit
b - airbag reinforcing frame
c - airbag cover

11-TYPE 2

TYPE 2: This is the passenger airbag fitted to the Vectra.

Both types of passenger airbag are situated behind the glove compartment.

STEP C1: Remove the glove box mounting screws, and pull it from the dashboard.

STEP C2: Remove the passenger side mixed air vent.

STEP C3: Unplug the airbag unit.

STEP C4A: CAVALIER AND CALIBRA: Unscrew the airbag-to-bracket nuts.
→ The airbag unit can now be lowered from under the dashboard.

STEP C4B: VECTRA: Remove the airbag cover screws and the cover.
→ Undo the airbag fixing screws and pull the unit out.

STEP C5: Refit as the reverse of removal.

Section D: Airbag control unit.

If the control unit is faulty and has to be replaced then the new unit must be programmed when connected. This work can only be carried out by an authorised Vauxhall dealer.

→ Before working on or near airbags, disconnect the battery earth terminal and allow one minute or preferably longer for the capacitor to discharge.
→ Before disconnecting the battery, make sure you have the radio security code.

☐ **STEP D1:** Remove the centre console.

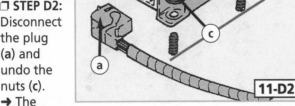

☐ **STEP D2:** Disconnect the plug (**a**) and undo the nuts (**c**).
➜ The control unit (**b**) can now be lifted out.

11-D2

Section E: Side airbags.

➜ Before working on or near airbags, disconnect the battery earth terminal and allow one minute or preferably longer for the capacitor to discharge.
➜ See **Chapter 10, Electrical, Dash, Instruments, Fact File: Disconnecting the Battery** BEFORE doing so!
➜ Before disconnecting the battery, make sure you have the radio security code.

IMPORTANT NOTE: If the side airbag deploys, renew the seat unit complete.

☐ **STEP E1:** The side airbags are situated in the seats. Detach the upholstery on the seat back and remove the backrest cover.

11-E1

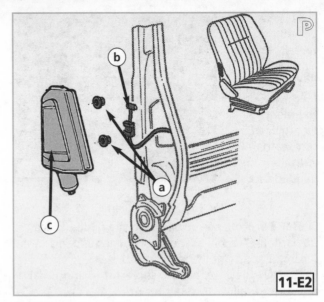

11-E2

☐ **STEP E2:** Remove the airbag retaining nuts (**a**), disconnect the wiring (**b**) and carefully remove the airbag (**c**).
➜ Store the airbag with the padded side up.

☐ **STEP E3:** To remove the side airbag sensor, first remove the door interior trim – see **Job 2**.

☐ **STEP E4:** Partly detach the water barrier to reveal the sensor.

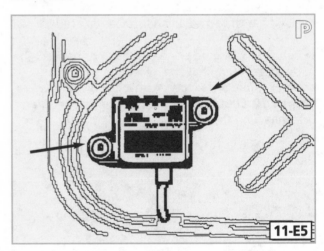

11-E5

☐ **STEP E5:** Release the sensor plug and remove the two fixing screws (arrowed).

☐ **STEP E6:** The sensor can now be removed.

☐ **STEP E7:** Replacement is the reverse of removal.
➜ Renew the fixing bolts when replacing sensors.
➜ NOTE: The sensors are 'handed' and the driver and passenger side sensors are NOT interchangeable.

CHAPTER 15: WIRING DIAGRAMS

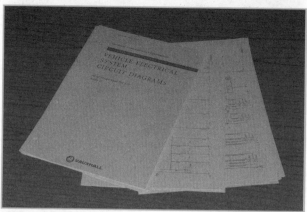

We don't pretend to include every wiring diagram for every model of Vauxhall covered in this manual. It simply can't be done in a single workshop manual.

· To show every wiring diagram for all of the vehicles covered by this manual would, we estimate, take well over a thousand pages!
· The wiring diagrams shown here are a small selection aimed at the (generally more complex) later vehicles.
· If you need a set of wiring diagrams specific to the vehicle you are working on, you should be able to purchase them, in loose-leaf form, from the Parts Department of your local Vauxhall or Opel dealership.
· We concentrate on providing you with a selection that might prove useful to you.
· We use these sample pages to show you how to use Vauxhall's own wiring diagrams - and it's *not* obvious if you don't know how!

HOW TO USE VAUXHALL WIRING DIAGRAMS

1. INDEX: At the start of each 'pack' of Vauxhall wiring diagrams is an index.
→ Against each component name is a number, which looks like a page number.
→ This 'page number' relates to the tracks shown on each page - Page **1** means, the page with **Tracks 100-on**; Page **2** means, the page with **Tracks 200-on**, and so on...

2. WIRE COLOURS AND SIZES: The colour codes and size of each wire are printed alongside the line for each wire.
→ Sometimes there is just one colour code, which is used for single-colour wire.
→ Usually there are two codes, such as **DGNWH**. When you look up the colour (see relevant list of colour codes) you'll see that **DGN** stands for **dark green** (the main colour) and **WH** stands for **white** (the secondary, or 'trace' colour).
→ The number, for example **0.35**, means that the wire cross section is **0.35 mm^2**.

3. 'TRACK' NUMBERS: These are numbers that run across the bottom of each page of diagrams.
→ The track numbers are organised in 'hundreds' - the first page shows Tracks 100-on; the second shows Tracks 200-on, and so on...
→ Track numbers do not relate to the vehicle - they are there to identify different parts of the wiring diagrams.

4. NUMBERS IN BOXES - CROSS REFERENCES: These numbers are cross-references to different parts ('Tracks') of the wiring diagrams.
→ When a wiring line ends with a number in a box - say **939**, for example - it means that the wiring continues on the page with Track No. **939** on it.
→ Unfortunately, on the samples we show here, some links to other track numbers will not be available. You would need a complete set of Vauxhall wiring diagrams for all of the cross-references to work properly.

5. COMPONENTS: Each component has a code.
→ In Vauxhall's wiring diagrams, the components included in the diagram - and their codes - are listed on the page facing the diagram, as shown here.
→ On the wiring diagram itself, the code number is printed next to the symbol for the component.
→ Where you see **AC** on the wiring diagram, for example, the list on the facing page explains that **AC** stands for **Air Conditioning Unit**.

Starting and charging systems (contd. on page 15-5)

Circuit Diagram Vectra-B

Circuits

Start & Charging	101 - 113
Voltage distribution	101 - 249

Grounding points

2	Engine, starter, alternator

Abbreviations

AB	Airbag
ABS	Anti-lock brake system
AC	Air conditioning
ASP	Outside mirror
AT	Automatic transmission
CRC	Cruise Control
DWA	Anti theft warning system
ECC	Electronic climate control
EMP	Radio
FH	Window lifters
HSH	Back window heating
HZG	Heating
IMO	Immobiliser
INS	Instrument
IRL	Interior lamp
LHD	Left-hand drive
LSW	Light switch
MID	Multi info display
MK	Engine cooling

MT	Manual transmission
MUT	Multitimer
NAV	Global positioning system
PBSL	Park brake shift lock
RFS	Back up lamps
RHD	Right-hand drive
SD	Sun roof
SH	Seat heating
SL	Stop lamps
SM	Engine control unit
TC	Traction control
TEL	Telephone
TFL	Day-time running light
TID	Triple info display
WI	Wiper
WS	Warning buzzer
ZIG	Cigarette lighter
ZV	Central door locking

Y21	Stroke magnet - Unlock ignition key

Component codes

FVx	Main fuse
Fx	Fuse
G1	Battery
G2	Alternator
M1	Starter
S1	Switch - Starter
X1	Instrument panel & Body rear
X5	Instrument panel & Engine
X7	Instrument panel & Engine

LIST OF COLOUR CODES

BN	brown
BU	blue
DBU	dark blue
DGN	dark green
YE	yellow
GN	green
GY	grey
LBU	light blue
LGN	light green
OC	ochre
OG	orange
PU	purple (violet)
PK	pink
RD	red
BK	black
WH	white
VT	violet

Starting and charging systems (contd. on page 15-5)

(contd. on page 15-5)

Starting and charging systems (contd. from page 15-2) and diagnostic connections.

Circuit Diagram Vectra-B

Circuits

Voltage distribution	101 - 249
Diagnostic link	211 - 234

Grounding points

1   Distribution engine compartment

Abbreviations

AB	Airbag
ABS	Anti-lock brake system
ASP	Outside mirror
AT	Automatic transmission
AZV	Trailer hitch
CD	CD-Changer
CRC	Cruise Control
D	Diesel
DWA	Anti theft warning system
ECC	Electronic climate control
EMP	Radio
FF	Trumpet Horn
FH	Window lifters
FI	Petrol
HSH	Back window heating
HZG	Heating
IMO	Immobiliser
INS	Instrument
LHD	Left-hand drive
MID	Multi info display
MK	Engine cooling
MUT	Multitimer
NAV	Global positioning system
NSL	Fog lamps, rear
NSW	Fog lamps, front
RHD	Right-hand drive
SD	Sun roof
SLP	Secondary air injection
SM	Engine control unit
SRA	Headlamp washer system
TC	Traction control
TEL	Telephone
TFL	Day-time running light
TID	Triple info display
TKS	Door contact switch
ZH	Add-on heater
ZV	Central door locking

Component codes

FVx	Main fuse
Fx	Fuse
X7	Instrument panel & Engine
X13	Diagnostic link

LIST OF COLOUR CODES

BN	brown
BU	blue
DBU	dark blue
DGN	dark green
YE	yellow
GN	green
GY	grey
LBU	light blue
LGN	light green
OC	ochre
OG	orange
PU	purple (violet)
PK	pink
RD	red
BK	black
WH	white
VT	violet

Starting and charging systems (contd. from page 15-2) and diagnostic connections.

Headlights, parking lights and turn indicators

Circuit Diagram Vectra-B

Circuits

High beam	300 - 311
Low beam	312 - 321
Parking lamps	324 - 339
Turn signal lamps	328 - 349

E8	High beam - Right
E9	Low beam - Left
E10	Low beam - Right
Fx	Fuse
K73	Relay - High beam
X1	Instrument panel & Body rear
X2	Instrument panel & Body front
X11	Instrument panel & Body rear
X12	Instrument panel & Body rear
X14	Body front & Body rear

Grounding points

1	Distribution engine compartment
6	Panel rear

Abbreviations

AZV	Trailer hitch
INS	Instrument
LHD	Left-hand drive
LWR	Headlamp levelling
MID	Multi info display
RHD	Right-hand drive
SRA	Headlamp washer system

Component codes

E1	Parking lamp - Left
E2	Tail lamp - Left
E4	Parking lamp - Right
E5	Tail lamp - Right
E7	High beam - Left

LIST OF COLOUR CODES

BN	brown
BU	blue
DBU	dark blue
DGN	dark green
YE	yellow
GN	green
GY	grey
LBU	light blue
LGN	light green
OC	ochre
OG	orange
PU	purple (violet)
PK	pink
RD	red
BK	black
WH	white
VT	violet

Headlights, parking lights and turn indicators

Stop lights and turn indicators

Circuit Diagram Vectra-B

Circuits

Stop lamps	402 - 427
Turn signal lamps	429 - 446

Grounding points

1	Distribution engine compartment
3	A - Pillar
6	Panel rear

Abbreviations

ABS	Anti-lock brake system
AT	Automatic transmission
AZV	Trailer hitch
CRC	Cruise Control
D	Diesel
HB	Hatchback
KW	Estate
LHD	Left-hand drive
MID	Multi info display
MUT	Multitimer
NB	Notchback
PBSL	Park brake shift lock
RHD	Right-hand drive
SL	Stop lamps
SM	Engine control unit
TC	Traction control

Component codes

H9	Stop lamp - Left
H10	Stop lamp - Right
H11	Turn signal lamp - Front, left
H12	Turn signal lamp - Rear, left
H13	Turn signal lamp - Front, right
H14	Turn signal lamp - Rear, right
H33	Auxiliary turn signal lamp - Left
H34	Auxiliary turn signal lamp - Right
H36	Stop lamp - Middle
K152	Relay - Turn signal lamp
K152.1	Relay - Turn signal lamp, left
K152.2	Relay - Turn signal lamp, right
S8	Switch - Stop lamp, single
S116	Switch - Stop lamp, double
X1	Instrument panel & Body rear
X11	Instrument panel & Body rear
X14	Body front & Body rear
X23	Body rear & Tail gate
X24	Body rear & Socket trailer
X35	Tail gate & Switch lamp

LIST OF COLOUR CODES

BN	brown
BU	blue
DBU	dark blue
DGN	dark green
YE	yellow
GN	green
GY	grey
LBU	light blue
LGN	light green
OC	ochre
OG	orange
PU	purple (violet)
PK	pink
RD	red
BK	black
WH	white
VT	violet

Stop lights and turn indicators

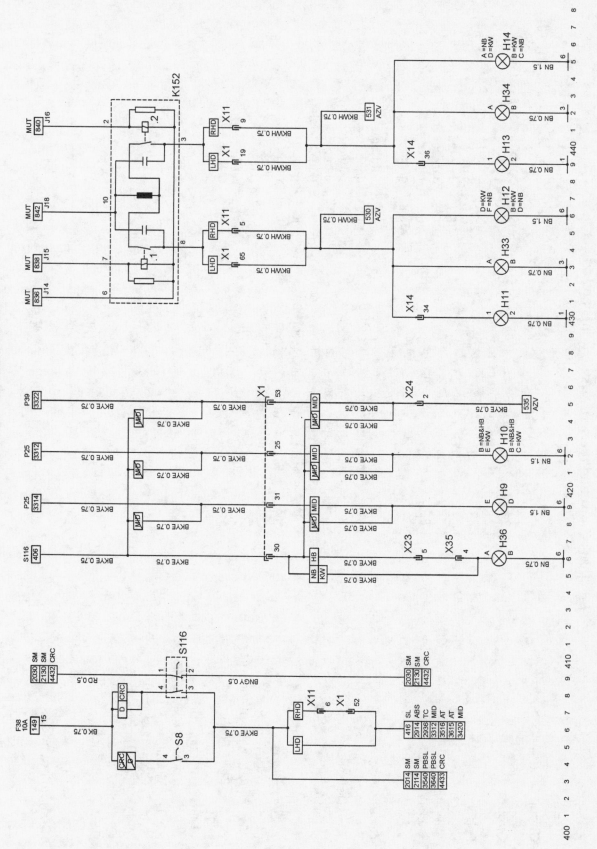

Fog lights (front and rear), reversing lights and trailer hitch

Circuit Diagram Vectra-B

Circuits

Fog lights, front	502 - 508
Fog lights, rear	513 - 520
Trailer hitch	522 - 537
Back up lamps	540 - 546

E20	Fog lamp - Front, left
E21	Fog lamp - Front, right
E24	Fog lamp - Rear, left
E39	Fog lamp - Rear, right
K5	Relay - Fog lamps, front
K89	Relay - Fog lamp, rear
S7	Switch - Back up lamp
S59	Switch - Socket trailer, fog lamp, rear
X1	Instrument panel & Body rear
X2	Instrument panel & Body front
X5	Instrument panel & Engine
X11	Instrument panel & Body rear
X12	Instrument panel & Body rear
X14	Body front & Body rear
X24	Body rear & Socket trailer
X27	Socket trailer
X33	Body rear & Socket trailer

LIST OF COLOUR CODES

BN	brown
BU	blue
DBU	dark blue
DGN	dark green
YE	yellow
GN	green
GY	grey
LBU	light blue
LGN	light green
OC	ochre
OG	orange
PU	purple (violet)
PK	pink
RD	red
BK	black
WH	white
VT	violet

Grounding points

1	Distribution engine compartment
6	Panel rear

Abbreviations

AT	Automatic transmission
AZV	Trailer hitch
INS	Instrument
LHD	Left-hand drive
MT	Manual transmission
MUT	Multitimer
NAV	Global positioning system
RHD	Right-hand drive
SL	Stop lamps
TEL	Telephone

Component codes

E17	Back up lamp - Left
E18	Back up lamp - Right

Fog lights (front and rear), reversing lights and trailer hitch

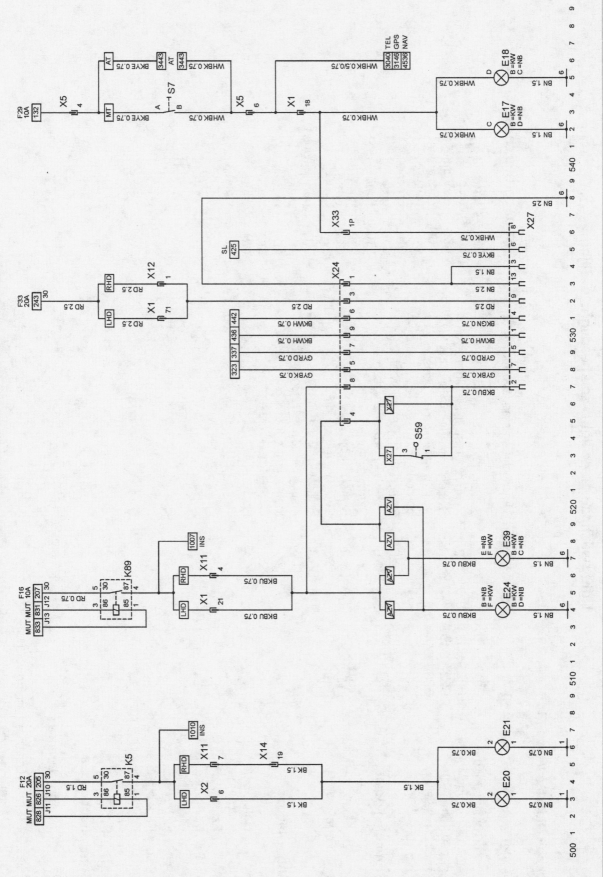

Light switch, no. plate lights and headlight levelling

Circuit Diagram Vectra-B

Circuits

Licence plate lamps	601 - 609
Day-time running lights	622 - 632
Light switch	622 - 649
Headlamp levelling	633 - 644

Grounding points

1	Distribution engine compartment
6	Panel rear

Abbreviations

HB	Hatchback
INS	Instrument
KW	Estate
MUT	Multitimer
NAV	Global positioning system
NB	Notchback
PBSL	Park brake shift lock
TFL	Day-time running light

Component codes

E3	Lamp - Licence plate
Fx	Fuse
K59	Relay - Daytime running light
M39	Motor - Headlamp levelling, left
M40	Motor - Headlamp levelling, right
S2	Switch - Light
S2.1	Switch - Light
S2.2	Dimmer - Instrument lights
S2.3	Switch - Headlamp levelling
S2.4	Switch - Illumination passenger compartment
S2.5	Switch - Fog lamps, front
S2.6	Switch - Fog lamps, rear
S5	Signal switch
S5.2	Switch - Low beam
S5.3	Switch - Turn signal
S5.4	Switch - Parking lamp
X1	Instrument panel & Body rear
X2	Instrument panel & Body front
X26	Code lamp license plate
X43	Body rear & Tail gate

LIST OF COLOUR CODES

BN	brown
BU	blue
DBU	dark blue
DGN	dark green
YE	yellow
GN	green
GY	grey
LBU	light blue
LGN	light green
OC	ochre
OG	orange
PU	purple (violet)
PK	pink
RD	red
BK	black
WH	white
VT	violet

Light switch, no. plate lights and headlight levelling

Circuit Diagram Vectra-B

Circuits

Headlamp washer	702 - 709
Window washer front / rear	710 - 714
Windscreen wiper	719 - 735
Back window wiper	738 - 749

Grounding points

1	Distribution engine compartment
6	Panel rear

Abbreviations

HB	Hatchback
KW	Estate
LHD	Left-hand drive
MUT	Multitimer
RHD	Right-hand drive

Component codes

K97	Relay - Time delay, washerpump, headlamps
K120	Relay - Wiper, back window
K124	Relay - Wiper, windscreen
M2	Motor - Wiper, windscreen
M8	Motor - Wiper, back window
M24	Pump - Washer, headlamps

Washers and wipers

M55	Pump - Windshield & back window
S9	Switch - Wiper unit
S9.1	Switch - Wiper, windshield
S9.2	Switch - Wiper, back window & washer unit
X1	Instrument panel & Body rear
X2	Instrument panel & Body front
X12	Instrument panel & Body rear
X14	Body front & Body rear
X23	Body rear & Tail gate
X42	Tail gate & Switch lamp
X44	Body rear & Tail gate

LIST OF COLOUR CODES

BN	brown
BU	blue
DBU	dark blue
DGN	dark green
YE	yellow
GN	green
GY	grey
LBU	light blue
LGN	light green
OC	ochre
OG	orange
PU	purple (violet)
PK	pink
RD	red
BK	black
WH	white
VT	violet

Washers and wipers

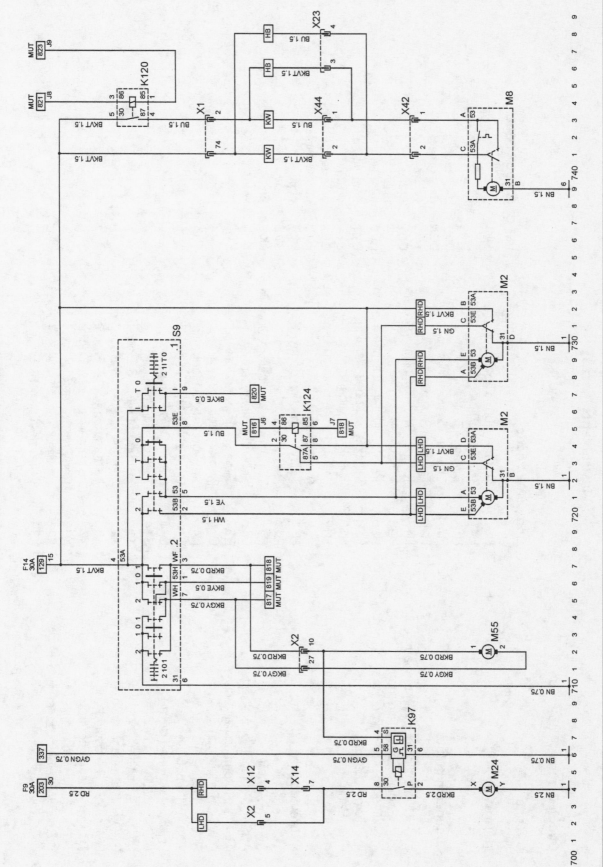

Circuit Diagram Vectra-B

Circuits

Instrument   900 - 1014

Grounding points

1   Distribution engine compartment
2   Engine, starter, alternator
4   Bracket steering column /
    crossmember
6   Panel rear

Abbreviations

AB   Airbag
ABS   Anti-lock brake system
AT   Automatic transmission
AZV   Trailer hitch
DIAG   Diagnostic link
IMO   Immobiliser
LHD   Left-hand drive
MID   Multi info display
MUT   Multitimer
RHD   Right-hand drive
SM   Engine control unit
TC   Traction control
TFL   Day-time running light
WS   Warning buzzer

Instruments (part 1)

Component codes

H3   Telltale - Turn signal lamp
H4   Telltale - Oil pressure
H5   Telltale - Brake system
H7   Charging indicator lamp
H16   Telltale - Glow time
H17   Telltale - Turn signal lamp trailer
H23   Telltale - Airbag
H26   Telltale - Anti-lock brake system
H28   Telltale - Seat belt warning
H30   Telltale - Engine
H42   Telltale - Automatic transmission,
        program power
H51   Telltale - Traction control
P1   Fuel indicator
P2   Temperature indicator - Coolant
P4   Sensor - Fuel reserve
P5   Sensor - Temperature, coolant
P7   Tachometer
P43   Speedometer - Electronic
S11   Control switch - Brake fluid
S13   Switch - Parking brake
S14   Switch - Oil pressure
X1   Instrument panel & Body rear
X2   Instrument panel & Body front
X5   Instrument panel & Engine
X7   Instrument panel & Engine
X21   Body rear & Tank fuel
X28   Engine & Injection valves - Fuel
X39   Instrument panel & Engine

LIST OF COLOUR CODES

BN   brown
BU   blue
DBU   dark blue
DGN   dark green
YE   yellow
GN   green
GY   grey
LBU   light blue
LGN   light green
OC   ochre
OG   orange
PU   purple (violet)
PK   pink
RD   red
BK   black
WH   white
VT   violet

Instruments (part 1)

Instruments (part 2), heater blower, cigar lighter and rear screen heater

Circuit Diagram Vectra-B

Circuits

Instrument	900 - 1014
Cigarette lighter	1014 - 1021
Rear screen heated	1021 - 1037
Blower heating	1037 - 1049

Component codes

E11	Lights - Instrument
E16	Lamp - Cigarette lighter
E19	Back window - Heated
H8	Telltale - High beam
H22	Telltale - Fog lamp, rear
H65	Telltale - Fog lamps, front
K1	Relay - Back window, heated
M3	Motor - Blower, heating
R3	Cigarette lighter
S3	Switch - Blower heating
S4	Switch - Back window & outside mirror, heating
S102	Switch - Recirculation, air
S131	Limit switch - Defroster lever
X1	Instrument panel & Body rear
X12	Instrument panel & Body rear
X23	Body rear & Tail gate
Y35	Solenoid valve - Recirculation air

Grounding points

1	Distribution engine compartment
4	Bracket steering column / crossmember
6	Panel rear

Abbreviations

ASP	Outside mirror
DWA	Anti theft warning system
ECC	Electronic climate control
HB	Hatchback
KW	Estate
LHD	Left-hand drive
MUT	Multitimer
NB	Notchback
NSL	Fog lamps, rear
NSW	Fog lamps, front
RHD	Right-hand drive

LIST OF COLOUR CODES

BN	brown
BU	blue
DBU	dark blue
DGN	dark green
YE	yellow
GN	green
GY	grey
LBU	light blue
LGN	light green
OC	ochre
OG	orange
PU	purple (violet)
PK	pink
RD	red
BK	black
WH	white
VT	violet

Instruments (part 2), heater blower, cigar lighter and rear screen heater

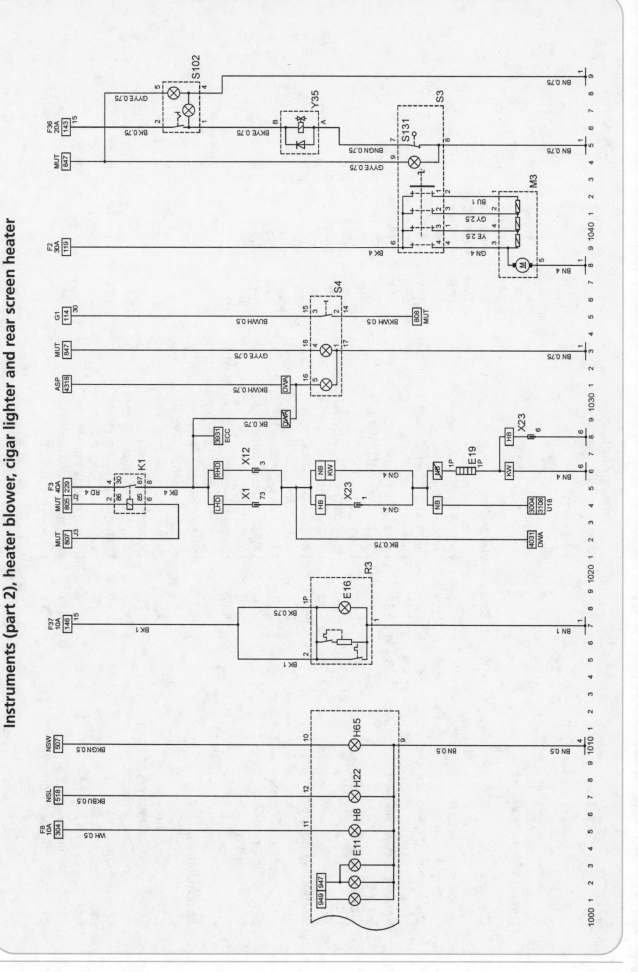

Interior lights

Circuit Diagram Vectra-B

Circuits

Door contact switch	1100 - 1117
Reading lamp, rear	1118 - 1126
Trunk lamp	1127 - 1132
Sunshade lamps	1133 - 1140
Passenger compartment lamp	1141 - 1146
Glove box lamp	1147 - 1149

Grounding points

1	Distribution engine compartment
3	A - Pillar
6	Panel rear

Abbreviations

CRP	Carphone
DWA	Anti theft warning system
FH	Window lifters
HB	Hatchback
KW	Estate
MUT	Multitimer
NB	Notchback
ZV	Central door locking

Component codes

E13	Lamp - Trunk
E15	Lamp - Glove box
E27	Reading lamp - Rear, left
E28	Reading lamp - Rear, right
E37	Lamp - Sunshade, left
E40	Lamp - Sunshade, right
E41	Lamp - Passenger compartment, disconnect delay
E50	Lamp - Driver door
E51	Lamp - Co-driver door
E52	Lamp - Door, rear, left
E53	Lamp - Door, rear, right
S15	Switch - Lamp, trunk
S16	Contact switch - Driver door
S17	Contact switch - Co-driver door
S31	Contact switch - Door, rear, left
S32	Contact switch - Door, rear, right
X1	Instrument panel & Body rear
X3	Instrument panel & Blower
X15	Body rear & Driver door
X16	Body rear & Co-driver door
X17	Body rear & Rear door, left
X18	Body rear & Rear door, right
X22	Body rear & Lid trunk
X35	Tail gate & Switch lamp
X43	Body rear & Tail gate

LIST OF COLOUR CODES

BN	brown
BU	blue
DBU	dark blue
DGN	dark green
YE	yellow
GN	green
GY	grey
LBU	light blue
LGN	light green
OC	ochre
OG	orange
PU	purple (violet)
PK	pink
RD	red
BK	black
WH	white
VT	violet

Interior lights

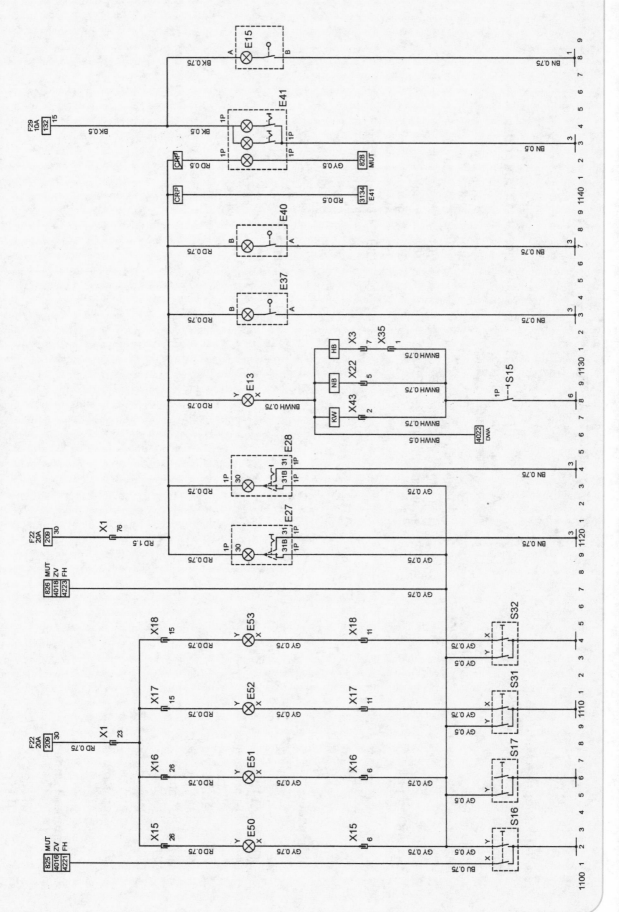

Radio/cassette and telephone

Circuit Diagram Vectra-B

Circuits

Radio	3000 - 3032
Telephone	3031 - 3050

Grounding points

4	Bracket steering column / crossmember

Abbreviations

AB	Airbag
CD	CD-Changer
DIAG	Diagnostic link
DWA	Anti theft warning system
HSH	Back window heating
MID	Multi info display
MUT	Multitimer
NB	Notchback
RC	Remote control
TC	Traction control
TID	Triple info display

Component codes

E19	Back window - Heated
H1	Radio
H56	Microphone - Phone

U17	Amplifier - Antenna, roof
U18	Amplifier - Antenna, back window
U29	Transmitter
X1	Instrument panel & Body rear
X38	Instrument panel & Phone

LIST OF COLOUR CODES

BN	brown
BU	blue
DBU	dark blue
DGN	dark green
YE	yellow
GN	green
GY	grey
LBU	light blue
LGN	light green
OC	ochre
OG	orange
PU	purple (violet)
PK	pink
RD	red
BK	black
WH	white
VT	violet

Radio/cassette and telephone

CD auto-changer and speakers

Circuit Diagram Vectra-B

Circuits

Speakers	3201 - 3237
CD-Changer	3238 - 3249

U24	CD - Changer
X1	Instrument panel & Body rear
X15	Body rear & Driver door
X16	Body rear & Co-driver door
X17	Body rear & Rear door, left
X18	Body rear & Rear door, right

LIST OF COLOUR CODES

BN	brown
BU	blue
DBU	dark blue
DGN	dark green
YE	yellow
GN	green
GY	grey
LBU	light blue
LGN	light green
OC	ochre
OG	orange
PU	purple (violet)
PK	pink
RD	red
BK	black
WH	white
VT	violet

Grounding points

4	Bracket steering column / crossmember

Abbreviations

EMP	Radio
LHD	Left-hand drive
MUT	Multitimer
RHD	Right-hand drive

Component codes

H37	Loudspeaker - Front, left
H38	Loudspeaker - Front, right
H39	Loudspeaker - Rear, left
H40	Loudspeaker - Rear, right
H52	Tweeter - Front, left
H53	Tweeter - Front, right
H57	Tweeter - Rear, left
H58	Tweeter - Rear, right
H66	Middle tone - Door, front, left
H67	Middle tone - Door, front, right

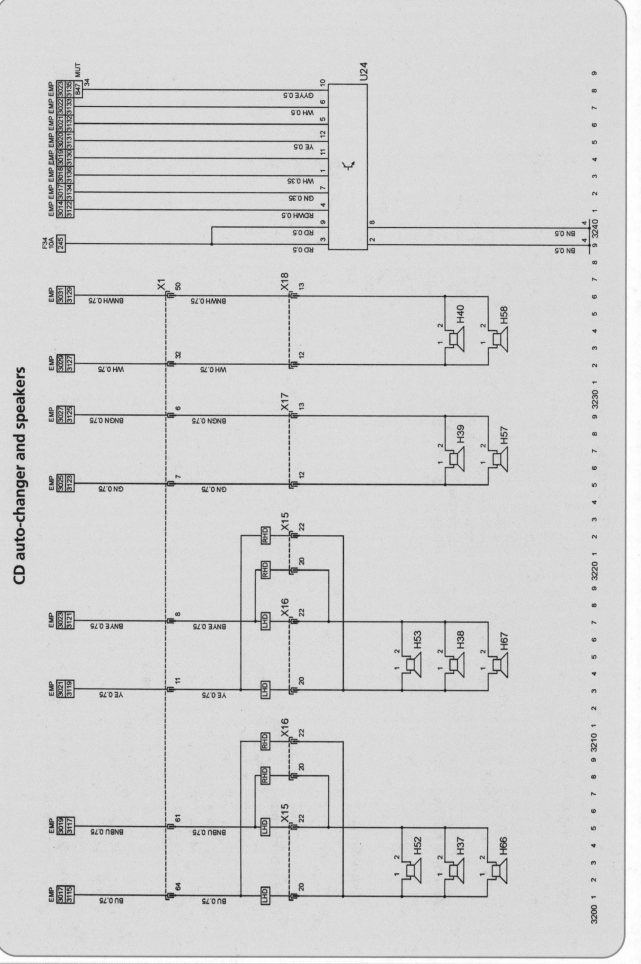

CD auto-changer and speakers

Door mirrors

Circuit Diagram Vectra-B

Circuits

Outside mirror	4300 - 4319
Outside mirror (Japan)	4320 - 4349

Grounding points

3   A - Pillar

Abbreviations

MUT   Multitimer

Component codes

K101	Relay - Mirror, parking position
K121	Relay - Mirror heating
M30	Outside mirror - Driver side
M30.1	Motor - Mirror adjustment
M30.2	Heating - Mirror
M30.3	Motor - Parking position
M31	Outside mirror - Co-driver side
M31.1	Motor - Mirror adjustment
M31.2	Heating - Mirror
M31.3	Motor - Parking position
S68	Switch - Outside mirror
S68.1	Switch - Outside mirror adjustment
S68.2	Switch - Outside mirror, left / right
S68.4	Switch - Outside mirror, parking position
X1	Instrument panel & Body rear
X15	Body rear & Driver door
X16	Body rear & Co-driver door

LIST OF COLOUR CODES

BN	brown
BU	blue
DBU	dark blue
DGN	dark green
YE	yellow
GN	green
GY	grey
LBU	light blue
LGN	light green
OC	ochre
OG	orange
PU	purple (violet)
PK	pink
RD	red
BK	black
WH	white
VT	violet

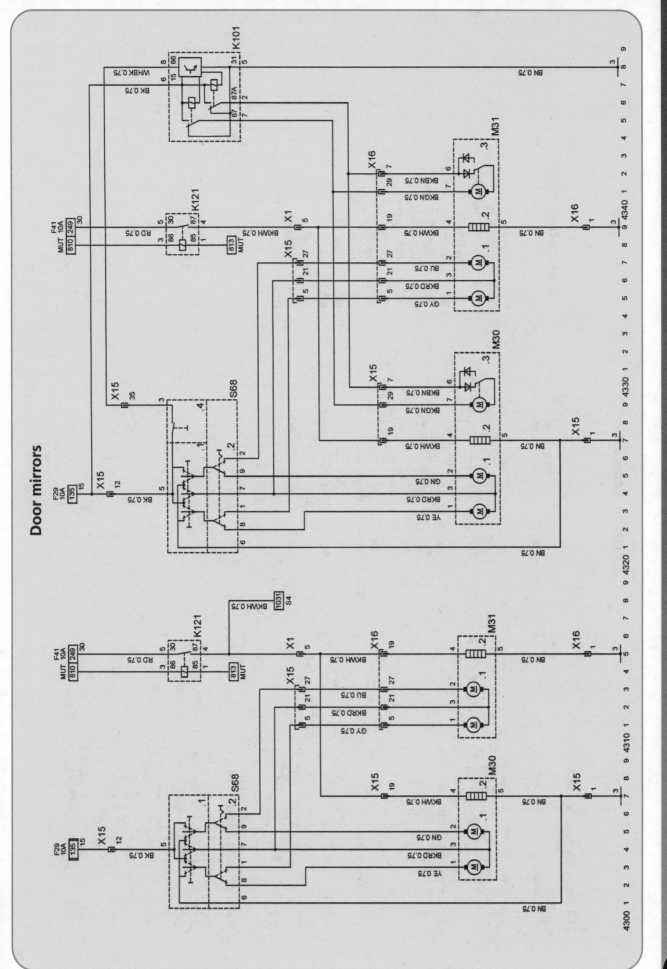

Door mirrors

Electric window lifters

Circuit Diagram Vectra-B

Circuits

Window lifters (Right-hand drive)	4200 - 4249

Grounding points

3	A - Pillar

Abbreviations

LHD	Left-hand drive
RHD	Right-hand drive
TKS	Door contact switch
ZV	Central door locking

Component codes

M47	Motor - Window lifter, driver door
M48	Motor - Window lifter, co-driver door
M49	Motor - Window lifter, door, rear, left
M50	Motor - Window lifter, door, rear, right
S37	Switch - Window lifter
S37.1	Switch - Window lifter, driver door
S37.2	Switch - Window lifter, co-driver door
S37.3	Switch - Window lifter, door, rear, left
S37.4	Switch - Window lifter, door, rear, right
S37.5	Safety switch
S37.7	Controller - Window lifter automatic
S79	Switch - Window lifter, door, rear, left
S80	Switch - Window lifter, door, rear, right
X1	Instrument panel & Body rear
X15	Body rear & Driver door
X16	Body rear & Co-driver door
X17	Body rear & Rear door, left
X18	Body rear & Rear door, right

LIST OF COLOUR CODES

BN	brown
BU	blue
DBU	dark blue
DGN	dark green
YE	yellow
GN	green
GY	grey
LBU	light blue
LGN	light green
OC	ochre
OG	orange
PU	purple (violet)
PK	pink
RD	red
BK	black
WH	white
VT	violet

Electric window lifters

Central door locking

Circuit Diagram Vectra-B

Circuits

Anti theft warning unit	4003 - 4150
Central door locking	4100 - 4150

Grounding points

3	A - Pillar

Abbreviations

HB	Hatchback
KW	Estate
LHD	Left-hand drive
MUT	Multitimer
NB	Notchback
RHD	Right-hand drive

Component codes

K37	Control unit - Central locking
K94	Control unit - Anti theft warning unit & central door locking
M18	Motor - Central locking, driver door
M19	Motor - Central locking, door, rear, left
M20	Motor - Central locking, door, rear, right
M32	Motor - Central locking, co-driver door
M37	Motor - Central locking, lid trunk/ tailgate
M41	Motor - Central locking, flap fuel filler
S41	Switch - Central locking, driver door
X15	Body rear & Driver door
X16	Body rear & Co-driver door
X17	Body rear & Rear door, left
X18	Body rear & Rear door, right
X22	Body rear & Lid trunk
X23	Body rear & Tail gate
X35	Tail gate & Switch lamp
X43	Body rear & Tail gate
X61	Body rear & Control unit - Anti theft warning unit & Central locking
X62	Body rear & Control unit - Anti theft warning unit & Central locking

LIST OF COLOUR CODES

BN	brown
BU	blue
DBU	dark blue
DGN	dark green
YE	yellow
GN	green
GY	grey
LBU	light blue
LGN	light green
OC	ochre
OG	orange
PU	purple (violet)
PK	pink
RD	red
BK	black
WH	white
VT	violet

Central door locking

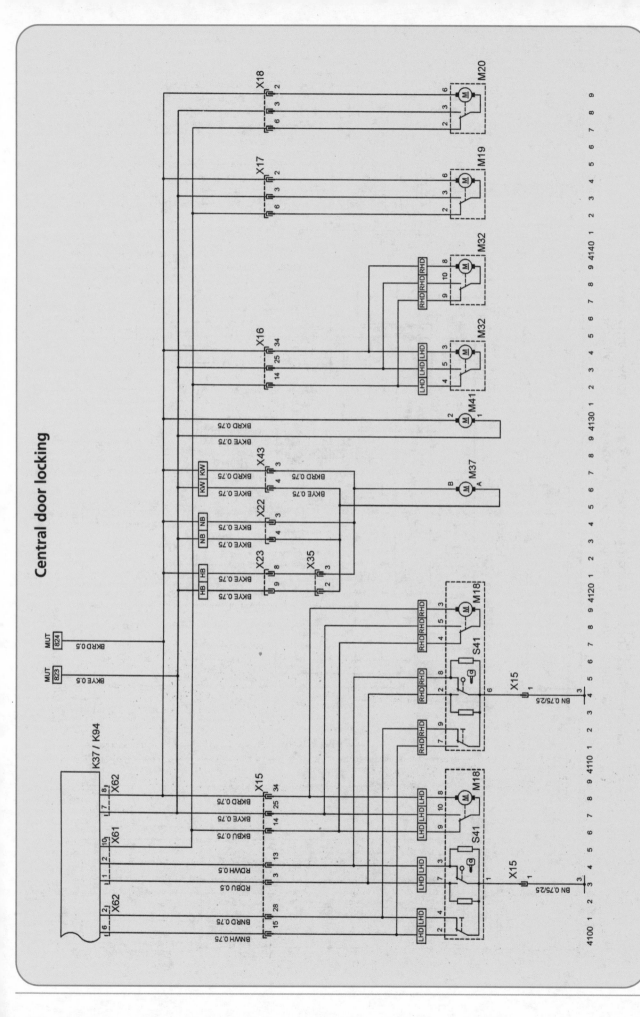

Cavalier and Calibra charging system, electric windows and mirrors
See relevant Vectra B wiring diagrams for component and wire codes.

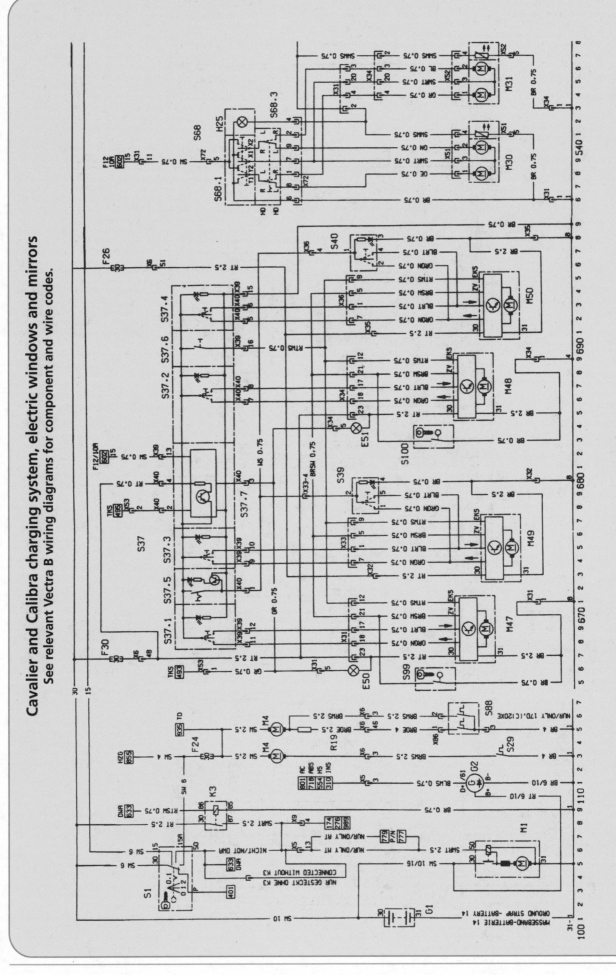

Cavalier and Calibra exterior lights

See relevant Vectra B wiring diagrams for component and wire codes.

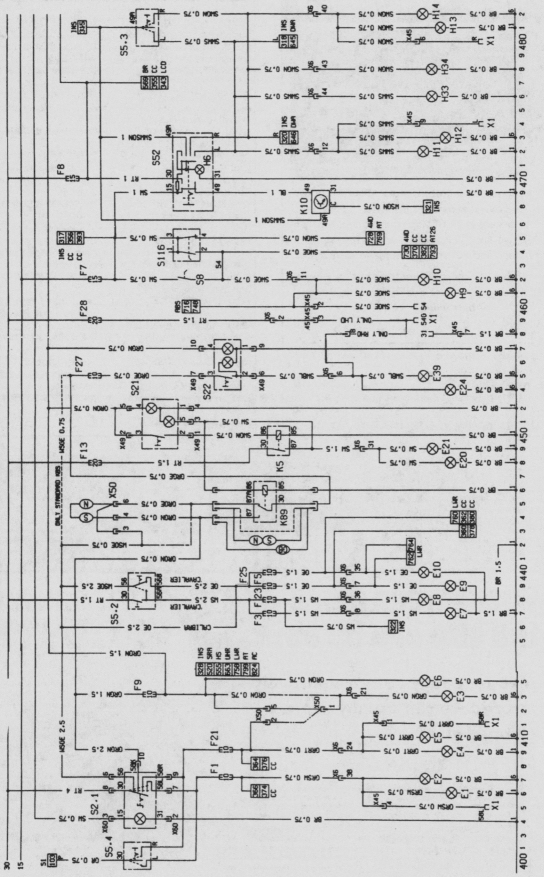

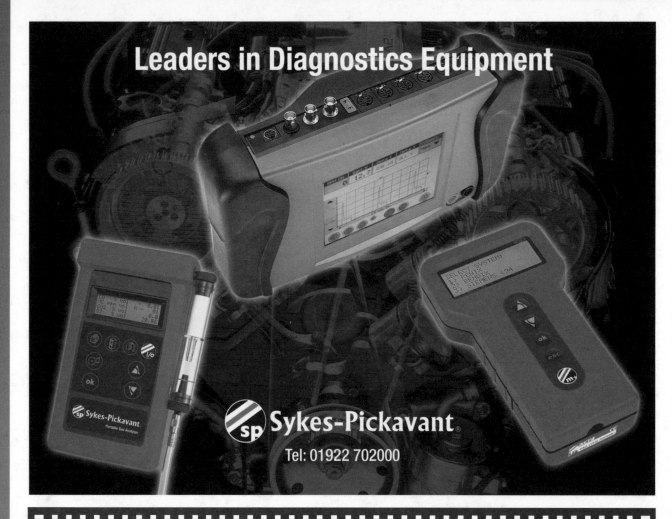

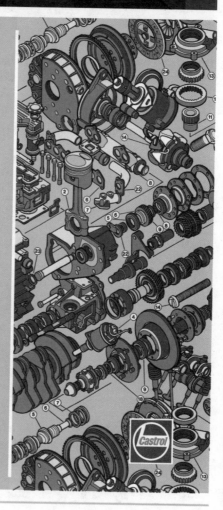

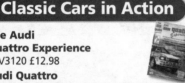

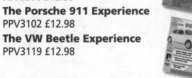

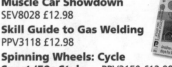